Nanostructure Design

METHODS IN MOLECULAR BIOLOGY™

John M. Walker, SERIES EDITOR

474. **Nanostructure Design:** *Methods and Protocols*, edited by *Ehud Gazit and Ruth Nussinov, 2008*
473. **Clinical Epidemiology:** *Practice and Methods*, edited by *Patrick Parfrey and Brendon Barrett, 2008*
472. **Cancer Epidemiology, Volume 2:** *Modifiable Factors*, edited by *Mukesh Verma, 2008*
471. **Cancer Epidemiology, Volume 1:** *Host Susceptibility Factors*, edited by *Mukesh Verma, 2008*
470. **Host-Pathogen Interactions:** *Methods and Protocols*, edited by *Steffen Rupp and Kai Sohn, 2008*
469. **Wnt Signaling, Volume 2:** *Pathway Models*, edited by *Elizabeth Vincan, 2008*
468. **Wnt Signaling, Volume 1:** *Pathway Methods and Mammalian Models*, edited by *Elizabeth Vincan, 2008*
467. **Angiogenesis Protocols:** *Second Edition*, edited by *Stewart Martin and Cliff Murray, 2008*
466. **Kidney Research:** *Experimental Protocols*, edited by *Tim D. Hewitson and Gavin J. Becker, 2008.*
465. **Mycobacteria,** *Second Edition*, edited by *Tanya Parish and Amanda Claire Brown, 2008*
464. **The Nucleus, Volume 2:** *Physical Properties and Imaging Methods*, edited by *Ronald Hancock, 2008*
463. **The Nucleus, Volume 1:** *Nuclei and Subnuclear Components*, edited by *Ronald Hancock, 2008*
462. **Lipid Signaling Protocols,** edited by *Banafshe Larijani, Rudiger Woscholski, and Colin A. Rosser, 2008*
461. **Molecular Embryology:** *Methods and Protocols, Second Edition*, edited by *Paul Sharpe and Ivor Mason, 2008*
460. **Essential Concepts in Toxicogenomics,** edited by *Donna L. Mendrick and William B. Mattes, 2008*
459. **Prion Protein Protocols**, edited by *Andrew F. Hill, 2008*
458. **Artificial Neural Networks:** *Methods and Applications*, edited by *David S. Livingstone, 2008*
457. **Membrane Trafficking**, edited by *Ales Vancura*, 2008
456. **Adipose Tissue Protocols,** *Second Edition*, edited by *Kaiping Yang, 2008*
455. **Osteoporosis**, edited by *Jennifer J. Westendorf, 2008*
454. **SARS- and Other Coronaviruses:** *Laboratory Protocols*, edited by *Dave Cavanagh, 2008*
453. **Bioinformatics, Volume 2:** *Structure, Function, and Applications*, edited by *Jonathan M. Keith, 2008*
452. **Bioinformatics, Volume 1:** *Data, Sequence Analysis, and Evolution*, edited by *Jonathan M. Keith, 2008*
451. **Plant Virology Protocols:** *From Viral Sequence to Protein Function*, edited by *Gary Foster, Elisabeth Johansen, Yiguo Hong, and Peter Nagy, 2008*
450. **Germline Stem Cells,** edited by *Steven X. Hou and Shree Ram Singh, 2008*
449. **Mesenchymal Stem Cells:** *Methods and Protocols*, edited by *Darwin J. Prockop, Douglas G. Phinney, and Bruce A. Brunnell, 2008*
448. **Pharmacogenomics in Drug Discovery and Development,** edited by *Qing Yan, 2008.*
447. **Alcohol:** *Methods and Protocols*, edited by *Laura E. Nagy, 2008*
446. **Post-translational Modifications of Proteins:** *Tools for Functional Proteomics, Second Edition*, edited by *Christoph Kannicht, 2008.*
445. **Autophagosome and Phagosome,** edited by *Vojo Deretic, 2008*
444. **Prenatal Diagnosis**, edited by *Sinhue Hahn and Laird G. Jackson, 2008.*
443. **Molecular Modeling of Proteins**, edited by *Andreas Kukol, 2008.*
442. **RNAi:** *Design and Application*, edited by *Sailen Barik, 2008.*
441. **Tissue Proteomics:** *Pathways, Biomarkers, and Drug Discovery*, edited by *Brian Liu, 2008*
440. **Exocytosis and Endocytosis**, edited by *Andrei I. Ivanov, 2008*
439. **Genomics Protocols,** *Second Edition*, edited by *Mike Starkey and Ramnanth Elaswarapu, 2008*
438. **Neural Stem Cells:** *Methods and Protocols, Second Edition*, edited by *Leslie P. Weiner, 2008*
437. **Drug Delivery Systems**, edited by *Kewal K. Jain, 2008*
436. **Avian Influenza Virus**, edited by *Erica Spackman, 2008*
435. **Chromosomal Mutagenesis**, edited by *Greg Davis and Kevin J. Kayser, 2008*
434. **Gene Therapy Protocols:** *Volume 2: Design and Characterization of Gene Transfer Vectors*, edited by *Joseph M. LeDoux, 2008*
433. **Gene Therapy Protocols:** *Volume 1: Production and In Vivo Applications of Gene Transfer Vectors*, edited by *Joseph M. LeDoux, 2008*
432. **Organelle Proteomics**, edited by *Delphine Pflieger and Jean Rossier, 2008*
431. **Bacterial Pathogenesis:** *Methods and Protocols*, edited by *Frank DeLeo and Michael Otto, 2008*
430. **Hematopoietic Stem Cell Protocols**, edited by *Kevin D. Bunting, 2008*
429. **Molecular Beacons:** *Signalling Nucleic Acid Probes, Methods and Protocols*, edited by *Andreas Marx and Oliver Seitz, 2008*
428. **Clinical Proteomics:** *Methods and Protocols*, edited by *Antonia Vlahou, 2008*
427. **Plant Embryogenesis**, edited by *Maria Fernanda Suarez and Peter Bozhkov, 2008*
426. **Structural Proteomics:** *High-Throughput Methods*, edited by *Bostjan Kobe, Mitchell Guss, and Thomas Huber, 2008*
425. **2D PAGE:** *Sample Preparation and Fractionation, Volume 2*, edited by *Anton Posch, 2008*
424. **2D PAGE:** *Sample Preparation and Fractionation, Volume 1*, edited by *Anton Posch, 2008*
423. **Electroporation Protocols:** *Preclinical and Clinical Gene Medicine*, edited by *Shulin Li, 2008*
422. **Phylogenomics**, edited by *William J. Murphy, 2008*
421. **Affinity Chromatography:** *Methods and Protocols, Second Edition*, edited by *Michael Zachariou, 2008*
420. **Drosophila:** *Methods and Protocols*, edited by *Christian Dahmann, 2008*
419. **Post-Transcriptional Gene Regulation**, edited by *Jeffrey Wilusz, 2008*
418. **Avidin–Biotin Interactions:** *Methods and Applications*, edited by *Robert J. McMahon, 2008*
417. **Tissue Engineering,** *Second Edition*, edited by *Hannsjörg Hauser and Martin Fussenegger, 2007*
416. **Gene Essentiality:** *Protocols and Bioinformatics*, edited by *Svetlana Gerdes and Andrei L. Osterman, 2008*
415. **Innate Immunity**, edited by *Jonathan Ewbank and Eric Vivier, 2007*
414. **Apoptosis in Cancer:** *Methods and Protocols*, edited by *Gil Mor and Ayesha Alvero, 2008*
413. **Protein Structure Prediction**, *Second Edition*, edited by *Mohammed Zaki and Chris Bystroff, 2008*
412. **Neutrophil Methods and Protocols**, edited by *Mark T. Quinn, Frank R. DeLeo, and Gary M. Bokoch, 2007*

METHODS IN MOLECULAR BIOLOGY™

Nanostructure Design

Methods and Protocols

Edited by

Ehud Gazit
Faculty of Life Science, Tel Aviv University
Tel Aviv, Israel

Ruth Nussinov
Center for Cancer Research Nanobiology Program
National Cancer Institute, Frederick, MD;
Medical School, Tel Aviv University
Tel Aviv, Israel

Humana Press

Editors
Ehud Gazit
Department of Molecular Biology
Faculty of Life Science
Tel Aviv University
Tel Aviv, Israel

Ruth Nussinov
Center for Cancer Research Nanobiology Program
SAIC-Frederick
National Cancer Institute
Frederick, MD
and
Department of Human Genetics
Medical School
Tel Aviv University
Tel Aviv, Israel

Series Editor
John M. Walker
School of Life Sciences
University of Hertfordshire
Hatfield, Hertfordshire AL10 9AB
UK

ISBN: 978-1-61737-932-1
e-ISBN: 978-1-59745-480-3
ISSN: 1064-3745
e-ISSN: 1940-6029
DOI: 10.1007/978-1-59745-480-3

Cover illustration: Provided by Aleksei Aksimentiev et al. (Chapter 11, Figures 4, 9a, 12, 13A)

Printed on acid-free paper

9 8 7 6 5 4 3 2 1

springer.com

Preface

We are delighted to present *Nanostructure Design: Methods and Protocols.* Nanotechnology is one of the fastest growing fields of research of the 21st century and will most likely have a huge impact on many aspects of our life. This book is part of the excellent *Methods in Molecular Biology*™ series as molecular biology offers novel and unique solutions for nanotechnology.

Nanostructure Design: Methods and Protocols is designed to serve as a major reference for theoretical and experimental considerations in the design of biological and bio-inspired building blocks, the physical characterization of the formed structures, and the development of their technological applications. It gives exposure to various biological and bio-inspired building blocks for the design and fabrication of nanostructures. These building blocks include proteins and peptides, nucleic acids, and lipids as well as various hybrid bioorganic molecular systems and conjugated bio-inspired entities. It provides information about the design of the building blocks both by experimental exploration of synthetic chemicals and biological prospects and by theoretical studies of the conformational space; the characterization of the formed nanostructures by various biophysical techniques, including spectroscopy (electromagnetic as well as nuclear magnetic resonance) together with electron and probe microscopy; and the application of bionanostructures in various fields, including biosensors, diagnostics, molecular imaging, and tissue engineering.

The book is divided into two sections; the first is experimental and the second computational. At the beginning of the book, Thomas Scheibel and coworkers describe the use of a natural biological self-assembled system, the spider silk, as an excellent source for the production of nano-ordered materials. Using recombinant DNA technology and bacterial expression, large-scale production of the unique silk-like protein is achieved.

In Chapter 2, by Anna Mitraki and coworkers in collaboration with Mark van Raaij, yet another fascinating biological system is explored for technological uses. The authors, inspired by biological fibrillar assemblies, studied a small trimerization motif from phage T4 fibritin. Hybrid proteins that are based on this motif are correctly folded nanorods that can withstand extreme conditions.

In Chapter 3, Maxim Ryadnov, Derek Woolfson, and David Papapostolou study yet another important self-assembly biological motif, the leucine zipper. Using this motif, the authors demonstrate the ability to form well-ordered fibrillar structures. In Chapter 4, Joseph Slocik and Rajesh Naik describe methodologies that exploit peptides for the synthesis of bimorphic nanostructures. Another demonstration of the use of peptides for self-assembled structures is described in Chapter 5 by Radhika P. Nagarkar and Joel P. Schneider. The authors use these peptides for the formation of hydrogel materials that may have many applications in diverse fields, including tissue engineering and regeneration.

In the last chapter of the book's experimental section (Chapter 6), Yingfu Li and coworkers describe a protocol for the preparation of a gold nanoparticle combined with a DNA scaffold on which nanospecies can be assembled in a periodical manner. This demonstrates the combination of biomolecules with inorganic nanoparticles for technological applications.

In Part II, on the computational approach, Bruce A. Shapiro and coauthors describe in Chapter 7 recent developments in applications of single-stranded RNA in the design of nanostructures. RNA nanobiology presents a relatively new approach for the development of RNA-based nanoparticles.

In Chapter 8, Idit Buch and coworkers describe self-assembly of fused homo-oligomers to create nanotubes. The authors present a protocol of fusing homo-oligomer proteins with a given three-dimensional structure to create new building blocks and provide examples of two nanotubes in atomistic model details.

The authors of Chapter 9, Joan-Emma Shea and colleagues, present a thorough discussion of the theoretical foundation of an enhanced sampling protocol to study self-assembly of peptides, with an example of a peptide cut from the Alzheimer Aβ protein. The self-assembly of Aβ peptides led to amyloid fibril formation. Thorough and efficient sampling is crucial for computational design of self-assembled systems.

In Chapter 10, Maarten G. Wolf, Jeroen van Gestel, and Simon W. de Leeuw also model amyloid fibril formation. The fibrillogenic properties of many proteins can be understood and thus predicted by taking the relevant free energies into account in an appropriate way. Their chapter gives an overview of existing simulation techniques that operate at a molecular level of detail.

Klaus Schulten and his coworkers provide an overview in Chapter 11 of the impressive array of computational methods and tools they have developed that should allow dramatic improvement of computer modeling in biotechnology. These include silicon bionanodevices, carbon nanotube-biomolecular systems, lipoprotein assemblies, and protein engineering of gas-binding proteins, such as hydrogenases.

In the final chapter (Chapter 12), Ugur Emekli and coauthors discuss the lessons that can be learned from highly connected β-rich structures for structural interface design. Identification of features that prevent polymerization of these proteins into fibrils should be useful as they can be incorporated in interface design.

Biology has already shown the merit of a nanostructure formation process; it is the essence of molecular recognition and self-assembly events in the organization of all biological systems. Biology offers a unique level of specificity and affinity that allows the fine tuning of nanoscale design and engineering. While much progress has been made, challenges are still ahead. We hope that *Nanostructure Design: Methods and Protocols*, which is based on biology and uses its principles and its vehicles toward design, will be useful for newcomers and experienced nanobiologists. It can also help scientists from other fields, such as chemistry and computer science, who would like to explore the prospects of nanobiotechnology.

Ehud Gazit
Ruth Nussinov

Contents

Contributors

CHRISTIAN ACKERSCHOTT • *TUM, Department Chemie, Lehrstuhl Biotechnologie, Garching, Germany*
ALEKSEI AKSIMENTIEV • *Beckman Institute for Advanced Science and Technology, University of Illinois at Urbana-Champaign, Urbana, IL*
GIOVANNI BELLESIA • *Department of Chemistry and Biochemistry, University of California, Santa Barbara, CA*
ECKART BINDEWALD • *Basic Research Program, SAIC-Frederick Inc., NCI-Frederick, Frederick, MD*
MICHAEL A. BROOK • *Department of Chemistry, McMaster University, Hamilton, Ontario, Canada*
ROBERT BRUNNER • *Beckman Institute for Advanced Science and Technology, University of Illinois at Urbana-Champaign, Urbana, IL*
IDIT BUCH • *Department of Human Genetics, Sackler Institute of Molecular Medicine, Sackler Faculty of Medicine, Tel Aviv University, Tel Aviv, Israel*
JORDI COHEN • *Beckman Institute for Advanced Science and Technology, University of Illinois at Urbana-Champaign, Urbana, IL*
JEFFREY COMER • *Beckman Institute for Advanced Science and Technology, University of Illinois at Urbana-Champaign, Urbana, IL*
EDUARDO CRUZ-CHU • *Beckman Institute for Advanced Science and Technology, University of Illinois at Urbana-Champaign, Urbana, IL*
SIMON W. DE LEEUW • *DelftChemTech, Delft University of Technology, Delft, The Netherlands*
UGUR EMEKLI • *Polymer Research Center and Chemical Engineering Department, Bogaziçi University, Istanbul, Turkey*
EHUD GAZIT • *Department of Molecular Biology, Faculty of Life Science, Tel Aviv University, Tel Aviv, Israel*
K. GUNASEKARAN • *Basic Research Program, SAIC-Frederick Inc., Center for Cancer Research Nanobiology Program, NCI-Frederick, Frederick, MD*
TURKAN HALILOGLU • *Polymer Research Center and Chemical Engineering Department, Bogaziçi University, Istanbul, Turkey*
DAVID HARDY • *Beckman Institute for Advanced Science and Technology, University of Illinois at Urbana-Champaign, Urbana, IL*
WOJCIECH KASPRZAK • *Basic Research Program, SAIC-Frederick Inc., NCI-Frederick, Frederick, MD*
SOTIRIA LAMPOUDI • *Department of Computer Science, University of California, Santa Barbara, CA*

YINGFU LI • *Departments of Chemistry and Biochemistry and Biomedical Sciences, McMaster University, Hamilton, Ontario, Canada*
ANNA MITRAKI • *Department of Materials Science and Technology, c/o Biology Department, University of Crete, Vassilika Vouton, Crete, Greece*
RADHIKA P. NAGARKAR • *Department of Chemistry and Biochemistry, University of Delaware, Newark, DE*
RAJESH R. NAIK • *Materials and Manufacturing Directorate, Air Force Research Lab, Wright-Patterson Air Force Base, OH*
RUTH NUSSINOV • *Center for Cancer Research Nanobiology Program, SAIC-Frederick, National Cancer Institute. Department of Human Genetics, Medical School, Tel Aviv University, Tel Aviv, Israel*
KATERINA PAPANIKOLOPOULOU • *Institute of Molecular Biology and Genetics, Vari 16672, Greece*
DAVID PAPAPOSTOLOU • *School of Chemistry, University of Bristol, Cantock's Close, Bristol, UK*
ARUNA RAJAN • *Beckman Institute for Advanced Science and Technology, University of Illinois at Urbana-Champaign, Urbana, IL*
LIN RÖMER • *Universität Bayreuth, Lehrstuhl Biomaterialien, 95440 Bayreuth, Germany*
MAXIM G. RYADNOV • *School of Chemistry, University of Bristol, Cantock's Close, Bristol, UK*
THOMAS SCHEIBEL • *Universität Bayreuth, Lehrstuhl Biomaterialien, 95440 Bayreuth, Germany*
JOEL P. SCHNEIDER • *Department of Chemistry and Biochemistry, University of Delaware, Newark, DE*
KLAUS SCHULTEN • *Beckman Institute for Advanced Science and Technology, University of Illinois at Urbana-Champaign, Urbana, IL*
BRUCE A. SHAPIRO • *Center for Cancer Research Nanobiology Program, National Cancer Institute, Frederick, MD*
JOAN-EMMA SHEA • *Department of Chemistry and Biochemistry, University of California, Santa Barbara, CA*
AMY SHIH • *Beckman Institute for Advanced Science and Technology, University of Illinois at Urbana-Champaign, Urbana, IL*
GRIGORI SIGALOV • *Beckman Institute for Advanced Science and Technology, University of Illinois at Urbana-Champaign, Urbana, IL*
JOSEPH M. SLOCIK • *Materials and Manufacturing Directorate, Air Force Research Lab, Wright-Patterson Air Force Base, OH*
CHUNG-JUNG TSAI • *SAIC-Frederick Inc., Center for Cancer Research Nanobiology Program, NCI-Frederick, Frederick, MD*
JEROEN VAN GESTEL • *DelftChemTech, Delft University of Technology, Delft, The Netherlands*

MARK J. VAN RAAIJ • *Institute of Molecular Biology of Barcelona (IBMB-CSIC); Parc Cientific de Barcelona, 08028 Barcelona, Spain*
CHARLOTTE VENDRELY • *TUM, Department Chemie, Lehrstuhl Biotechnologie, Garching, Germany*
MAARTEN G. WOLF • *DelftChemTech, Delft University of Technology, Delft, The Netherlands*
HAIM J. WOLFSON • *School of Computer Science, Tel Aviv University, Tel Aviv, Israel*
DEREK N. WOOLFSON • *School of Chemistry, University of Bristol, Cantock's Close, Bristol, UK; Department of Biochemistry, School of Medical Sciences, University Walk, Bristol, UK*
YING YIN • *Beckman Institute for Advanced Science and Technology, University of Illinois at Urbana-Champaign, Urbana, IL*
YAROSLAVA YINGLING • *Center for Cancer Research Nanobiology Program, National Cancer Institute, Frederick, MD*
WEIAN ZHAO • *Department of Chemistry, McMaster University, Hamilton, Ontario, Canada*

I

Experimental Approach

1

Molecular Design of Performance Proteins With Repetitive Sequences

Recombinant Flagelliform Spider Silk as Basis for Biomaterials

Charlotte Vendrely, Christian Ackerschott, Lin Römer, and Thomas Scheibel

Summary

Most performance proteins responsible for the mechanical stability of cells and organisms reveal highly repetitive sequences. Mimicking such performance proteins is of high interest for the design of nanostructured biomaterials. In this article, flagelliform silk is exemplary introduced to describe a general principle for designing genes of repetitive performance proteins for recombinant expression in *Escherichia coli*. In the first step, repeating amino acid sequence motifs are reversely transcripted into DNA cassettes, which can in a second step be seamlessly ligated, yielding a designed gene. Recombinant expression thereof leads to proteins mimicking the natural ones. The recombinant proteins can be assembled into nanostructured materials in a controlled manner, allowing their use in several applications.

Key Words: Biomaterials; recombinant production; repetitive sequence; spider silk proteins.

1. Introduction

Proteins with repetitive sequences often have specific structural properties and functions in nature. Such proteins comprise transcription factors, developmental proteins *(1)*, or structural biomaterials like elastin *(2)*, collagen *(3)*, and silk *(4)*.

Spider silks, for instance, possess outstanding mechanical properties *(5–7)*, which are highly important for the stability, for a spider's web. Among the diversity of silks produced by an individual spider, major ampullate silk forms the frame of the web and is responsible for its strength. In contrast, flagelliform silk building the capture spiral provides the elasticity necessary for dissipating

From: *Methods in Molecular Biology, vol. 474: Nanostructure Design: Methods and Protocols*
Edited by: E. Gazit and R. Nussinov © Humana Press, Totowa, NJ

the energy of prey flying into the web. Typically, all spider silks are composed of proteins that have a highly repetitive core sequence flanked by short, nonrepetitive sequences at the amino and carboxy termini (**Fig. 1**) *(8,9)*.

Sequence comparison of common spider silk proteins reveals four oligopeptide motifs that are repeated several times in each individual protein: (1) $(GA)_n/(A)_n$, (2) GPGGX/GPGQQ, (3) GGX, and (4) "spacer" sequences that contain charged amino acids *(4,10–14)*. Previously, distinct secondary structure contents (i.e., nanostructures) have been detected for silk proteins, depending on these amino acid sequences. The structural investigation of the motifs has often been performed using either entire silk fibers or short, nonassembled peptides mimicking the described oligopeptide sequences. Methods like Fourier transform infrared (FTIR), X-ray diffraction, and nuclear magnetic resonance (NMR) revealed that oligopeptides with the sequence $(GA)_n/(A)_n$ tend to form α-helices in solution but β-sheet structures in assembled fibers *(15–22)*. Such β-sheets presumably assemble the crystalline domains found within the natural silk fiber *(19,23–25)*.

In contrast, the structures adopted by GPGGX/GPGQQ and GGX repeats remain unclear. Based on X-ray diffraction studies, these regions have been described to resemble amorphous "rubber" *(26,27)*, and NMR studies suggested that they form 3_1-helical structures or can be incorporated into β-sheets *(17,19)*. Flagelliform silk, which is rich in GPGGX and GGX motifs (**Fig. 1**), likely

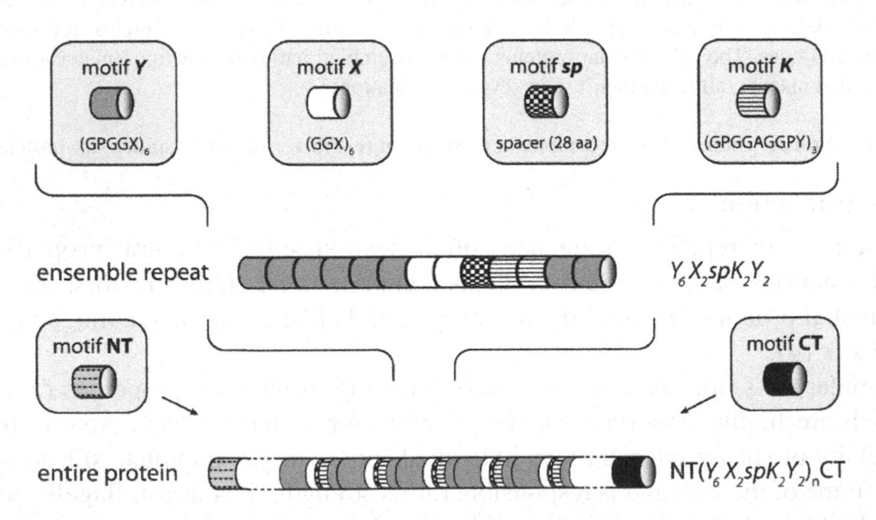

Fig. 1. Repetitive nature of the flagelliform silk protein sequence. The core sequence consists of 11 ensemble repeats that contain four consensus motifs: *Y*, *X*, *sp*, and *K*. Sfl, the recombinant protein mimicking the core domain of natural flagelliform protein, is composed of $Y_6X_2spK_2Y_2$. In the natural protein, the repetitive core sequence is flanked by nonrepeated sequences at the amino terminus (NT) and the carboxy terminus (CT).

folds into β-turn structures ***(28,29)***, which yield a right-handed helix termed *β-spiral* on stacking, similar to structural elements of elastin ***(13,14,30)***.

The outstanding properties of silk materials together with the modular nature of the underlying proteins have prompted researchers to design proteins mimicking natural silk in a modular approach. This design strategy employs synthetic DNA modules that are reversely transcripted from oligopeptide motifs characteristic for spider silk proteins. The DNA modules are assembled step by step, yielding a synthetic gene, which can be recombinantly expressed in hosts such as bacteria. Designed recombinant silk proteins allow controlled assembly of nanostructures and morphologies for various applications and therefore reflect a fascinating new generation of biomaterials.

2. Materials

1. *Escherichia coli* strains DH10B and BLR [DE3] (Novagen, Merck Biosciences Ltd., Darmstadt, Germany).
2. Plasmids: pFastBac1 (Invitrogen, Carlsbad, CA).
3. Oligonucleotide primers (MWG Biotech AG, Ebersberg, Germany).
4. Restriction enzymes: *Alw*NI, *Bam*HI, *Bgl*II, *Bse*RI, *Bsg*I, *Eco*RI, *Hin*dIII, and *Nco*I (New England Biolabs, Beverly, MA).
5. T4 ligase (Promega Biosciences Inc., San Luis Obispo, CA).
6. Agarose, polymerase chain reaction (PCR), and DNA sequencing equipment.
7. LB (Luria Bertani) medium.
8. Appropriate antibiotics: Ampicillin stock solution (100 mg/mL).
9. Isopropyl-β-D-galactopyranoside (IPTG) 1*M* stock solution.
10. Lysis buffer: 20 m*M* HEPES, 5 m*M* NaCl, pH 7.5.
11. Lysozyme (Sigma-Aldrich, St. Louis, MO).
12. $MgCl_2$ 2 M solution.
13. Proteinase-free deoxyribonuclease (DNase) I (Roche, Mannheim, Germany).
14. Protease inhibitors (Serva, Heidelberg, Germany).
15. Inclusion bodies washing buffer: 100 m*M* Tris-HCl, 20 m*M* ethylenediaminetetraacetic acid (EDTA), pH 7.0.
16. Q Sepharose (Amersham Biosciences, Piscataway, NJ).
17. Fast protein liquid chromatographic (FPLC) equipment.
18. Binding buffer: 20 m*M* HEPES, 5 m*M* NaCl, 8*M* urea, pH 7.5.
19. Elution buffer: 20 m*M* HEPES, 1*M* NaCl, 8*M* urea, pH 7.5.
20. 4*M* ammonium sulfate.
21. 1*M* ammonium carbonate.
22. Hexafluoroisopropanol (HFIP): Toxic solution; handle with care.

3. Methods

3.1. Design of a Cloning Vector

We developed a gene design method to recombinantly produce spider silk proteins in bacteria ***(31–33)***. The commercially available vector pFastbac1, featuring an origin of replication and a cassette for antibiotic resistance for selection, was equipped with a specific multiple-cloning site (MCS). The MCS was generated by two complementary synthetic oligonucleotides, which were annealed by decreasing the temperature from 95°C to 20°C with an increment of 0.1 K/s. Mismatched double strands were denatured at 70°C, and again the temperature was decreased to 20°C. The denaturing and annealing cycle was repeated 10 times, and 10 additional cycles were performed with a denaturing step at 65°C (instead of 70°C). The resulting double-stranded DNA fragment exhibited sticky ends for ligation with the vector pFastbac1 digested with *Bgl*II and *Hin*dIII. Both recognition sites were destroyed after ligation using T4 ligase. The resulting cloning vector pAZL contains recognition sites for the restriction enzymes *Bse*RI, *Bsg*I, *Bam*HI, *Nco*I, *Eco*RI, and *Hin*dIII, which can individually be employed for various steps of the cloning procedure.

3.1.1. Cloning Strategy

The amino acid sequence of a repetitive protein is divided into distinct characteristic oligopeptide motifs. The amino acid sequences of these motifs are backtranslated into DNA sequences. To obtain double-stranded DNA cassettes, both sense and antisense strands are synthesized (*see* **Notes 1** and **2**), which are annealed as described in case of the MCS.

The repetitive core sequence of flagelliform silk contains mainly four amino acid motifs (**Fig. 1**), which have been backtranslated into DNA sequences using the codon usage of *Escherichia coli*. For each construct, two complementary synthetic oligonucleotides were designed in a way that each 3′ end has two additional bases for direct cloning into linearized pAZL digested with either *Bse*RI or *Bsg*I (**Fig. 2A**). Multimerization or combination of the DNA cassettes was performed by digesting (1) pAZL containing one desired DNA cassette with *Alw*NI and *Bsg*I and (2) pAZL containing another cassette with *Alw*NI and *Bse*RI (**Fig. 2B**). After ligation using T4 ligase, pAZL was reconstituted, now containing both DNA cassettes. Since the recognition sequences of *Bsg*I and *Bse*RI are situated 14 and 8 basepairs away from the respective restriction site, all restriction sites are omitted between both DNA cassettes, and the cloning system allows direct ligation of the two cassettes without additional linker or spacer regions (**Fig. 2B**).

3.1.2. Cassettes With Specific Flagelliform Silk Sequences

The flagelliform silk protein of *Nephila clavipes* contains nonrepetitive amino- and carboxyterminal regions and 11 ensemble repeats in the core domain. Each reflects subrepeats with distinct recurring oligopeptide motifs (**Fig. 1**). From there, a spacer (sp) and three repeating motifs (*Y, X, K*) have been selected for backtranslation into oligonucleotides, which were then annealed as described.

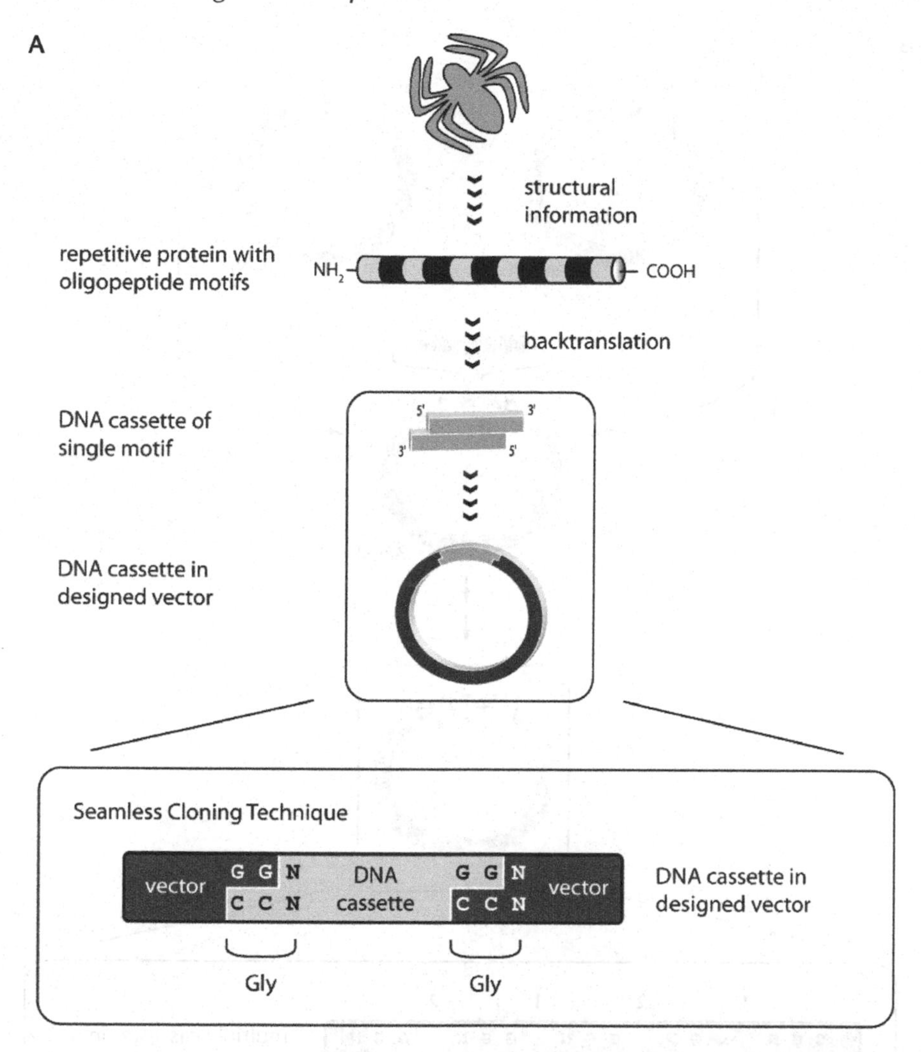

Fig. 2. Cloning strategy for the production of proteins with repeated sequences. (**A**) The protein of interest is analyzed and its amino acid sequence is backtranslated into DNA cassettes corresponding to single-oligopeptide motifs. Single DNA cassettes are incorporated into a predesigned vector. In the chosen example, the seamless cloning technique leads to the incorporation of a codon for a glycine residue between two cassettes. This connecting glycine residue is the natural linker in flagelliform silk but would also be a perfect linker for other peptide motifs since glycines do not significantly perturbate the protein structure. (**B**) Motifs 1 and 2 are joined in one plasmid by seamless ligation using restriction enzymes *Bsg*I and *Bse*RI. By repeatedly digesting/ligating the respective plasmid, it is possible to obtain vectors containing a defined number and composition of motifs separated by glycine residues.

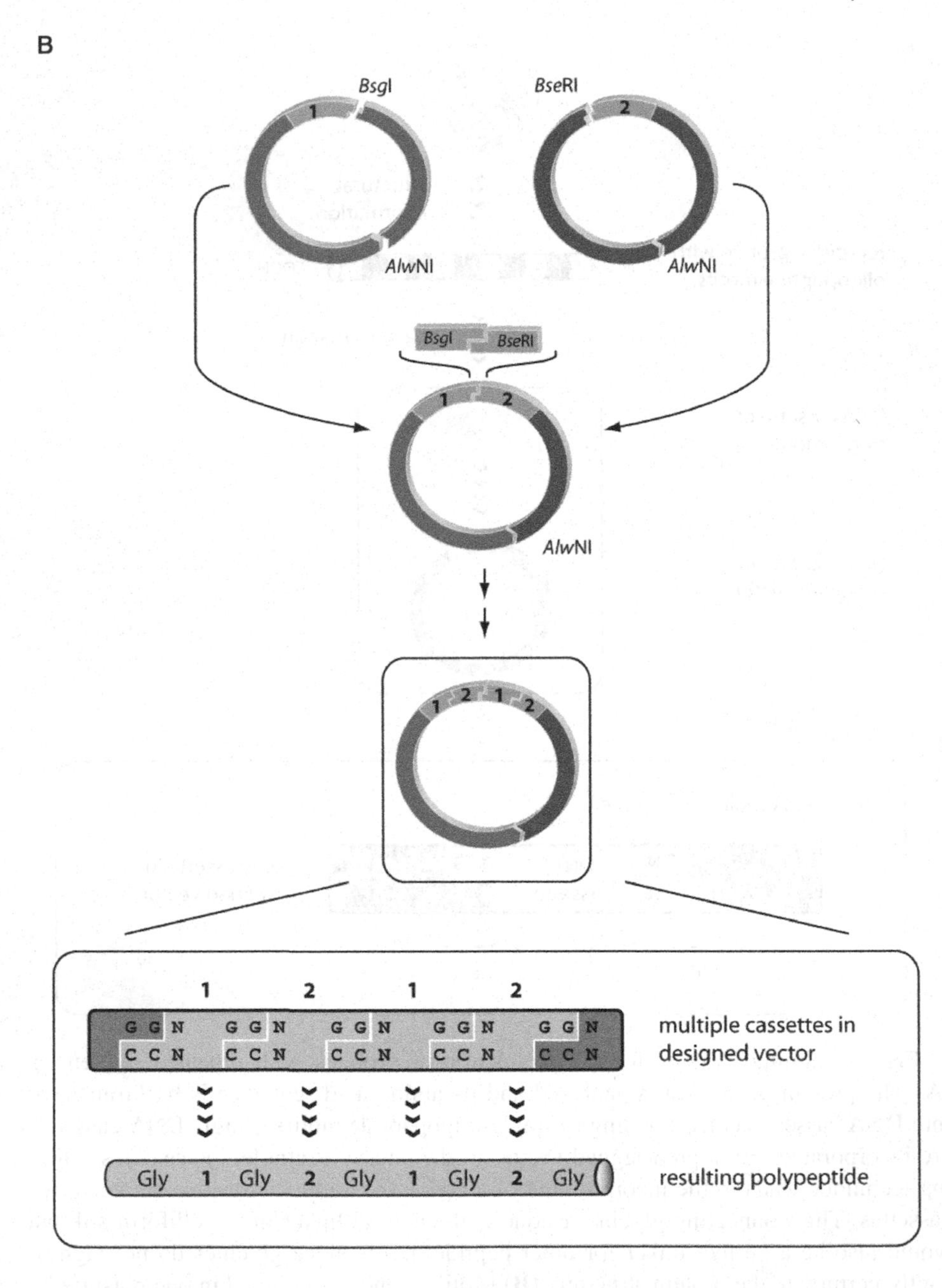

The gene sequences of the native aminoterminal (NT) and carboxyterminal (CT) regions were amplified by PCR and inserted into pAZL like the other synthetic DNA sequences.

The DNA cassettes *Y, X, K*, and *sp* were ligated to mimic one ensemble repeat of the native flagelliform protein. A consensus sequence of a single ensemble repeat is reflected by the sequence *sfl:* $Y_6X_2spK_2Y_2$. Successively, starting with a single *sfl* module, various constructs have been designed, leading to the exemplary proteins Sfl_3, Sfl-CT, Sfl_3-CT, NT-Sfl, and NT-Sfl-CT useful for studying structure-function relationships of individual silk motifs.

3.2. Recombinant Production of Sfl Proteins

After engineering various artificial flagelliform genes, they were subcloned into expression vectors pET21 or pET28 using the restriction enzymes *Bam*HI and *Hin*dIII or *Nco*I and *Hin*dIII, respectively. *Escherichia coli* BLR [DE3] was transformed with the corresponding plasmid (*see* **Note 3**), and single clones were incubated in a 4-mL preculture at 37°C overnight. After inoculation of a 2-L culture of LB medium, expression was initiated at OD_{600} 0.8 using 1 m*M* IPTG at 30°C.

Escherichia coli cells were harvested 3–4 h after induction, and the cell pellet was resuspended in lysis buffer at 4°C (5 mL per gram of cells). On addition of 0.2 mg lysozyme per milliliter, the suspension was incubated at 4°C for 30 min until becoming viscous. Protease inhibitor was added before the cells were ultrasonicated. DNA was digested with 3 m*M* $MgCl_2$, 10 μg/mL DNase, followed by incubation at room temperature for 30 min. Then, 0.5 volumes of 60 m*M* EDTA, 2–3% Triton X-100 (v/v), 1.5*M* NaCl, pH 7.0, were added, and the suspension was incubated at 4°C for another 30 min. Recombinant flagelliform proteins are entirely found in inclusion bodies, which were sedimented at 20,000*g* at 6°C for 30 min. The inclusion bodies were resuspended in 100 m*M* Tris-HCl, 20 m*M* EDTA, pH 7.0, using an ultraturax. These steps were repeated one or two additional times to wash the inclusion bodies. After a final centrifugation step, the inclusion bodies were frozen in liquid nitrogen and stored at −20°C.

3.3. Purification of Flagelliform Proteins From Inclusion Bodies

Frozen inclusion bodies were dissolved in binding buffer, and the solution was applied to an equilibrated Q Sepharose Fast Flow column (20 mL self-packed, flow 1–1.5 mL/min), which was eluted by a linear sodium chloride gradient. Flagelliform silk proteins were eluted between 200 and 250 m*M* NaCl. Pooled protein fractions were precipitated at 30% w/v ammonium sulfate (final concentration 1.2*M* ammonium sulfate). After sedimentation, the protein pellet was dissolved in 20 m*M* HEPES, 5 m*M* NaCl, 8*M* urea, pH 7.5m and dialyzed against 10 m*M* ammonium hydrogen carbonate. The protein (purity > 98%) (*see* **Note 4**) was aliquoted, frozen in liquid nitrogen, lyophilized overnight, and stored at −20°C.

3.4. Assembly of Recombinant Proteins Developing New Materials

Over the past few years, various studies have explored the potential of insect and spider silks as new materials. Regenerated or recombinant silks can be assembled in various forms, like threads, micro- or nanofibers, hydrogels, porous sponges, and films *(34)*. Such biomaterials could be employed in biomedical, cosmetic, and technical applications.

Here, the example of a silk film is presented. The properties of those films are mediated by the employed protein dissolved in an appropriate solvent *(35,36)*. Exemplarily, lyophilized Sfl is dissolved in HFIP and cast on a surface like polyethylene (PE). After evaporation of HFIP, the remaining Sfl film can be peeled off the surface (**Fig. 3**). The Sfl film reveals mainly β-sheet structure. The thickness of this silk film can be easily controlled by the amount of the protein and the size of the area where the organic solution is cast.

3.5. Design of Novel Proteins

The polymeric nature of spider silk inspires the design of novel proteins with defined nanostructures and desired properties. Specific motifs can be integrated to improve the solubility of the protein, to control its assembly process, and to control thermal, chemical, biological, and mechanical properties. For example, motifs have been incorporated into silk protein sequences to control assembly *(37–40)*. Side-specific functionalization is also feasible by incorporating amino acids with chemically active side groups, such as lysine or cysteine *(32,41,42)*. Conceiving the addition of larger peptide motifs with specific functionalities or structures will lead to novel chimeric proteins *(43,44)*.

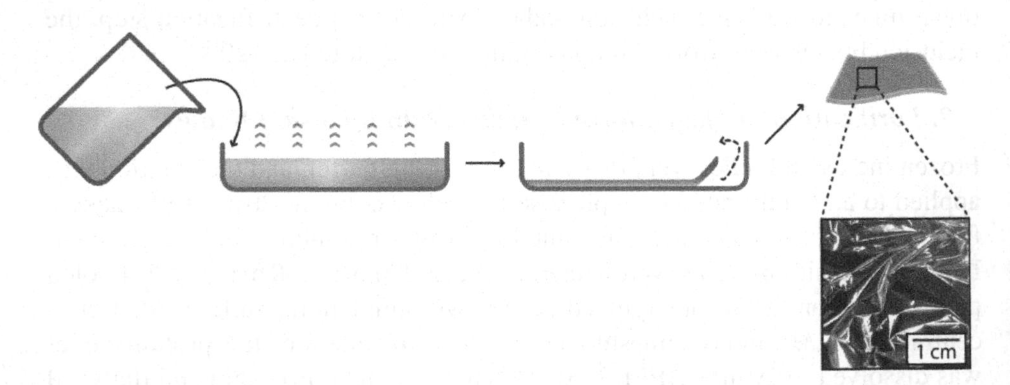

Fig. 3. Film casting using engineered flagelliform silk proteins. Lyophilized protein is dissolved in hexafluoro-2-propanol. The protein solution is cast on a surface, and the film is peeled off after evaporation of the solvent.

Based on such technology, chimera combining a spider silk domain and an elastin peptide or a dentin matrix protein have been successfully engineered ***(43,45)***. Chimeric silk proteins are capable of providing a wide variety of functions or structures based on their peptide motifs, including chemically active sites, enzyme activity, receptor-binding sites, and so on ***(42,46–48)***. The combination of such potential with repetitive sequences will allow the design of new performance proteins with defined nanostructures and chosen functionalities.

4. Notes

1. The length of the oligonucleotides is generally between 30 and 120 bases, depending on technical limitations during synthesis.
2. Screening of bacterial clones can be facilitated by adjusting the codon usage to incorporate a restriction site for a defined enzyme within the DNA cassette.
3. An appropriate bacterial strain is important for the production of proteins with repetitive sequences. *Escherichia coli* BLR [DE3] does not contain recombinase activity, preventing homolog recombination and subsequent shortening of the repetitive genetic information.
4. Since the employed spider silk proteins do not comprise tryptophan residues, fluorescence measurements of the purified protein will reveal a maximum at 305 nm after excitation of tyrosine residues at 280 nm, but no tryptophan fluorescence maximum (350 nm) on excitation at 295 nm ***(31)***. Therefore, protein purity can easily be checked and quantified.

Acknowledgments

We thank members of the Fiberlab and Lasse Reefschläger for critical comments on the manuscript. This work was supported by grants from DFG (SCHE 603/4-2) and ARO (W911NF-06-1-0451).

References

1. Faux NG, Bottomley SP, Lesk AM, et al. (2005) Functional insights from the distribution and role of homopeptide repeat-containing proteins. *Genome Res.* **154**, 537–551.
2. Foster JA, Bruenger E, Gray WR, Sandberg LB. (1973) Isolation and amino acid sequences of tropoelastin peptides. *J. Biol. Chem.* **248**, 2876–2879.
3. Fietzek PP, Kuhn K. (1975) Information contained in the amino acid sequence of the alpha1(I)-chain of collagen and its consequences upon the formation of the triple helix, of fibrils and crosslinks. *Mol. Cell. Biochem.* **8**, 141–157.
4. Xu M, Lewis RV. (1990) Structure of a protein superfiber: spider dragline silk. *Proc. Natl. Acad. Sci. U. S. A.* **87**, 7120–7124.
5. Gosline JM, Guerette PA, Ortlepp CS, Savage KN. (1999) The mechanical design of spider silks: from fibroin sequence to mechanical function. *J. Exp. Biol.* **202**, 3295–3303.

6. Vollrath F. (2000) Strength and structure of spiders' silks. *J. Biotechnol.* **74**, 67–83.
7. Gosline JM, Lillie M, Carrington E, Guerette P, Ortlepp C, Savage K. (2002) Elastic proteins: biological roles and mechanical properties *Phil. Trans. R. Soc. Lond. B.* **357**, 121–132.
8. Bini E, Knight DP, Kaplan DL. (2004) Mapping domain structures in silks from insects and spiders related to protein assembly. *J. Mol. Biol.* **335**, 27–40.
9. Hu X, Vasanthavada K, Kohler K, et al. (2006) Molecular mechanisms of spider silk. *Cell. Mol. Life. Sci.* **63**, 1986–1999.
10. Anderson SO. (1970) Amino acid composition of spider silks. *Comp. Biochem. Physiol.* **35**, 705–711.
11. Work RW, Young CT. (1987) The amino acid compositions of major and minor ampullate silks of certain orb-web-building spiders (Araneae, Araneidea) *J. Arachnol.* **15**, 65–80.
12. Lombardi SL, Kaplan DL. (1990) The amino acid composition of major ampullate gland silk (dragline) of *Nephila clavipes* (Araneae, Tetragnathidae). *J. Arachnol.* **18**, 297–306.
13. Hayashi CY, Lewis RV. (1998) Evidence from flagelliform silk cDNA for the structural basis of elasticity and modular nature of spider silks. *J. Mol. Biol.* **275**, 773–784.
14. Hayashi CY, Shipley NH, Lewis RV. (1999) Hypotheses that correlate the sequence, structure, and mechanical properties of spider silk proteins. *Int. J. Biol. Macrom.* **24**, 271–275.
15. Simmons A, Ray E, Jelinski LW. (1994) Solid state ^{13}C NMR of *N. clavipes* dragline silk establishes structure and identity of crystalline regions. *Macromolecules* **27**, 5235–5237.
16. Hijirida DH, Do KG, Michal C, Wong S, Zax D, Jelinski LW. (1996) ^{13}C NMR of *Nephila clavipes* major ampullate silk gland. *Biophys. J.* **71**, 3442–3447.
17. Kümmerlen J, van Beek JD, Vollrath F, Meier BH. (1996) Local structure in spider dragline silk investigated by two-dimensional spin-diffusion NMR. *Macromolecules* **29**, 2820–2928.
18. Seidel A, Liivak O, Calve S, et al. (2000) Regenerated spider silk: processing, properties, and structure. *Macromolecules* **33**, 775–780.
19. van Beek JD, Hess S, Vollrath F, Meier BH. (2002) The molecular structure of spider dragline silk: folding and orientation of the protein backbone. *Proc. Natl. Acad. Sci. U. S. A.* **99**, 10266–10271.
20. Lawrence BA, Vierra CA, Moore AMF. (2004) Molecular and mechanical properties of major ampullate silk of the black widow spider, *Latrodectus hesperus. Biomacromolecules* **5**, 689–685.
21. Yang M, Nakazawa Y, Yamauchi K, Knight D, Asakura T. (2005) Structure of model peptides based on *Nephila clavipes* dragline silk spidroin (MaSp1) studied by ^{13}C cross polarization/magic angle spinning NMR. *Biomacromolecules* **6**, 3220–3226.
22. Zhou C, Leng B, Yao J, et al. (2006) Synthesis and characterization of multiblock copolymers based on spider dragline silk proteins. *Biomacromolecules* **7**, 2415–2419.

23. Simmons A, Michal C, Jelinski L. (1996) Molecular orientation and two-component nature of the crystalline fraction of spider dragline silk. *Science* **271**, 84–87.
24. Jelinski LW, Blye A, Liivak O, et al. (1999) Orientation, structure, wet-spinning, and molecular basis for supercontraction of spider dragline silk. *Int. J. Biol. Macromol.* **24**, 197–201.
25. Riekel C, Bränden C, Craig C, Ferrero C, Heidelbach F, Müller M. (1999) Aspects of X-ray diffraction on single spider fibers. *Int. J. Biol. Macromol.* **24**, 179–186.
26. Gosline JM, Denny MW, DeMont ME. (1984) Spider silk as rubber. *Nature* **309**, 551–552.
27. Termonia Y. (1994) Monte Carlo diffusion model of polymer coagulation. *Macromolecules* **27**, 7378–7381.
28. Zhou Y, Wu S, Conticello VP. (2001) Genetically directed synthesis and spectroscopic analysis of a protein polymer derived from a flagelliform silk sequence. *Biomacromolecules* **2**, 111–125.
29. Ohgo K, Kawase T, Ashida J, Asakura T. (2006) Solid-state NMR analysis of a peptide (Gly-Pro-Gly-Gly-Ala)6-Gly derived from a flagelliform silk sequence of *Nephila clavipes. Biomacromolecules* **7**, 1210–1214.
30. Becker N, Oroudjev E, Mutz S, et al. (2003) Molecular nanosprings in spider capture-silk threads. *Nat. Mater.* **2**, 278–283.
31. Huemmerich D, Helsen CW, Quedzuweit S, Oschman J, Rudolph R, Scheibel T. (2004) Primary structure elements of spider dragline silks and their contribution to protein solubility. *Biochemistry* **43**, 13604–13612.
32. Scheibel T. (2004) Spider silks: recombinant synthesis, assembly, spinning, and engineering of synthetic proteins. *Microb. Cell. Fact.* **3**, 14–23.
33. Vendrely C, Scheibel T. (2007) Biotechnological production of spider silk proteins enables new applications. *Macromol. Biosci.* **7**, 401–409.
34. Altman GH, Diaz F, Jakuba C, et al. (2003) Silk-based biomaterials. *Biomaterials* **24**, 401–416.
35. Huemmerich D, Slotta U, Scheibel T. (2006) Processing and modification of films made from recombinant spider silk proteins. *Appl. Phys.* A **82**, 219–222.
36. Slotta U, Tammer M, Kremer F, Kölsch P, Scheibel T. (2006) Structural analysis of spider silk films. *Supramol. Chem.* **18**, 465–472.
37. Szela S, Avtges P, Valluzzi R, et al. (2000) Reduction-oxidation control of beta-sheet assembly in genetically engineered silk. *Biomacromolecules* **1**, 534–542.
38. Winkler S, Wilson D, Kaplan DL. (2000) Controlling β-sheet assembly in genetically engineered silk by enzymatic phosphorylation/dephosphorylation. *Biochemistry* **39**, 12739–12746.
39. Valluzzi R, Winkler S, Wilson D, Kaplan DL. (2002) Silk: molecular organization and control of assembly. *Philos. Trans. R. Soc. Lond. B Biol. Sci.* **357**, 165–167.
40. Wong Po Foo C, Bini E, Huang J, Lee SY, Kaplan DL. (2006) Solution behaviour of synthetic silk peptides and modified recombinant silk proteins. *Appl. Phys.* A **82**, 193–203.

41. Scheibel T, Parthasarathy R, Sawicki G, Lin X-M, Jaeger H, Lindquist S. (2003) Conducting nanowires built by controlled self-assembly of amyloid fibers and selective metal deposition. *Proc. Natl. Acad. Sci. U. S. A.* **100**, 4527–4532.
42. Scheibel T. (2005) Protein fibers as performance proteins: new technologies and applications. *Curr. Opin. Biotechnol.* **16**, 427–433.
43. Scheller J, Henggeler D, Viviani A, Conrad U. (2004) Purification of spider silk-elastin from transgenic plants and application for human chondrocyte proliferation. *Transgenic Res.* **13**, 51–57.
44. Wong Po Foo C, Patwardhan SV, Belton DJ, et al. (2006) Novel nanocomposites from spider silk-silica fusion (chimeric) proteins *Proc. Natl. Acad. Sci. U. S. A.* **103**, 9428–9433.
45. Huang J, Wong C, George A, Kaplan DL. (2007) The effect of genetically engineered spider silk-dentin matrix protein 1 chimeric protein on hydroxyapatite nucleation. *Biomaterials* **28**, 2358–2367.
46. Cappello J, Crissman J, Dorman M, et al. (1990) Genetic engineering of structural protein polymers. *Biotechnol. Prog.* **6**, 198–202.
47. Yang M, Asakura T. (2005) Design, expression, and solid-state NMR characterization of silk-like materials constructed from sequences of spider silk, *Samia cynthia ricini* and *Bombyx mori* silk fibroins. *J. Biochem.* **137**, 721–729.
48. Bini E, Wong Po Foo C, Huang J, Karageorgiou V, Kitchel B, Kaplan DL. (2006) RGD-functionalized bioengineered spider silk dragline silk biomaterial. *Biomacromolecules* **7**, 3139–3145.

2

Creation of Hybrid Nanorods From Sequences of Natural Trimeric Fibrous Proteins Using the Fibritin Trimerization Motif

Katerina Papanikolopoulou, Mark J. van Raaij, and Anna Mitraki

Summary

Stable, artificial fibrous proteins that can be functionalized open new avenues in fields such as bionanomaterials design and fiber engineering. An important source of inspiration for the creation of such proteins are natural fibrous proteins such as collagen, elastin, insect silks, and fibers from phages and viruses. The fibrous parts of this last class of proteins usually adopt trimeric, β-stranded structural folds and are appended to globular, receptor-binding domains. It has been recently shown that the globular domains are essential for correct folding and trimerization and can be successfully substituted by a very small (27-amino acid) trimerization motif from phage T4 fibritin. The hybrid proteins are correctly folded nanorods that can withstand extreme conditions. When the fibrous part derives from the adenovirus fiber shaft, different tissue-targeting specificities can be engineered into the hybrid proteins, which therefore can be used as gene therapy vectors. The integration of such stable nanorods in devices is also a big challenge in the field of biomechanical design. The fibritin foldon domain is a versatile trimerization motif and can be combined with a variety of fibrous motifs, such as coiled-coil, collagenous, and triple β-stranded motifs, provided the appropriate linkers are used. The combination of different motifs within the same fibrous molecule to create stable rods with multiple functions can even be envisioned. We provide a comprehensive overview of the experimental procedures used for designing, creating, and characterizing hybrid fibrous nanorods using the fibritin trimerization motif.

Key Words: Fibritin; fibrous proteins; fusion proteins; nanorods; trimerization.

1. Introduction

1.1. Fibrous Proteins in Nature and Their Possible Use in Applications

Fibrous proteins such as collagens, elastins, silkworm and insect silks, and viral fibers are mainly designed for providing mechanical functions and structural support in nature *(1–5)*. Their primary sequences consist of repetitive sequences

From: *Methods in Molecular Biology, vol. 474: Nanostructure Design: Methods and Protocols*
Edited by: E. Gazit and R. Nussinov © Humana Press, Totowa, NJ

that serve as building blocks for their bottom-up, controlled self-assembly, leading to complex molecular architectures. Furthermore, site-specific changes can be easily introduced at the sequence level to achieve their functionalization. This possibility of chemical and structural control at the nanoscale confers considerable advantages to fibrous biomaterials compared to conventional, nonbiological fibrous materials *(6–8)*.

Fibrous proteins from viruses are a distinctive family of extracellular proteins, usually entirely composed of β-structure *(9)*. Their fibrous parts fold into triple β-structured folds and are joined to globular domains, usually located at their C-termini. These globular domains are essential for the trimerization of the fibrous parts *(10,11)*. On top of excellent mechanical properties, this family of proteins, as well as amyloid-forming peptides derived from them, is exceptionally resistant to extreme conditions (temperature, detergents, denaturants) *(12,13)*. This exceptional resistance offers the possibility of interfacing with the inorganic world, that is, to use biological nanofibers and nanotubes in nanodevices *(6,14–16)*.

In adenoviruses, the C-terminal globular domain is the cell receptor domain, essential for attachment specificity *(17,18)*. Adenoviruses are used as gene therapy vectors, and new generations of vectors seek to selectively target tissues. If an attaching specificity different from the one conferred by the natural C-terminal is desired, the C-terminal domain has to be replaced by another motif/domain. This domain has to be a trimerization motif to support trimerization of the fibrous part *(19)*. This kind of "knobless" construct/vector, further derivatized with the desired tissue-targeting motifs, can be used for enabling gene therapy and tissue engineering applications *(20–22)*.

1.2. Methodology for Creating and Studying Chimeric Proteins Between Fibritin and Triple-Stranded Segments of Fibrous Proteins

The methodology for creating and studying chimeric proteins was first developed from fundamental studies aimed at structural understanding of fibrous proteins. In phage T4 fibritin, a 27-amino acid (aa) domain (amino acids 457–483 of fibritin) forms a trimeric globular, β-propellor-like structure located at the C-terminal end of a triple coiled-coil motif *(23)*. This small domain (termed *foldon*) can fold and trimerize autonomously *(24)*. N-terminal deletion mutants with an intact foldon domain trimerize successfully, whereas fibritins with a deleted or mutated foldon domain fail to fold correctly. It has therefore been proposed that the foldon domain serves as a registration motif for the segmented triple coiled-coil motif of the fibritin *(25)*. It has been subsequently shown that chimeric proteins between the foldon domain and fibrous sequences from collagen, phage T4 short-tail fibers, and HIV glycoproteins can be created and can fold successfully *(26–28)*.

We have recently created chimeric proteins by replacing the head of the adenovirus fiber by the fibritin foldon to gain insight into the trimerization mechanisms of the fiber. The previously reported structure of a stable fragment (residues 319–582 of the fiber) ***(29)*** served as the basis for the chimeric protein design. Three chimeric proteins were constructed, two comprising the shaft segment (residues 319–392) with the foldon domain in its C-terminal end (replacing the natural head domain) and one with the foldon domain in the N-terminal end of the shaft segment. In one of the chimeric proteins with the foldon at the C-terminal end, the natural linker sequence (Asn-Lys-Asn-Asp-Asp-Lys, residues 393–398) that connects the globular head to the shaft was used to connect the shaft sequences to the foldon domain. In the second, as well as in the protein with the foldon at its N-terminal end, the two domains were joined without incorporating the linker sequence. The chimeric proteins with the foldon domain appended to the C-terminus of the fiber shaft sequences fold into highly stable, sodium dodecyl sulfate (SDS)-resistant trimers, indicative of correct folding and assembly ***(30)***. The crystal structure of these two chimeras was subsequently solved, showing that the individual domains retain their native fold (*see* **Fig. 3**) ***(31)***. The results suggested that the foldon domain not only ensures correct trimerization of the shaft sequences but also allows them to assume their triple β-spiral fold. This result combined with the versatility of the foldon domain, suggests that its fusion to longer adenovirus shaft segments as well as segments from other trimeric, β-structured fiber proteins should be feasible.

The experimental methodology described can be applied to the following areas:

Fundamental studies of new fibrous folds. Although several novel fibrous folds emerged during the last decade, many still remain unresolved. The asymmetric nature (coexistence of globular and fibrous parts; large differences in relative dimensions) and natural flexibility in trimeric fibrous proteins are major barriers in crystallogenesis. Even when crystals can be obtained, they often suffer from local disorder. Replacement of big globular terminal domains with the fibritin foldon, allowing the creation of stable crystallizable fragments, can become a general strategy leading to solving the fibrous part structures.

Construction of gene therapy vectors. When the fibritin foldon replaces the C-terminal globular head of adenovirus, it enables correct trimerization of shaft repeats, but the construct is devoid of biological activity. However, it can be derivatized with tissue-targeting motifs that can offer different attaching specificities and therefore could be used as gene therapy vectors.

Rational design of fibrous constructs with controlled dimensions. Adding a desired number of building blocks derived from β-structured fibrous motifs to the fibritin foldon can create stable nanorods that could be used for integration in devices.

2. Materials

All reagents are analytical grade.

2.1. Cloning

1. Oligonucleotides have been synthesized by MWG-biotech; reconstitute the lyophilized powder at 100 pmol/μL.
2. Perform polymerase chain reaction (PCR) using Pwo polymerase (Roche).
3. Bacteriophage T4 genomic DNA is obtained from Sigma.
4. Restriction enzymes are purchased from Roche.
5. For preparative gel electrophoresis of DNA fragments less than 1 kb, use Nusieve GTG agarose. The isolated fragments can be purified using the QIAquick Gel Extraction kit (Qiagen).
6. For the ligation reactions, the Rapid DNA Ligation Kit is supplied by Roche.
7. The amplified DNA fragments are cloned in the PT7-7 vector *(32)*.
8. Plasmid DNA production is performed in the strain DH5α (Invitrogen).
9. For plasmid DNA purification, the Plasmid Miniprep Kit (Qiagen) can be used.

2.2. Culture and Lysis of Escherichia Coli

1. Protein expression is performed in the strain JM109(DE3) (Promega).
2. Prepare 1 L of LB (Luria Bertani) medium supplemented with sorbitol (330 m*M*), betaine hydrochloride (2.5 m*M*), and ampicillin (100 μg/mL).
3. Dissolve ampicillin (Sigma) at 100 mg/mL in water, aliquot, and store at −80°C.
4. Isopropyl-β-D-thiogalactopyranoside (IPTG) is dissolved in water at 0.5*M* and stored at −80°C in aliquots.
5. Ethylenediaminetetraacetic acid (EDTA) stock solution: 0.5*M* in water. Dissolve 18.6 g EDTA (disodium salt, dihydrate, M = 372.2) into 70 mL water, titrate to pH 8.0, and make up to 100 mL. Store at room temperature or at 4°C.
6. Lysis buffer: 50 m*M* Tris-HCl at pH 8.0, 2 m*M* EDTA, 20 m*M* NaCl. Add a tablet of Roche Complete™ protease inhibitors to lysis buffer just before use.
7. Streptomycin sulfate is purchased from Sigma.
8. For cell lysis, use a French press.

2.3. Purification (see Note 1)

1. Anion exchange chromatography:
   a. Column: Resource Q column (Pharmacia).
   b. Buffer A: 10 m*M* Tris-HCl buffer at pH 8.5, 1 m*M* EDTA.
   c. Buffer B: 10m*M* Tris-HCl buffer at pH 8.5, 1 m*M* EDTA, 200 m*M* NaCl.
2. Hydrophobic interaction chromatography:
   a. Column: Phenyl superose 5/5 column (Pharmacia).
   b. Buffer 1: 25 m*M* Na_2HPO_4, 25 m*M* NaH_2PO_4, 1 m*M* EDTA, 1.7*M* ammonium sulfate, pH 6.5.
   c. Buffer 2: 25 m*M* Na_2HPO_4, 25 m*M* NaH_2PO_4, 1 m*M* EDTA, pH 6.5.
3. For the fractional precipitation, ammonium sulfate is purchased from Sigma.

2.4. Sodium Dodecyl Sulfate Polyacrylamide Gel Electrophoresis

Sodium dodecyl sulfate polyacrylamide gel electrophoresis (SDS-PAGE) is considered a standard procedure known to all biochemists and molecular biologists; therefore, we do not describe it in detail here. However, all the recipes for preparing buffers and setting up gels according to the original protocols *(33)* can be found on the Jonathan King lab Web site, http://web.mit.edu/king-lab/www/cookbook/cookbook.htm.

2.5. Crystallogenesis

For crystallization of proteins in general, several aspects are important. First, reagents should be crystallization grade if available or otherwise of the highest purity that can be obtained. The experiments should be set up using glassware or high-quality plastics resistant to common organic solvents and clear enough for convenient visualization of the experiments afterward. Provision of a reliable fixed-temperature room or fixed-temperature incubators free of excessive vibrations is necessary for storing the potentially long-term crystallization experiments. Finally, a stereomicroscope is needed for observation of the crystallization trials, if possible with magnification greater than 50-fold to observe microcrystals and to judge if precipitates appear crystalline. A camera for recording crystallization results is also useful.

Crystallization plates can be obtained from many sources. First, tissue culture plates are available from general laboratory suppliers; these plates (e.g., Linbro plates and Terasaki plates) can be adapted for crystallization use. Plates developed and marketed especially for crystallization purposes are also available and, although somewhat more expensive, can be recommended for their ease of use. Our favorites are ready-to-use sitting-drop vapor diffusion plates, for example, CrysChem plates (Hampton Research, Aliso Viejo, CA, http://www.hamptonresearch.com/) and CompactClover plates (Jena Bioscience, Jena, Germany, http://www.jenabioscience.com/); these are to be covered with extraclear tape and should be resistant to organic solvent if these are used in the crystallization screen. There are several worldwide suppliers of crystallization reagents, ready-made crystallization screens, crystallization plates, and other materials useful for protein crystallography. Hampton Research was the first company on the market and still has a leading position. More recently, companies like Molecular Dimensions (Apopka, FL, http://www.moleculardimensions.com/) and Jena Bioscience have come onto the market, also providing a full catalogue of crystallization reagents and consumables. These companies often also supply material for data collection, such as goniometer heads, adaptors, capillaries, and loops for mounting crystals.

3. Methods

3.1. Cloning

1. One critical parameter for successful amplification in a PCR is the correct design of oligonucleotide primers. While constructing chimeric proteins, the aim of primer design is not only to obtain a balance between specificity and efficiency of amplification but also to ensure structural compatibility of the joining parts. If crystal structures are available, their inspection is necessary to provide guidelines for the incorporation or not of appropriate linker sequences. In the case study described here, it was estimated that the joining of the shaft domain residues 319–392 and the fibritin foldon domain residues 457–483 should not introduce structural conflicts because the three Gly392 residues of the fiber shaft lie on a triangle with sides of 13.5 Å and the three Gly457 residues of the foldon domain lie on a triangle with sides of 12.5 Å. In most of the known triple-stranded folds from viruses, hinges exist between globular and fibrous parts ***(9)***. If the fold is unknown, hinges can still be predicted from inspection of sequences, breaking of sequence repeat patterns, and so on. When constructing a chimeric protein between fibrous parts of unknown fold and the foldon domain, incorporating the hinge sequences as linker sequences between the two parts is a good initial strategy for avoiding structural conflicts.
2. The fragment coding for the foldon domain, spanning residues Gly457 to Leu483 of fibritin can be obtained by PCR using the bacteriophage T4 genomic DNA as a template. The forward primer I is designed to contain the *BamHI* restriction site that results in the addition of two residues, Gly and Ser at the N-ter of the foldon (**Table 1**). The reverse primer II contains the *ClaI* site for cloning into the expression vector PT7.7. For the construction of the chimeras that comprise the foldon domain in replacement of the natural head domain, fiber shaft fragments starting from Val319 can be joined to the N-ter of the foldon by incorporating or not the 6-aa linker of the fiber protein (Asn-Lys-Asn-Asp-Asp-Lys, residues 393–398)

Table 1
Primers Used in Constructing Foldon-Adenovirus Shaft Chimeras

N°	SEQUENCE (5′→3′)	Restriction site
I	CAAGCTCTCCAAGGATCCGGTTATATT	*Bam*HI
II	CTTGCGGCCCCATCGATTGCTGGTTATAAAAAGGTAGA	*Cla*I
III	GAAGGAGATATACATATGGTTAGCATAAAAA	*Nde*I
IV	GGTAAGTTTGTCATCATTGGATCCTCCTATTGTAAT	*Bam*HI
V	TTGTCCACAGGGATCCTTTGTCATCATTTTTGT	*Bam*HI
VI	CAAGCTCTCCAACATATGGGTTATATTCCTGAA	*Nde*I
VII	CTTGCGGCCCCATGTTATGCGGATCCTAAAAAGGTAGA	*Bam*HI
VIII	CAAACAATACTAAAGGATCCGGAGTTAGCATAAAAA	*Bam*HI
IX	CAGGGTAAGTTTGTCATCGATTTTTATCCTATTGTAA	*Cla*I

that connects the globular head to the shaft (**Fig. 1**). The DNA fragments are amplified between primers III–IV and III–V subsequently from the complementary DNA of the protein containing the original stable fragment (Val319–Glu582) cloned in the pT7.7 vector.

3. For adenoviruses, as for most of the triple-stranded fibers from viruses, the globular trimerization domains are located at their C-termini *(34)*. Therefore, the most natural design is to place the foldon at the C-terminus of the chimeric proteins. In the framework of the original study, the authors created also a construct that connects the foldon domain to the N-terminal end of the shaft segment (**Fig. 1**). Although the resulting proteins fail to fold into SDS-resistant trimers, the construction strategy is also mentioned. For this construct, the gene encoding the foldon (Gly457–Leu483) is amplified between primers VI and VII and the shaft segment, spanning residues Val319–Glu582, is amplified between primer VIII and primer IX (containing stop codon taa). The generation of the *BamHI* site results in the introduction of three residues, Gly-Ser-Gly, between the two segments.
4. Set up the PCR reactions according to the product instructions provided with the polymerase. Place tubes into the preheated thermal cycler and perform each amplification for 30 cycles according to the following schedule: 30-s denaturation at 95°C, 30-s annealing at 60°C, and 45-s extension at 72°C.
5. Purify the PCR reactions by preparative gel electrophoresis on a 4% Nusieve GTG agarose gel, cut the bands of interest with a sharp blade, and extract the amplified DNA fragments using the QIAquick Gel Extraction kit.
6. Digest 1 μg of the purified bands and pT7.7 vector with the corresponding restriction enzymes for 1 h 30 min at 37°C and purify the DNA using the QIAquick Gel Extraction kit.
7. Set up 20-μL ligation reactions according to the instructions of the Roche Rapid DNA Ligation Kit. Start with 35–50 ng of vector while the insert to vector ratio is kept at 3:1.
8. Aliquot 100 μL of competent DH5α cells into an Eppendorf tube and add 4 μL of the ligation mixture. Swirl the contents of the tube gently and incubate on ice for 30 min. Heat pulse each transformation reaction in a 42°C water bath for 2 min. Add 900 μL of LB medium to each tube and incubate at 37°C for 1 h with shaking at 220 rpm. Centrifuge for 3 min at 1300 g discard 800 μL, and resuspend the pelleted cells in the remaining 200 μL. Use a sterile spreader to plate 200 μL of the transformed bacteria onto LB agar plates that contain ampicillin (100 μg/mL). Colonies will appear following overnight incubation at 37°C.
9. Culture single colonies overnight in 5 mL LB medium supplemented with ampicillin (100 μg/mL). Harvest cells and isolate the plasmid DNA using the Plasmid Miniprep Kit. Positive clones are identified by restriction enzyme digestion.

3.2. Protein Expression

1. Pick a single colony from a freshly streaked plate, inoculate 10 mL LB supplemented with ampicillin (100 μg/mL), and grow overnight with shaking at 37°C.
2. Inoculate 1 L of LB medium containing sorbitol (330 m*M*), betaine hydrochloride (2.5 m*M*), and ampicillin (100 μg/mL) with 10 mL of an overnight culture and grow at 37°C until the OD_{600} reaches 0.4. Cool culture to 22°C.

fiber shaft residues 319-392

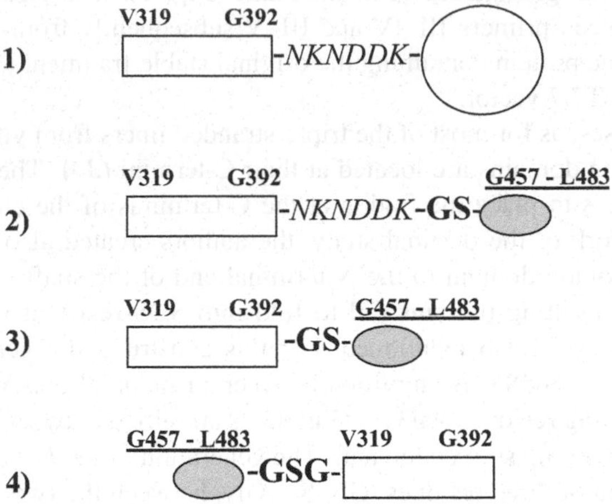

B

18	**319 V S I K K S S G L N F D N -**
19	**T A I A I N A G K G L E F D T N T S E S P D I -**
20	**N P I K T K I G S G I D Y N E N -**
21	**G A M I T K L G S G L S F D N S -**
22	**G A I T I G 392**

fibritin foldon residues 457-483

457 G Y I P E A P R D G Q A Y V R K D G E W V L L S T F L 483

Fig. 1. **(A)** Schematic representation of the domain structure of the chimeric proteins: **(1)** the stable adenovirus fiber fragment (fiber residues 319–582). Residues belonging to the shaft domain (V319 to G392) are symbolized with a rectangle, and residues belonging to the globular head (L399–E582) are symbolized with a circle. Residues 393–398 (NKNDDK) form the linker that connects the fibrous and globular parts and are drawn in italics. **(2)** The chimeric protein that comprises the fibritin foldon domain (fibritin residues G457 to L483, oval shape) fused to the C-terminus of the shaft domain with use of the natural linker between the two domains. For the sake of clarity, the numbers corresponding to the fibritin residues are underlined. The residues GS, highlighted in bold, are not part of the coding sequence and are introduced as a result of the cloning strategy. **(3)** The chimeric protein with the foldon domain appended at the C-terminal end of the shaft domain without the use of the natural linker sequence. **(4)** The chimeric protein with the foldon domain appended at the N-terminal end of the shaft domain. The residues GSG, highlighted in bold, do not belong to the coding sequence and are introduced as a result of the cloning strategy. **(B)** Amino acid sequences of the fiber shaft residues 319–392 and of the fibritin foldon residues 457–483. The fiber shaft sequence repeat numbers (repeats 18–22 according to ***29***) are indicated to the left. The repeats are not aligned. (From **ref.** ***30*** with permission.)

3. When the temperature of the culture reaches 22°C, add IPTG to 0.5 m*M* final concentration to induce protein expression. Continue the incubation for 14 h at 22°C.
4. Harvest the bacterial cells by centrifugation at 14,000 g at 4°C for 10 min and pour off the supernatant.
5. Add approximately 20 mL of lysis buffer (50 m*M* Tris-HCl pH 8.0, 2 m*M* EDTA, 20 m*M* NaCl) containing a tablet of Roche Complete protease inhibitors.
6. After cell lysis using a French press, remove cell debris by centrifugation at 43,000 g at 4°C for 20 min.
7. Recuperate the supernatant and add streptomycin sulfate (Sigma) to a final concentration of 1% (w/v). Stir the suspension for a further 15 min in the cold room to remove the viscous nucleic acid. Centrifuge for 15 min at 19,000 rpm and 4°C and discard the pellet.

3.3. Protein Purification (see Note 1)

1. Measure the volume of the supernatant after streptomycin sulfate treatment and pour it into a glass beaker.
2. Weigh 0.361 g of solid ammonium sulfate for every 1 mL of protein solution to reach a final concentration of 60% saturation.
3. Place the beaker on ice and stir with a magnetic stirrer. Add the ammonium sulfate to the protein solution slowly and in small batches.
4. After addition is complete, incubate for 15 min on ice and then remove the precipitated protein by centrifugation at 43,000 g at 4°C for 20 min.
5. Decant off the supernatant into a measuring cylinder and determine the total volume.
6. Add 0.129 g of solid ammonium sulfate per milliliter of protein solution to take the concentration from 60% to 80% saturation as described above.
7. After centrifugation, discard supernatant and dissolve the protein precipitate in 1 mL of 10 m*M* Tris-HCl buffer pH 8.5, 1 m*M* EDTA.
8. Transfer the protein suspension into a dialysis bag and dialyze against 10 m*M* Tris-HCl buffer pH 8.5, 1 m*M* EDTA, overnight in the cold room.
9. The next day, equilibrate the Resource Q column with 10 m*M* Tris-HCl buffer pH 8.5, 1 m*M* EDTA (buffer A) at a flow rate of 3 mL/min. Load the sample onto the column and elute the protein with a gradient of 0–200 m*M* NaCl (buffer B).
10. Pool the fractions containing the protein, bring them to 1.7 *M* ammonium sulfate, and dialyze against a phosphate buffer (25 m*M* Na_2HPO_4, 25 m*M* NaH_2PO_4, 1 m*M* EDTA, 1.7 *M* ammonium sulfate, pH 6.5) overnight in the cold room.
11. Apply the sample to a Pharmacia phenyl superose 5/5 column equilibrated with buffer 1 (25 m*M* Na_2HPO_4, 25 m*M* NaH_2PO_4, 1 m*M* EDTA, 1.7 *M* ammonium sulfate, pH 6.5). Elute with a linear gradient of 1.7–0.0 *M* ammonium sulfate (buffer 2).
12. The chimeric protein elutes at about 1.5 *M* ammonium sulfate. Collect the fraction and precipitate the purified protein by adding ammonium sulfate to 80% saturation. Store the precipitated protein at 4°C.

3.4. Characterization of Chimeric Proteins With Denaturing and Nondenaturing SDS-PAGE

A hallmark of well-folded, trimeric β-structured fibers is their SDS resistance. In the standard Laemmli SDS buffer, which contains 2% final SDS, all or most of these fibers are stable at 4°C. For the trimers to be completely dissociated, boiling for 3 min in sample buffer is recommended. This SDS resistance is a precious biochemical tool that allows easy assessment of native, trimeric states of proteins from nonnative and even intermediate states. In standard SDS gels, trimers do not bind SDS efficiently and migrate slowly in the gel; the denatured, misfolded, or aggregated forms are completely dissociated by SDS and migrate in the monomer position. Since this methodology can be applied to cell lysates, it allows rapid screening of various chimeric constructs before purification and selects the ones that fold successfully into trimers for further purification and characterization.

The following procedure is recommended:

1. Mix the lysate or protein solution with Laemmli SDS sample buffer and split in two tubes.
2. Place one tube on ice.
3. Place the second tube in a heating block for 3 min at 100°C, then put the tube on ice and let it cool.
4. Run the two samples in adjacent wells in the SDS gel and compare the running positions.

If the nonboiled band migrates with an apparent mass compatible with a trimer that chases to the monomer band after boiling, it is a good indication that the chimeric protein folds into a trimer (**Fig. 2**). It is very important for the gel to be refrigerated since it has been observed that above room temperature partial unfolding of native trimers can occur, leading to "open" forms that migrate slower than the native trimer *(12)*. If the SDS gel is not refrigerated, partial unfolding induced by the combination of SDS and temperature may occur in situ and lead to formation of slower migrating bands.

3.5. Crystallization

For crystallization, several aspects of the protein preparation have to be considered. A high degree of purity (better than 99%) is important, although preliminary experiments with somewhat less-pure preparation can give some useful initial information about solubility and in some cases even yield crystals. As important as "chemical" purity is conformational homogeneity or, in other words, absence of flexibility. At this point appropriate design of the expression vector comes into play, as does the presence of a not too flexible linker between the fused domains. If purification aids such as histidine tags are to be introduced,

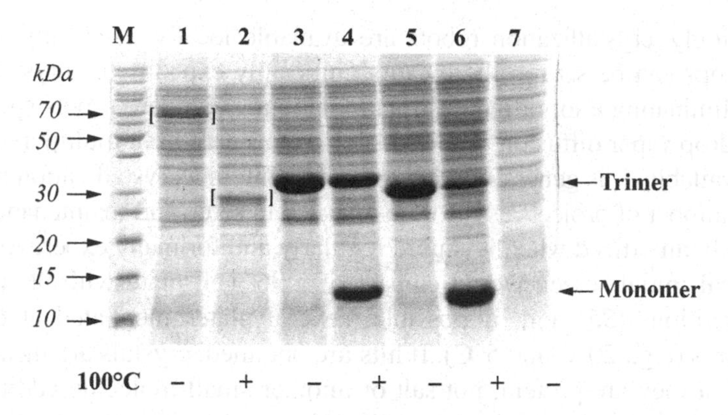

Fig. 2. Expression of the chimeric proteins with the foldon domain at their C-terminus. After a pellet supernatant fractionation of *Escherichia coli* lysates, supernatants were electrophoresed on a 12.5% sodium dodecyl sulfate (SDS) polyacrylamide gel and visualized with Coomassie blue staining. Electrophoresis was carried at 4°C. The + symbol indicates boiling in loading buffer containing 2% SDS for 3 min prior to loading in the gel. Lane 7, lysate of noninduced bacteria. Lanes 1 and 2, supernatants of lysates of the original fiber stable fragment, nonboiled and boiled, respectively, are shown to allow comparison with the chimeric proteins. The trimer (lane 1) and monomer (lane 2) positions are marked with brackets. Lanes 3 and 4, chimeric protein with linker, nonboiled and boiled, respectively. Lanes 5 and 6, chimeric protein without the linker, nonboiled and boiled, respectively. The trimer and monomer positions for the chimeric proteins are marked with arrows. Lane M, molecular mass markers. (From **ref. *30*** with permission.)

it is preferable to include a protease cleavage site between the tags and the protein to be crystallized as the purification tags lead to undesirable flexibility. Having said that, there are examples of successful crystallization of proteins including these purification tags, especially if these are relatively small.

The fibrous fusion proteins discussed here are in general not expected to be air sensitive, so vapor diffusion is the method of choice. Sitting-drop vapor diffusion can be recommended for ease of setup, visualization, and crystal harvesting. These can be sealed with extra-clear tape, which permits opening individual wells by carefully removing the tape only from that well and resealing with a piece of the same tape. If the proteins are found or expected to be oxidation sensitive, vapor diffusion experiments can be set up under a nitrogen atmosphere, or more easily, microbatch experiments can be performed. In microbatch experiments, protein solution is directly mixed with precipitant solution and incubated under a layer of mineral oil, allowing for slow evaporation of aqueous solvent through the oil layer. A percentage of silicon oil can be mixed with the mineral oil if faster evaporation is desired.

Increasingly, crystallization robots are available locally, especially if small-volume drops can be set up; these can significantly expedite the crystallization process, eliminating a lot of tedious manipulations. There are robots specialized in sitting-drop vapor diffusion or microbatch experiments, but multipurpose ones are also available. It is generally not worth investing in a crystallization robot for a limited number of projects as the time invested in setup and maintenance of the robot is only amortized when it is used regularly and for many experiments.

A typical initial screen would consist of a 96-well plate with 96 very different conditions ***(35)*** and, if possible, several plates incubated at different temperatures (e.g., 20°C and 5°C). If hits are obtained, crystals are measured to confirm that they are protein, not salt or another small-molecule additive, and to assess their diffraction limit and quality. However, in many cases, no crystals are obtained in the first screen. If crystalline precipitates are obtained, further screens are performed around these conditions to see if crystals can be obtained. At the same time, it is worth carefully examining the cloning strategy and the expression and purification procedure to see if improvements in protein purity and conformational homogeneity can be obtained. In addition to these initial more-or-less random screens (and if enough material is available), it is worth screening common precipitants like ammonium sulfate and polyethylene glycol 6000 at different concentrations, pH, and temperatures.

Crystallization trials should be regularly examined, with the results noted in a notebook or spreadsheet system and photographically documented if possible. A suitable regime would be a quick examination straight after setup, then more extensive ones after a day, after 3 d, after a week, after 2 wk, after a month, and so forth until suitable crystals have been obtained or the drops have dried. For more complete information on protein crystallization, several textbooks are available ***(36–38)***; a special issue of the *Journal of Structural Biology* about protein crystallization methods is also very useful ***(39)***.

3.6. Structure Determination

3.6.1. Choice of Method

Structure solution by crystallography is in principle feasible for molecules of almost any size if, of course, crystals can be obtained. If the protein is not too large, structure solution by nuclear magnetic resonance (NMR) spectroscopic methods may be considered ***(40)***. This has the major advantage of not having to crystallize the protein, although depending on the protein size different labeling techniques will be necessary. For up to around 30 kDa (trimeric size), labeling with carbon-13 and nitrogen-15 is likely to be sufficient, while with additional deuterium labeling structures of size up to 50–60 kDa may be tractable. Given the trimeric foldon size of just over 9 kDa, this would permit solving unknown trimeric nanorod structures of just over 20 or 40–50 kDa, respectively (7 or

13–17 kDa per monomer, respectively). Introduction of a protease cleavage site between the fibrous and foldon domains would allow removal of the foldon domain and the study of the fibrous domain on its own, allowing structure solution of trimeric nanorods of up to 30 and 50–60 kDa trimeric size by NMR spectroscopic methods.

3.6.2. Data Collection

The first step of data collection is the recovery of crystals from the crystallization setup. As protein crystals are generally fragile, they will either have to be carefully transferred to a quartz capillary and mounted in conditions in which the crystal will not dry up or attract moisture from the surrounding atmosphere and dissolve. They can be picked up with a nylon or plastic microloop slightly larger than the crystal. If the loop is then covered with a plastic hood filled with a drop of mother liquor, data collection can proceed at room temperature. To prolong crystal life, a crystal can also be briefly incubated in a suitable cryoprotectant; in this case, they can either be flash frozen at 100 K or frozen in liquid nitrogen ***(41)***. If data collection is then performed at 90–120 K, significant increase in crystal lifetime can be obtained (radiation damage decreases at lower temperature; ***42***).

Depending on the space group of the crystals obtained and the structure solution method that is to be used, somewhat different data collection procedures will need to be employed. In all cases, complete datasets are necessary and, if the anomalous signal is to be exploited, high multiplicity. For high-symmetry space groups, a relatively small fraction of reciprocal space needs to be explored, while for lower-symmetry space groups, a larger fraction of reciprocal space will need to be covered (i.e., more images per dataset will have to be collected). For structure solution by molecular replacement or isomorphous replacement methods (see **Subheading 3.6.3.**), high multiplicity is not a necessity, while for anomalous dispersion methods it is. High-multiplicity datasets will require longer data collection times; at the same time, radiation damage will have to be avoided. Therefore, to allow successful structure solution, at times resolution will have to be sacrificed for data completeness or multiplicity.

3.6.3. Structure Solution

Given that the foldon structure is known, structure solution by a molecular replacement technique ***(43)*** will be possible if the foldon is a significant fraction of the total protein, say 25–30%. If molecular replacement is not successful, heavy atom derivatives will have to be produced for structure solution by multiple isomorphous replacement (MIR), single isomorphous replacement using anomalous signal (SIRAS), multiwavelength anomalous dispersion (MAD; ***44***), or single-wavelength anomalous dispersion (SAD; ***45***). Common derivatives are mercury compounds, which bind to cysteine residues and are especially useful for MIR or SIR(AS), or seleno-methionine derivatives, especially useful for the MAD method *(46)*.

Heavy atoms are generally introduced into preformed protein crystals by soaking techniques *(47)*, although cocrystallization is also a possibility. Selenomethionine can be introduced into proteins instead of methionine by growing methionine-auxotroph bacteria in expression cultures in the presence of selenomethionine *(48)* or by inhibition of the methionine synthesis pathway and provision of the necessary amino acids and seleno-methionine in expression cultures *(49)*. If no cysteines or methionines are present in the natural sequence, these can be introduced by site-directed mutagenesis. A discussion and explanation of macromolecular phasing methods is available in **ref. *50*** and in several textbooks and compilations *(51–56)*.

3.6.4. Model Building, Refinement, Validation, and Analysis

Once interpretable electron density maps have been obtained, a model for the protein will have to be built either "by hand" using molecular graphics programs or, if the map is of sufficient quality (resolution better than 2.3 Å), in combination with automated building procedures like Arp-Warp *(57)*. Once a complete protein model, including ordered solvent molecules, has been built, the structure should be refined using appropriate geometric restraints and the best-available dataset with respect to completeness and resolution. Refmac is the program of choice for refinement as it uses maximum likelihood targets *(58)*. Validation of the structure is always necessary as important errors in model building and refinement may have gone unnoticed. Molprobity is the software of choice for this purpose *(59)*. Validation judges parameters used in refinement such as bond distances and angles, planarity of aromatic groups, and parameters not used in refinement such as whether all amino acids are in suitable environments respective to their nature (polar, apolar, charged), whether the Ramachandran plot of the structure looks reasonable, and so on.

Once the structure has been solved, and preferably refined to completion, the structure will have to be analyzed, first judging whether a new fibrous fold has been discovered or whether the structure is similar to other known structures. The program DALI can perform similarity searches against the protein structure database automatically *(60)*. Further analysis will concern the biological interest of the structure. In the case of viral fibers, examine whether the structure may contain regions implicated in receptor binding or interaction with other biomolecules.

The structure is also likely to be of interest for materials science, and inspection may reveal the presence of surface loops that may be modified without affecting the structure. These modified surface loops may then be used to bind small molecules or other proteins to function as sensors, metals in an attempt to make the fibers conductive, and the like. As many biological fibers contain sequence repeats, inspection of the structure will also likely reveal the start and end of the structural repeat, which is important for design of longer fibers made up of repeating sequences.

3.6.5. Verification and Application of the Foldon Fusion Strategy for Structure Solution of Trimeric Fibrous Domains

Papanikolopoulou et al. *(31)* have shown that the C-terminal four adenovirus type 2 fiber shaft repeats have the same structure when fused to a C-terminal foldon domain (**Fig. 3A**) as the native fold *(29)*, showing that the foldon fusion strategy is viable and valid for solving crystal structures of unknown fibrous

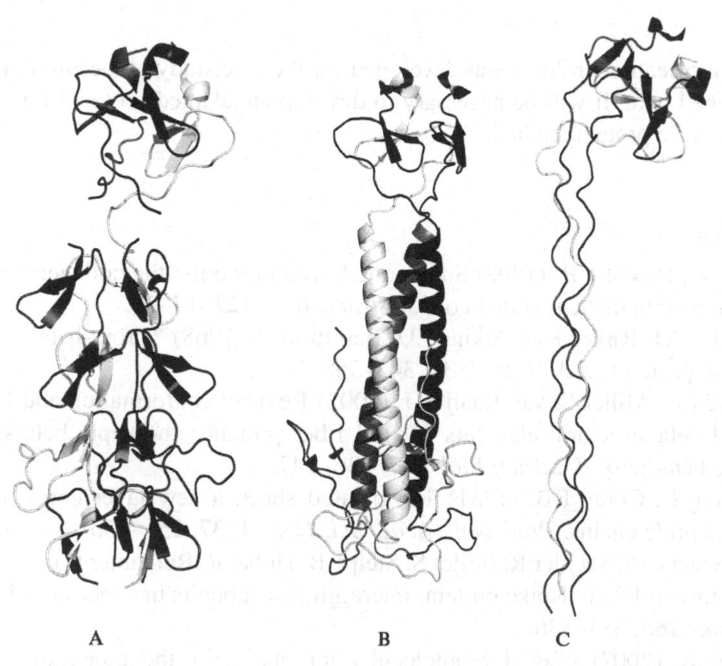

Fig. 3. Crystal structures of foldon fusion proteins. (**A**) Fusion construct consisting of human adenovirus type 2 fiber shaft residues 319–392 (bottom), a Gly-Ser linker, and bacteriophage T4 fibritin residues 457–483 (top). Note the partially disordered linker *(31)*. (**B**) Structure of "minifibritin," a fusion construct consisting the N-terminal domain of the bacteriophage T4 fibritin with the C-terminal foldon *(61)*. (**C**) Fusion construct consisting of synthetic collagen sequence (GPP) repeats and the foldon domain (top). Note the pronounced angle between the two domains, caused by the stagger of the collagen triple helix *(27)*. These figures were prepared using the deposited coordinates (pdb-codes 1V1H, 1OX3, and 1NAY, respectively) and the Pymol program *(62)*.

domains. Also, it showed that a flexible linker between the two domains is not necessarily an obstacle to crystallization and structure solution. With regard to applications, the foldon fusion strategy has been used to solve the structure of the N-terminal domain of the bacteriophage fibritin itself (**Fig. 3B**; *61*), a structure that could not be solved before due to flexibility of the intermediate domains. A third example is the structure of a collagen model peptide, a Gly-Pro-Pro repeating sequence, in which the staggered collagen triple helix leads to a rather sharp angle between the collagen and foldon domains (**Fig. 3C**; *27*). However, this was not an insurmountable problem for crystallization and structure solution. What to our knowledge has not been tried with success is incorporating a protease site between the trimeric fibrous protein domain and the foldon domain, so that after correct trimerization and partial purification, the foldon domain can be removed and the fibrous domain further purified and crystallized.

4. Note

1. This purification protocol was developed for the case study of the chimeric protein described here. It will be necessary to develop an adapted protocol for each case of chimeric protein studied.

References

1. Beck K, Brodsky B. (1998) Supercoiled protein motifs: the collagen triple-helix and the alpha-helical coiled coil. *J. Struct. Biol.* **122**, 17–29.
2. Geddes AJ, Parker KD, Atkins ED, Beighton E. (1968) “Cross-beta” conformation in proteins. *J. Mol. Biol.* **32**, 343–358.
3. Mitraki A, Miller S, van Raaij MJ. (2002) Review: conformation and folding of novel beta-structural elements in viral fiber proteins: the triple beta-spiral and triple beta-helix. *J. Struct. Biol.* **137**, 236–247.
4. Pauling L, Corey RB. (1951) The pleated sheet, a new layer configuration of polypeptide chains. *Proc. Natl. Acad. Sci. U. S. A.* **37**, 251–256.
5. Steinbacher S, Seckler R, Miller S, Steipe B, Huber R, Reinemer P. (1994) Crystal structure of P22 tailspike protein: interdigitated subunits in a thermostable trimer. *Science* **265**, 383–386.
6. Gazit E. (2007) Use of biomolecular templates for the fabrication of metal nanowires. *FEBS J.* **274**, 317–322.
7. Rajagopal K, Schneider JP. (2004) Self-assembling peptides and proteins for nanotechnological applications. *Curr. Opin. Struct. Biol.* **14**, 480–486.
8. Woolfson DN, Ryadnov MG. (2006). Peptide-based fibrous biomaterials: some things old, new and borrowed. *Curr. Opin. Chem. Biol.* **10**, 559–567.
9. Mitraki A, Papanikolopoulou K, Van Raaij MJ. (2006) Natural triple beta-stranded fibrous folds. *Adv. Protein Chem.* **73**, 97–124.
10. Hong JS, Engler JA. (1996) Domains required for assembly of adenovirus type 2 fiber trimers. *J. Virol.* **70**, 7071–7078.

11. Novelli A, Boulanger PA. (1991) Deletion analysis of functional domains in baculovirus-expressed adenovirus type 2 fiber. *Virology* **185**, 365–376.
12. Mitraki A, Barge A, Chroboczek J, Andrieu JP, Gagnon J, Ruigrok RW. (1999) Unfolding studies of human adenovirus type 2 fibre trimers. Evidence for a stable domain. *Eur. J. Biochem.* **264**, 599–606.
13. Papanikolopoulou K, Schoehn G, Forge V, et al. (2005) Amyloid fibril formation from sequences of a natural beta-structured fibrous protein, the adenovirus fiber. *J. Biol. Chem.* **280**, 2481–2490.
14. Reches M, Gazit E. (2003) Casting metal nanowires within discrete self-assembled peptide nanotubes. *Science* **300**, 625–627.
15. van Raaij MJ, Mitraki A. (2004) Beta-structured viral fibres: assembly, structure and implications for materials design. *Curr. Opin. Solid State Mater. Sci.* **8**, 151–156.
16. Yemini M, Reches M, Rishpon J, Gazit E. (2005) Novel electrochemical biosensing platform using self-assembled peptide nanotubes. *Nano. Lett.* **5**, 183–186.
17. Bewley MC, Springer K, Zhang YB, Freimuth P, Flanagan JM. (1999) Structural analysis of the mechanism of adenovirus binding to its human cellular receptor, CAR. *Science* **286**, 1579–1583.
18. Burmeister WP, Guilligay D, Cusack S, Wadell G, Arnberg N. (2004) Crystal structure of species D adenovirus fiber knobs and their sialic acid binding sites. *J. Virol.* **78**, 7727–7736.
19. Krasnykh V, Belousova N, Korokhov N, Mikheeva G, Curiel DT. (2001) Genetic targeting of an adenovirus vector via replacement of the fiber protein with the phage T4 fibritin. *J. Virol.* **75**, 4176–4183.
20. Glasgow JN, Everts M, Curiel DT. (2006) Transductional targeting of adenovirus vectors for gene therapy. *Cancer Gene Ther.* **13**, 830–844.
21. Nicklin SA, Wu E, Nemerow GR, Baker AH. (2005) The influence of adenovirus fiber structure and function on vector development for gene therapy. *Mol. Ther.* **12**, 384–393.
22. Noureddini SC, Curiel DT. (2005) Genetic targeting strategies for adenovirus. *Mol. Pharm.* **2**, 341–347.
23. Tao Y, Strelkov SV, Mesyanzhinov VV, Rossmann MG. (1997) Structure of bacteriophage T4 fibritin: a segmented coiled coil and the role of the C-terminal domain. *Structure* **5**, 789–798.
24. Frank S, Kammerer RA, Mechling D, et al. (2001) Stabilization of short collagen-like triple helices by protein engineering. *J. Mol. Biol.* **308**, 1081–1089.
25. Letarov AV, Londer YY, Boudko SP, Mesyanzhinov VV. (1999) The carboxy-terminal domain initiates trimerization of bacteriophage T4 fibritin. *Biochemistry (Mosc.)* **64**, 817–823.
26. Miroshnikov KA, Marusich EI, Cerritelli ME, et al. (1998) Engineering trimeric fibrous proteins based on bacteriophage T4 adhesins. *Protein Eng.* **11**, 329–332.
27. Stetefeld J, Frank S, Jenny M, et al. (2003) Collagen stabilization at atomic level. Crystal structure of designed (GlyProPro)(10)foldon. *Structure (Camb.)* **11**, 339–346.
28. Yang X, Lee J, Mahony EM, Kwong PD, Wyatt R, Sodroski J. (2002) Highly stable trimers formed by human immunodeficiency virus type 1 envelope

glycoproteins fused with the trimeric motif of T4 bacteriophage fibritin. *J. Virol.* **76**, 4634–4642.
29. van Raaij MJ, Mitraki A, Lavigne G, Cusack S. (1999) A triple beta-spiral in the adenovirus fibre shaft reveals a new structural motif for a fibrous protein. *Nature* **401**, 935–938.
30. Papanikolopoulou K, Forge V, Goeltz P, Mitraki A. (2004) Formation of highly stable chimeric trimers by fusion of an adenovirus fiber shaft fragment with the foldon domain of bacteriophage t4 fibritin. *J. Biol. Chem.* **279**, 8991–8998.
31. Papanikolopoulou K, Teixeira S, Belrhali H, Forsyth VT, Mitraki A, van Raaij MJ. (2004) Adenovirus fibre shaft sequences fold into the native triple beta-spiral fold when N-terminally fused to the bacteriophage T4 fibritin foldon trimerisation motif. *J. Mol. Biol.* **342**, 219–227.
32. Tabor S. (1990) Expression using the T7 RNA polymerase/promoter system. In: Ausubel FA, Brent R, Kingston RE, Moore DD, Seidman JG, Smith JA, and Struhl K, eds. *Current Protocols in Molecular Biology*. New York: Greene and Wiley-Interscience.
33. King J, Laemmli UK. (1971) Polypeptides of the tail fibres of bacteriophage T4. *J. Mol. Biol.* **62**, 465–477.
34. Schwarzer D, Stummeyer K, Gerardy-Schahn R, Muhlenhoff M. (2007) Characterization of a novel intramolecular chaperone domain conserved in endosialidases and other bacteriophage tail spike and fiber proteins. *J. Biol. Chem.* **282**, 2821–2831.
35. Jancarik J, Kim SH. (1991) Sparse matrix sampling: a screening method for crystallization of proteins. *J. Appl. Cryst.* **24**, 409–411.
36. Bergfors T. (1999) *Protein Crystallization Techniques*. International University Line, La Jolla, CA.
37. McPherson A. (1989) *Preparation and Analysis of Protein. Crystals*. Krieger, Malabar, FL.
38. McPherson A. (2004) Introduction to protein crystallization. *Methods* **34**, 254–265.
39. McPherson A. (2003) Macromolecular crystallization in the structural genomics era. *J. Struct. Biol.* **142**, 1–2.
40. Foster MP, McElroy CA, Amero CD. (2007) Solution NMR of large molecules and assemblies. *Biochemistry* **46**, 331–340.
41. Hope H. (1990) Crystallography of biological macromolecules at ultra-low temperature. *Annu. Rev. Biophys. Biophys. Chem.* **19**, 107–126.
42. Massover WH. (2007) Radiation damage to protein specimens from electron beam imaging and diffraction: a mini-review of anti-damage approaches, with special reference to synchrotron X-ray crystallography. *J. Synchrotron Radiat.* **14**, 116–127.
43. Rossmann MG. (2001) Molecular replacement–historical background. *Acta Crystallogr.* **57**, 1360–1366.
44. Ealick SE. (200) Advances in multiple wavelength anomalous diffraction crystallography. *Curr. Opin. Chem. Biol.* **4**, 495–499.
45. Dodson E. (2003) Is it jolly SAD? *Acta Crystallogr.* **59**, 1958–1965.

46. Hendrickson W. (1999) Maturation of MAD phasing for the determination of macromolecular structures. *J. Synchrotron Radiat.* **6**, 845–851.
47. Garman E, Murray JW. (2003) Heavy-atom derivatization. *Acta Crystallogr.* **59**, 1903–1913.
48. Hendrickson WA, Horton JR, LeMaster DM. (1990) Selenomethionyl proteins produced for analysis by multiwavelength anomalous diffraction (MAD): a vehicle for direct determination of three-dimensional structure. *EMBO J.* **9**, 1665–1672.
49. Doublie S, Carter C. (1992) *Preparation of Selenomethionyl Protein Crystals.* Oxford University Press, New York.
50. Taylor G. (2003) The phase problem. *Acta Crystallogr.* **59**, 1881–1890.
51. Blow D. (2002) *Outline of Crystallography for Biologists.* Oxford University Press, New York.
52. Carter C. (2003) *Methods in Enzymology Parts C and D.* Elsevier, New York.
53. Drenth J. (1999) *Principles of Protein X-ray Crystallography.* Springer-Verlag, New York.
54. McPherson A. (2002) *Introduction to Macromolecular Crystallography.* Wiley, New York.
55. McRee D. (1999) *Practical Protein Crystallography.* Academic Press, New York.
56. Rhodes G. (1993) *Crystallography Made Crystal Clear.* Academic Press, New York.
57. Morris RJ, Perrakis A, Lamzin VS. (2003) ARP/wARP and automatic interpretation of protein electron density maps. *Methods Enzymol.* **374**, 229–244.
58. Murshudov GN, Vagin AA, Dodson EJ. (1997) Refinement of macromolecular structures by the maximum-likelihood method. *Acta Crystallogr.* **53**, 240–255.
59. Lovell SC, Davis IW, Arendall WB 3rd, et al. (2003) Structure validation by C-alpha geometry: phi,psi and Cbeta deviation. *Proteins* **50**, 437–450.
60. Holm L, Sander C. (1999) Protein folds and families: sequence and structure alignments. *Nucleic Acids Res.* **27**, 244–247.
61. Boudko SP, Strelkov SV, Engel J, Stetefeld J. (2004) Design and crystal structure of bacteriophage T4 fibritin NCCF. *J. Mol. Biol.* **339**, 927–935.
62. DeLano WL. (2002) *The PyMOL Molecular Graphics System.* DeLano Scientific, San Carlos, CA. Available at: http://www.pymol.org.

3

The Leucine Zipper as a Building Block for Self-Assembled Protein Fibers

Maxim G. Ryadnov, David Papapostolou, and Derek N. Woolfson

Summary

Nanostructured materials are receiving increased attention from both academia and industry. For example, the fundamental understanding of fiber formation by peptides and proteins both is of interest in itself and may lead to a range of applications. A key idea here is that the folding and subsequent supramolecular assembly of the monomers can be programmed within polypeptide chains. Thus, with an understanding of so-called sequence-to-structure relationships for these peptide assemblies, it may be possible to design novel nanostructures from the bottom up that exhibit properties determined by, but not characteristic of, their component building blocks. In this respect, the α-helical leucine zipper presents an excellent place to start in the rational design of ordered nanostructures that span several length scales. Indeed, such systems have been put forward and developed to different degrees. Despite their apparent diversity, they employ similar assembly routes that can be compiled into one basic methodology. This chapter gives examples and provides methods of what can be achieved through leucine zipper-based assembly of fibrous structures.

Key Words: Fibers; α-helical leucine zipper; hierarchical self-assembly; nanostructures; peptide design; supramolecular chemistry.

1. Introduction

Fundamental studies in designing novel nanostructures, such as protein fibers, advance our understanding of protein folding, assembly, and chemistry. Potential applications in this area include the fabrication of scaffolds for cell growth in culture and templates for the controlled and directed assembly of inorganic materials *(1)*. Peptide-based fibers and matrices can be, and indeed have been, assembled from a variety of protein-folding motifs as well as from artificial peptide amphiphiles *(2)*. The underpinning concept here is that rules and guides for assembly processes are programmed into such motifs, and that

From: *Methods in Molecular Biology, vol. 474: Nanostructure Design: Methods and Protocols*
Edited by: E. Gazit and R. Nussinov © Humana Press, Totowa, NJ

this potential introduces nanoscale features often reflecting the chemistry of the motif into the targeted higher-order assemblies *(3,4)*.

This chapter focuses on peptide-based fibrous assemblies *(5)* and the use of one peptide-folding motif in particular. This is the leucine zipper (LZ), which is commonly found in nature and is reasonably well understood *(6)*. The LZ is one of the most straightforward elements for protein–protein interactions known *(7)*. It comprises two polypeptide chains that, by wrapping around one another, form a rope-like helical bundle. There are excellent sequence-to-structure rules *(8)* that guide the folding, supramolecular assembly, and stability of LZs. For instance, a key feature of LZ sequences is the heptad repeat in which hydrophobic (H) and polar (P) amino acids are ordered into an HPPHPPP pattern. This pattern dictates both the folding of LZ strands and their subsequent association into bundles. LZ peptides that fold to stable structures are usually required to be about three to five heptads long, which is readily accessible to modern-day peptide synthesis. Finally, LZ peptides have proved to be ideal candidates for engineering nanostructures *(9–12)* because of a direct relationship between sequence and scale, namely, one folded heptad repeat meters out about 1 nm of structure (**Fig. 1**). Combined, these features of the LZ facilitate the rational design of peptide- and protein-based nanoscale fibers *(3)*. These include straight and kinked fibers *(13)*; branched fibers *(14)* and polygonal matrices *(12,15)*; fibers decorated with functional peptides *(16)*, inorganic materials, dyes *(17)*, and even whole proteins *(16)*; and finally fibers that respond to changes in their environment *(18)*.

As this is a practical guide to designing and fabricating fibrous nanostructures, we focus necessarily on our own experiences with a system that we refer to as the self-assembled peptide fiber (SAF) system *(19)*.

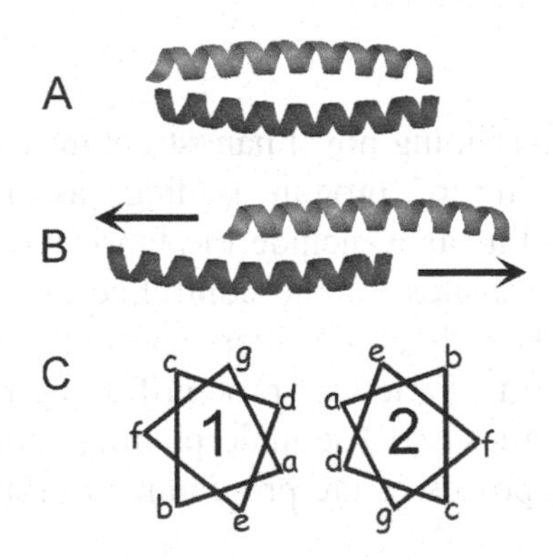

Fig. 1. (**A**) The two-heptad leucine zipper peptides. (**B**) An axially staggered heterodimer with "sticky ends" to promote longitudinal assembly into a contiguous superhelix (indicated by arrows). (**C**) The heptad repeat, ***abcdefg***, signature of LZ sequences configured onto helical wheels. (The helices for **A** and **B** were taken from PDB entry 2ZTA.)

2. Materials

2.1. Peptide Synthesis

2.1.1. 9-Fluorenylmethoxycarbonyl Solid-Phase Peptide Synthesis

1. Solvents: Anhydrous dimethylformamide (DMF), piperidine, pyridine, diisopropylethylamine (DIPEA), *N*-methylpyrrolidone (NMP), dichloromethane (DCM), trifluoroacetic acid (TFA), triisopropyl silane (TIS), 1,2-ethanedithiol (EDT) (Rathburn or Fluka).
2. 9-Fluorenylmethoxycarbonyl (Fmoc) amino acid derivatives and hydroxybenzotriazole (HOBt) uronium salts (*O*-benzotriazole-*N,N,N′,N′*-tetramethyluronium tetrafluoroborate [TBTU], *O*-benzotriazole-*N,N,N′,N′*-tetramethyluronium hexafluorophosphate [HBTU], or *O*-(7-azabenzotriazole-1-yl)-*N,N,N′N′*-tetramethyluronium hexafluorophosphate [HATU]) from Merck Biosciences (Nottingham, UK) or Applied Biosystems (Warrington, UK); Pd(0) from Sigma.
3. Resins (100–200 mesh, 0.2–0.6 mmol/g): Wang or polyethylene glycol-polystyrene (PEG-PS) and peptide amide linker-polystyrene (PAL-PS) or Rink Amide 4-methylbenzhydrylamine (MBHA) for carboxyl-free and amide peptides, respectively (Merck Biosciences or Applied Biosystems).
4. Deprotection reagent: 20% piperidine in DMF.
5. Coupling mixture: 0.5*M* HOBt uronium salt and 1*M* DIPEA in DMF.
6. Cleavage cocktails: TFA/TIS/water (95:2:5:2.5) (cocktail 1) and TFA/TIS/EDT/water (94.5:1:2:5:2.5) (cocktail 2) for cysteine-containing peptides.
7. Methyltrityl (Mtt) deprotection reagent: DCM/TFA/TIS (94:1:5) (reagent M) for Fmoc-Lys(Mtt)-OH.
8. Allyl deprotection reagent: Pd(0) (3 Eq) in $CHCl_3$/AcOH/NMM (37:2:1) under Ar.
9. Capping mixture: 5% acetic anhydride and 6% pyridine in DMF.

2.1.2. Conjugation and Ligation

1. Fluorescein-5-maleimide, tetramethylrhodamine-5-maleimide, succinimidyl esters of 5- (and 6-) carboxyfluorescein and 5- (and 6-) carboxytetramethylrhodamine from Molecular Probes. α-bromoacetic acid, dithiothreitol (DTT), and tris(2-carboxyethyl)phosphine (TCEP) from Sigma.
2. Buffer 1 (*see* **Note 1**): 0.1*M* 3-(*N*-morpholino)propanesulfonic acid (MOPS; pH 7.5–8.2) or 0.1*M* *N*-2-hydroxyethylpropane sulfonic acid (EPPS; pH 7.9–8.2), 2 m*M* ethylenediaminetetraacetic acid (EDTA).
3. Buffer 2: 8*M* urea, 0.6*M* Tris-HCl, 5 m*M* EDTA (pH 8.4–8.6).
4. Buffer 3: 50 m*M* Tris-HCl, 50 m*M* NaCl (pH 7.0–7.4).
5. PD-10 columns (Sephadex G-25 M) from Amersham Biosciences.

2.1.3. Reversed-Phase High-Performance Liquid Chromatography and Mass Spectrometry

1. Mobile phase: 5% (buffer A) and 95% (buffer B) aqueous CH_3CN containing 0.1% TFA.

2. Stationary phase: Vydac C18 and C8 reversed-phase columns. Analytical (5 μm, 4.6 mm internal diameter [id] × 250 mm) and semipreparative (5 and 10 μm, 10 mm id × 250 mm).
3. High-performance liquid chromatographic (HPLC) system: Models PU-980, PU-2086 (Jasco, Japan).
4. TofSpec E MALDI (matrix-assisted laser desorption/ionization) spectrometer (Micromass Ltd., UK), 4700 Proteomics Discovery System (Applied Biosystems).

2.2. Fiber Assembly

For the incubation buffer, use 10 m*M* MOPS (pH 7.4–7.6).

2.3. Fiber Decoration

1. Incubation buffer: 10 m*M* MOPS (pH 7.4–7.6).
2. Streptavidin 5 and 10 nm colloidal gold-labeled (streptavidin-gold *Streptomyces avidinii*, 0.4 mL in 0.01*M* phosphate-buffered saline [PBS], pH 7.4, containing 0.02% polyethylene glycol [PEG], 20% glycerol, and 15 m*M* sodium azide) from Sigma.
3. Anti-FLAG BioM2 (1 mg/mL in 50% glycerol, 10 m*M* sodium phosphate, pH 7.4, 150 m*M* NaCl containing 0.02% sodium azide) from Sigma.
4. Biotin from Sigma.

2.4. Polar Assembly

1. Incubation buffer: 10 m*M* MOPS (pH 7.4–7.6).
2. 5- (and 6-) carboxyfluorescein and 5- (and 6-) carboxytetramethylrhodamine from Molecular Probes.

2.5. Spectroscopy

2.5.1. Circular Dichroism

1. Quartz cuvettes (0.1–1 mm; Starna, UK).
2. Jasco-J 715, 810 spectropolarimeters fitted with Peltier temperature controllers.

2.5.2. Fourier Infrared

For Fourier infrared, use a thermostated Bruker Tensor 27 spectrometer fitted with a BioATR-II cell with a ZnSe crystal.

2.6. Microscopy

2.6.1. Confocal Fluorescence Microscopy

1. Glass slides and coverslips from Fisher.
2. Bio-Rad MRC600 confocal microscope fitted with a ×60 oil immersion lens, a krypton/argon mixed-gas laser, a dual-excitation filter, and K1 (520 nm, fluorescein) and K2 (585 nm, rhodamine) filter-block set.
3. The COMOS software (Bio-Rad) for image collection.

2.6.2. Wide-Field Fluorescence Microscopy

1. Glass slides and coverslips from Fisher.
2. Carl Zeiss Vision wide-field microscope fitted with ×10 and ×40 LD lenses, a krypton/argon mixed-gas laser, a dichromic mirror, and two filters: 520 nm for fluorescein and 585 nm for rhodamine.
3. ORCA ER camera for data collection, Carl Zeiss Vision image software for data analysis.

2.6.3. Transmission Electron Microscopy

1. Specimen grids: Carbon film on 3.05-mm copper grid, 400 lines/inch (Agar Scientific, UK).
2. Self-closing fine-tip Dumont tweezers NOC (Agar Scientific).
3. Stains: Uranyl acetate, ammonium molybdate (Agar Scientific); 1% (w/v) aqueous solutions, filtered through a 0.2-μm pore size Minisart units (Sartorius, UK) and stored at 5°C.
4. JEOL JEM 1200 EX transmission electron microscope (tungsten filament operated at 120 kV), fitted with a MegaViewII digital camera, using Soft Imaging Systems GmbH analySIS 3.0 image analysis software.
5. The ImageJ software for image analysis and measurements of fiber dimensions (http://rsb.info.nih.gov/ij/index.html).

2.7. Fiber Diffraction With Partially Aligned Samples

2.7.1. Partial Alignment of Fibers: Stretched Frame Procedure

1. Borosilicate thin-walled capillaries (1.5-mm outer diameter, 1.17-mm inner diameter) (Harvard Apparatus, UK).
2. Cutting stone (Hampton Research, USA).
3. Standard beeswax: Break into pieces, transfer the pieces into a glass beaker.
4. Mounting clay (Hampton Research).
5. Petri dishes, 90-mm diameter, Parafilm® (Fisher Scientific, UK).

2.7.2. Fiber Diffraction

1. Rigaku CuKα rotating anode X-ray source (wavelength 1.5418 Å) and R-AXIS IV image-plate detector (Rigaku, Japan).
2. Instrument control and data collection through the CrystalClear™ software (Rigaku).
3. Data handled and converted to image files with MOSFLM (http://www.mrc-lmb.cam.ac.uk/harry/mosflm/).

3. Methods

3.1. Basic Design Rules for LZ Sequences

Leucine zippers comprise two helices that pack intertwined or bundled in a rope-like fashion (**Fig. 1A,B**). The hallmark of LZ sequences is the heptad repeat—$(abcdefg)_n$—in which the first (***a***) and fourth (***d***) sites are usually hydrophobic (**Fig. 1C**). The others are normally occupied by polar or small amino acids to provide water-soluble surfaces. To specify the dimer in LZ designs, the (***a***) and (***d***) residues are preferably made isoleucine and leucine, respectively. These form the core of the bundle. Residues at (***e***) and (***g***) can be used to further cement the core by electrostatic interactions and hence are often made oppositely charged; lysines and glutamates are commonly used amino acids at the sites. The other sites are taken by polar or small amino acids, usually glutamine and alanine, which have high α-helical propensities. Given these rules, sequences as short as three to six heptads can be designed to yield stable LZs. Inclusions of tyrosines or tryptophans at one of the (*f*) positions are often used as chromophores for accurate concentration measurements.

3.2. LZs as Nanoscale Building Blocks

Morphological and functional properties of LZ-based nanostructures can be rationally programmed into LZ sequences ***(2,3)***. The sequences are also highly accommodating of various molecular topologies that are often required by supramolecular hierarchies ***(15)***. Such an architectural flexibility is particularly beneficial for responsive materials or in applications for which nanometer precision is the key requirement. For example, nanoscale fibrillar networks able to self-arrange in three-dimensional (3D) cell culture are attractive for tissue engineering ***(20)***, whereas networks of metal nanoparticles separated at distances of as few as several nanometers may find use in molecular electronics ***(21)***. Intrinsically reversible, LZ assemblies can also be used as self-assembling particles ***(9)***, reactors, ***(10)*** or switches ***(22)***. Altogether, this considerably extends the assortment of functional nanostructures and places high demands on synthesis.

3.3. Synthesis of LZ Blocks

The synthesis of LZ sequences is relatively straightforward and can be achieved by solid-phase peptide synthesis (SPPS). Automated SPPS has been developed as a technique of choice for both peptide chemists and nonspecialists. Some cases, however, require SPPS to be supplemented with subsidiary capabilities based on more specific and efficient synthetic procedures. Examples include the need for long peptides to avoid structuring of peptide chains synthesized on resin or engineering specific molecular topologies that cannot be tackled using conventional SPPS protocols ***(23)***. Semisynthetic methods developed to date, such as conjugation, fragment condensation, enzymatic coupling, and chemoselective ligation, have been proposed as approaches complementary to SPPS *(24)*. Chemoselective ligation, which is typically performed in aqueous buffers, has proved remarkably effective for postsynthetic assembly of geometrically varied constructs. Compatibility of such molecular grafting with folding and assembly of peptides has been supported by a number of LZ designs ***(10,13,14,16,25,26)***. Although the majority of the reported examples are empirical, the properties are unique in peptide assembly and have not been demonstrated for other peptide elements to the same extent *(2)*.

3.4. Self-Assembled Fibers

Natural LZ structures fold to give in-register assemblies. Thus, it can be envisaged that the staggering of LZ strands would lead to longitudinal filamentous structures reminiscent of natural fibrin, vimentin, or tropomyosin *(27)*. Indeed, LZ-based fibers were engineered using an axially staggered *(28)* or sticky-ended *(19)* assembly (**Fig. 1B**). These effectively shift the adjacent helices, which can be rationally designed. The shift can vary from a few amino acids *(28)* to some heptads *(19,29,30)* as in SAF *(19)*, one of the first examples of the "shift-based" assembly (**Fig. 1B**). Axially staggered helices are mainly arranged by placing complementary charged residues at ***e*** and ***g*** positions in heptad repeats along the sequence *(19,28)*. Specifically, the SAF arises from the coassembly of two 28-mers (standards) that form a staggered heterodimer with oppositely charged sticky ends (**Fig. 1**). The resulting dimers, which are 2 nm wide and 4 nm long, propagate longitudinal assembly and lateral thickening to yield fibers 70 ± 20 nm thick and tens of microns long *(13,29)* (**Fig. 2A,B**).The fibers exhibit a conserved nanoscale order as judged by wide-angle X-ray fiber diffraction and positive-stain transmission electron microscopy (TEM), which reveals regularly striated surface ultrastructure of the fibers (**Fig. 2D**). The striations run perpendicular to the long fiber axis and are separated at the lengths of the individual peptides. The SAF system works as a binary mixture; that is, each peptide is individually inactive (unfolded) and assembles only with its companion *(19)* (**Fig. 1B,2**). The binary design allows for the incorporation of additional partners, *specials*, that can introduce a functionality not displayed by the standard assemblies. Given that LZs are rope-like, the termini of the helices are easily accessible for derivatization or fusions; that is, the assembly of staggered helices can be directed by nonlinear LZ conjugates *(2)*. The special can be a modified standard or a hybrid construct complementary to standards. Specials incorporated into the SAF system in this way can be used to follow the assembly on the nanoscale *(17)* and to control *(13–15)* or decorate *(16)* the mesoscopic architecture of the fiber (**Fig. 3**).

3.5. Peptide Synthesis

3.5.1. Fmoc Solid-Phase Synthesis

All peptides are synthesized on solid phase on 0.1-m*M* scale using standard Fmoc/^{t}Bu, Fmoc/^{t}Bu/Allyl, or Fmoc/^{t}Bu/Mtt solid-phase protocols with a hydroxybenzotriazole uronium salt/DIPEA as coupling reagents (*see* **Notes 2** and **3**). Fourfold excess of amino acids and coupling reagents are used to ensure efficient coupling.

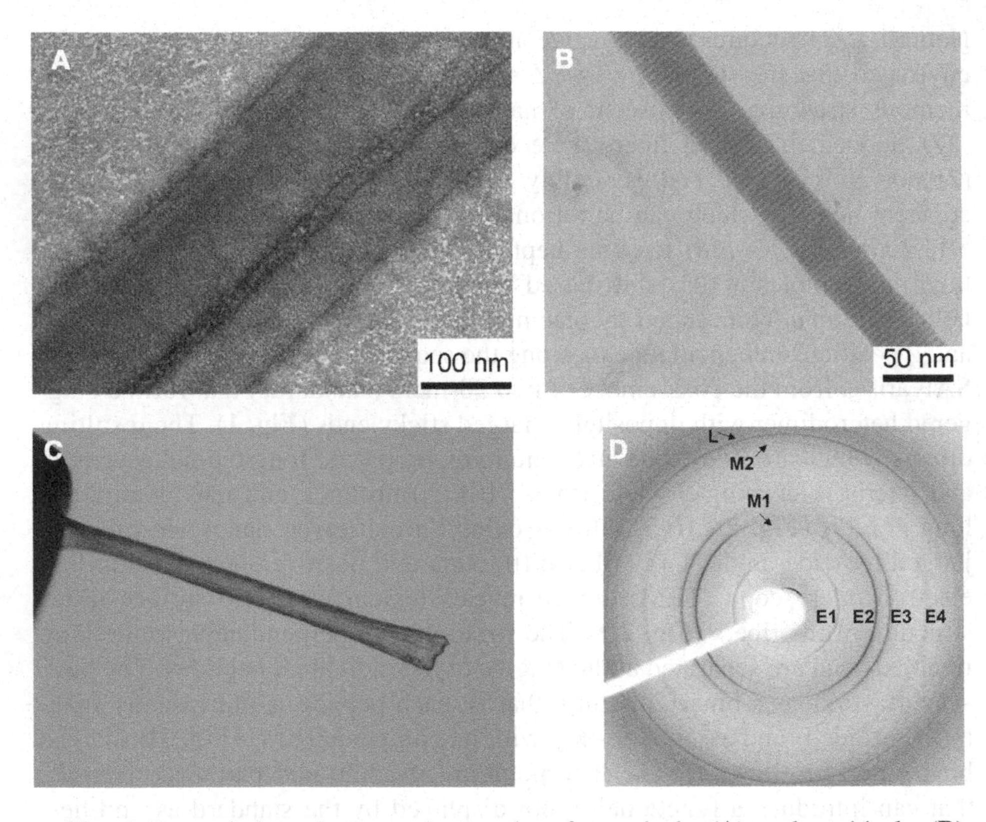

Fig. 2. Transmission electron micrographs of negatively (**A**) and positively (**B**) stained standard self-assembled peptide fibers (SAFs). (**C**) Partially dried and aligned fiber stem prepared using the stretched-frame procedure. (**D**) Fiber diffraction pattern of a partially aligned fiber sample; the meridional diffractions M1 and M2 correspond to the heptad repeat length (1.03 nm) and the helical rise per turn (5.14 nm), respectively. This indicates the assembly parallel to the fiber main axis. The equatorial diffractions E1–E4 indicate that the leucine zipper (LZ) superhelices are packed hexagonally across the width of the fiber (the corresponding real-space distances are E1, 15.6 nm; E2, 9.8 nm; E3, 7.8 nm; E4, 5.93nm) *(33)*.

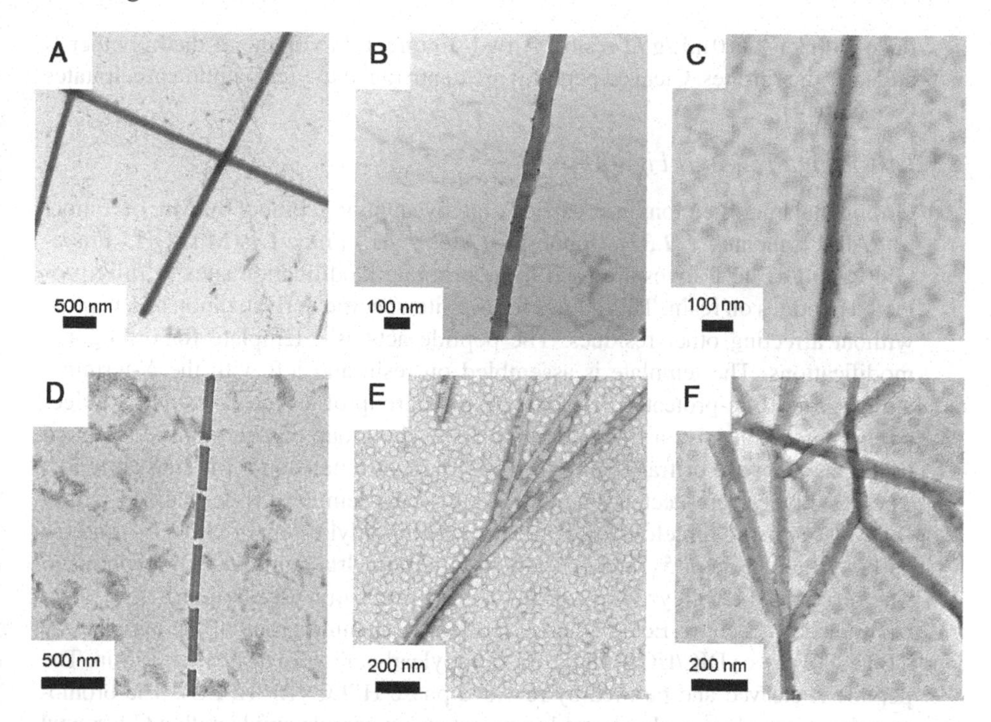

Fig. 3. Transmission electron micrographs of standard (**A**) and special (**B–F**) self-assembled peptide fibers (SAFs). (**B**) Biotinylated SAFs coated with streptavidin labeled with 10-nm gold particles. (**C**) FLAG-modified SAFs coated with biotinylated anti-FLAG antibodies and subsequently with streptavidin labeled with 5-nm gold particles. Segmented (**D**), branched (**E**) fibers and fibrillar matrices (**F**) assembled from standard and special SAFs. (Reproduced from **refs.** ***15*** and ***16*** with permission from American Chemical Society.

1. Mtt deprotection: Removal of Mtt groups is performed by washing with reagent M (10 mL/g of resin) three times for 3 min followed by washing three times with DMF or 5% pyridine in DMF and then DMF.
2. OAl/Aloc-deprotection: Removal of allyl-based groups is catalyzed by Pd(0) (15 mL/g of resin) for 2 h with gentle agitation followed by washing with 0.5% DIPEA in DMF, 0.5% sodium diethyldithiocarbamate (DEDC) in DMF, and DMF two times each.
3. Capping: Acetylation and amidation of peptides are usually referred to as *capping*. Partial capping, in which only one of the two is performed, is equally common. Amidation is done using Rink Amide MBHA or PAL-PS resins. Acetylation is the final step in the synthesis: The resin is kept under the capping mixture (10 mL/g of resin) for 30 h, which is followed by washing three times with DMF.
4. Postsynthetic TFA cleavage: The resin is kept under cleavage cocktail 1 or 2 (15 mL/g of resin) for 2–3 h, then removed by filtration under vacuum and washed

once with TFA (10 mL/g of resin). A two- to threefold volume of diethyl ether is added to the filtrates. Cleaved peptides are centrifuged to yield peptide precipitates (*see* **Note 4**).

3.5.2. Conjugation and Ligation

1. Orthogonal modifications are carried out by applying Fmoc/^{t}Bu/Mtt or Fmoc/^{t}Bu/Allyl schemes ***(31,32)***. Fmoc-Lys(Aloc)-OH, Fmoc-Lys(Mtt)-OH, Fmoc-Glu(OAl)-OH derivatives are used to incorporate modification sites in fully protected peptides on resin. This allows for the site-specific derivatization of a peptide without affecting other residues. The peptide acts as a template for orthogonal modifications. The template is assembled on resin and left with the *N*-terminal amino group Boc-protected. The Mtt or Aloc group of a lysinyl residue is selectively removed to give a free ε-amino group. The group can be modified or used to initiate stepwise or fragment orthogonal peptide extensions ***(14)*** (*see* **Note 5**).
2. The selectively deprotected ε-amino group of the template is derivatized with a fluorophore using threefold excess of the succinimidyl esters of either 5- (and 6-) carboxyfluorescein or 5- (and 6-) carboxytetramethylrhodamine in DMF on resin. The reactions are catalyzed by DIPEA (6 Eq) for 30 min (*see* **Note 6**).
3. α-Bromoacetic acid (4 Eq) is coupled to the free ε-amino group of the template by TBTU/DIPEA or DIC/HOBt (DIC, diisopropylcarbodiimide) (4 Eq) for 30 min. The peptide is cleaved and purified by reversed-phase HPLC (RP-HPLC). The bromoacetylated peptide is further ligated to an extension peptide amide with a C-terminal cysteine (*see* **steps 4** and **5**).
4. The ligation is carried out with threefold excess of the extension peptide for 2–6 h in buffer 1 containing TCEP (10 Eq) at 0.5 m*M* peptides (1 mL) and monitored by MALDI-TOF (time of flight) (2-μL aliquots).
5. The extension peptide (1 mg) is dissolved in the denaturing urea buffer 3 (0.5 mL), to which 2-mercaptoethanol (5 μL) is added, and incubated for 30–120 min at 40°C. The obtained solution is mixed with the bromoacetylated peptide (3 mg in 1 mL water) and incubated for 30 min (*see* **Note 7**).
6. The use of fluorescein-5-maleimide and tetramethylrhodamine-5-maleimide allows for the derivatization of the template peptide in water, in which case the peptide is mutated to have a single cysteine residue to provide an orthogonal site. A maleimide sample is dissolved in buffer 3 at 0.5–1 mg/mL. Mix 0.8 mL of the solution with 0.2 mL of the template peptide (1 m*M*) in buffer 3 or 2 containing TCEP (5 Eq) for 1 h (*see* **Notes 8** and **9**).
7. Conjugates are eluted with water (pH 7.0, **Subheading 3.5.3.**) or 1*N* acetic acid (**Subheading 3.5.3.**) through pre-equilibrated PD-10 columns.

3.5.3. RP HPLC and Mass Spectrometry

Peptides are purified using 45–60 min linear 20–60% or 40–80% buffer B gradients (flow rate of 4.5 mL/min). The same gradients are used for analytical HPLC with a flow rate of 1 mL/min. Eluted peptides are lyophilized and analyzed by MALDI-TOF.

3.6. Fiber Assembly

1. Standard: Stock solutions of both peptides are prepared in water at 1.5–3 m*M*. Aliquots are diluted to final concentrations of 50–200 μ*M* (200–300 μL) in each peptide by 10 m*M* MOPS (pH 7.0–7.4) (*see* **Note 10**). The obtained solutions are incubated at a set temperature (5, 20, 36°C) over 6–20 h before TEM analysis.
2. Specialist: Aliquots of specialist peptides stocks (1.5–3 m*M*) are mixed with those of standard peptides in 0.001–1 ratios and incubated over 6–20 h in 10 m*M* MOPS (pH 7.0–7.4) prior to TEM analysis.

3.7. Fiber Decoration

3.7.1. Protein-Binding Assays

1. The 200 μL fiber solutions prepared from either SAF peptide with its biotinylated, FLAG-modified, or streptag-modified (orthogonal extension) companion, both biotinylated peptides (each at 100 μ*M*) are incubated for 16–24 h at 20°C in MOPS (pH 7.0–7.4).
2. Streptavidin-gold conjugates (SGCs) are diluted 10 times with 10 m*M* MOPS (pH 7.0–7.4) containing phosphorocholine chloride (10 m*M*) to optimal concentrations of stock solutions of 0.05 m*M* (calculated from protein absorption at 280 nm and the extinction coefficient of 41,820 M^{-1} cm^{-1} [http://ca.expasy.org/tools/protparam.html]). Diluted conjugates are equilibrated for 30 min at 20°C before adding to appropriate fiber preparations preincubated for 30–45 min with 0.1% polyoxyethylenesorbitan monolaureate (Tween-20®). This is followed by either of the following routes:
   a. A designated aliquot (2–100 μL) of the diluted SGCs is mixed with a fiber preparation *in situ* to incubate for 30–60 min (*see* **Note 11**).
   b. Add 2 μL of anti-FLAG antibody to FLAG-modified fibers and incubate at 4°C for 8–16 h. Add SGCs (*see* **step 2a**).

3.7.2. Binding Competition Assays

1. Biotin washing: Preparations of naked and SGC-decorated fibers in SGCs (5–20 μ*M*) are washed with MOPS buffer containing biotin (0.3–0.5 m*M*) (*see* **Note 12**).
2. Biotin preincubation: Biotinylated and naked fibers are incubated in biotin-containing MOPS buffer (0.3–0.5 m*M*) prior to adding SGCs (5–20 μ*M*) (*see* **Note 13**).
3. Streptag-fused fibers: Streptag (FSHPQNT) has a lower affinity for streptavidin (~78 m*M*) than biotin (~1 f*M*). Recruitment of SGCs to streptag-coated fibers is similar to that for naked fibers.

3.7.3. Preparation for TEM Visualization

1. Following incubation, an 8-μL drop of a peptide solution is applied to a carbon-coated copper specimen grid held by a pair of self-closing tweezers and dried with a filter paper after 30–60 s (*see* **Note 14**).

2. To wash off unspecifically bound SGCs, the grid is placed upside down on a 20-μL drop of the MOPS buffer for 3 min. This is repeated three times. When higher concentrations of SGCs are used, an additional three washings (3 min each) can be performed. The grid is dried and examined under the microscope.

3.8. Polar Assembly

1. Mix 100-μL aliquots of unlabeled peptides (100 μ*M* each) with a fluorescein-labeled peptide (0.1 μ*M*) in MOPS (pH 7.0) to mature for 16 h at 20°C. Rhodamine-labeled peptide (0.1 μ*M*) is added to the mixture, and the obtained solution is left for another 24 h before visualization using fluorescence microscopy.
2. A 100-μL aliquot of unlabeled peptides SAF-p1 and -p2 (100 μ*M* each) is initially incubated for 24 h at 20°C. The sample is then incubated for another 24 h with a rhodamine-labeled SAF-p2 peptide (1 μ*M*), followed by a further 24 h incubation with fluorescein-labeled SAF-p1 peptide (1 μ*M*). The resulting preparation is examined by fluorescence microscopy. The experiment is also done in the opposite order of adding labeled peptides.
3. Free 5- (and 6-) carboxyfluorescein and 5- (and 6-) carboxytetramethylrhodamine are used as controls.

3.9. Spectroscopy

3.9.1. Circular Dichroism Spectroscopy

1. Circular dichroism (CD) spectra: Recorded for 50–200 μ*M* peptide solutions at pH 7.0–7.4 (MOPS, HEPES, or phosphate buffer), at 5–20°C. Data points are recorded at 1-nm intervals using a 1-nm bandwidth and 4- to 16-s response times. Following baseline correction, ellipticities in millidegrees are converted to molar ellipticities (deg cm^2 dmol res^{-1}) by normalizing for the concentration of peptide bonds. Any contribution to the CD spectra from special peptides is neglected; for example, the concentration of peptide bonds for 100 μ*M* is taken as 54 μ*M* ($2 \times (28 - 1) \times 100$) for all spectra.
2. Thermal denaturation: Data points for thermal unfolding curves are collected through 1°C/min ramps using a 2-nm bandwidth, averaging the signal for 8–16 s at 1°C intervals.

3.9.2. Fourier Transform Infrared Spectroscopy

Spectra (200–400 interferograms) are recorded in the range 400–4000 cm^{-1} with a spectral resolution of 2 nm^{-1}. Acquired data are processed using proprietary software (OPUS).

3.10. Microscopy

3.10.1. Confocal Fluorescence Microscopy

1. The matured fluorescein-labeled fibers are visualized directly or after being mixed with a rhodamine-labeled peptide using the dual-excitation filter.

2. A 6-μL drop of a peptide solution is deposited on a microscope slide; the slide is covered by a glass slip. Fibers are visualized with a ×60 oil immersion lens using either a set of dual filters (blue and yellow for images if both fluorophores are present) or the relevant filter individually if only one fluorophore is present.
3. Collect 15 images of the same view and average to 1 by Kalman averaging. The averaged images are converted to the red-green-blue (RGB) TIFF (tagged imaged file format) files. For dual-filter images, left-hand 384 × 512 pixels are adjusted to allow the contrast to cover the full range and then changed to green. Right-hand 384 × 512 pixels are adjusted to the full-contrast range and changed to red. The two halves of the image are overlaid and mixed to create the final, false-color image. For single-filter images, the entire image is adjusted to cover the full range and changed to a relevant color.

3.10.2. Wide-Field Fluorescence Microscopy

1. The matured fluorescein-labeled fibers are visualized directly or after mixing with a rhodamine-labeled peptide.
2. A 6-μL drop of a peptide solution is deposited on a microscope slide; the slide is covered by a glass slip. Fibers are visualized with a ×40 lens; the DIC imaging mode is used to optimize the focus. Fluorescence data are collected with the relevant excitation wavelength.
3. The images are processed, false-color red (rhodamine) or green (fluorescein), using Carl Zeiss Vision imaging software before overlaying.

3.10.3. Transmission Electron Microscopy

1. An 8-μL drop of the fiber samples is applied onto a specimen grid held by a pair of self-closing tweezers and dried with a filter paper after 30–60 s (*see* **Notes 14** and **15**).
2. An 8-μL drop of a stain solution (1% uranyl acetate or 2% ammonium molybdate) is deposited onto the specimen grid and dried with a filter paper after 30–60 sec (*see* **Note 16**).
3. Specimen grids are allowed to air dry for 15 min before observation.
4. Digital images are collected with the MegaViewII digital camera as RGB TIFF files.
5. Image files are opened in ImageJ. The software for distance measurements is calibrated with the image scale bar as a reference. The fast Fourier transform (FFT) is calculated to characterize any repeated motif, such as orthogonal striations (**Fig. 2**).

3.11. Fiber Diffraction With Partially Aligned Samples

3.11.1. Partial Alignment of Fibers: Stretched-Frame Procedure

1. Capillaries are cut in 2.5-cm sections using the cutting stone.
2. The beeswax is melted in the beaker using a benchtop hot plate (marks 3–4 are usually enough with most stir plate/hot plate units). One extremity (typically 1–2 nm) of each capillary section is dipped into the melted wax and cooled. The operation is repeated to create a flat surface on the end of the capillary. The capillary is stored at ambient temperature.

3. The waxed capillaries are mounted on small pieces of the clay at the bottom of a Petri dish; the waxed ends should face each other, 1–2 mm apart, with both capillaries precisely aligned.
4. A 10-μL drop of the fiber solution is placed between the waxed ends of aligned capillaries. Depending on the system, either fresh peptide mixture or grown fiber solutions can be used.
5. A Petri dish is sealed with a strip of Parafilm and stored in an incubator set at the required temperature. The fiber solution is air dried within 48 h to yield a stem of partially dried fibers.
6. If the resulting dried stem links both capillaries, one of the clay-mounted capillaries is carefully moved back. The stem will break at the point of contact between sample and wax to give a partially dried fiber stem pointing perpendicularly from the wax support.

3.11.2. Fiber Diffraction

1. The sample prepared as described in **Subheading 3.11.1**. is mounted onto a goniometer head. The dried fiber stem is aligned and focused with the camera so that the fiber axis is oriented vertically. Only use the cryostream with samples that have been prepared at 5°C.
2. The image plate is set to a 300-mm sample to the detector distance. One single frame is collected with a 0.5° width over 10 min.
3. The intensity of diffraction as a function of the angle 2Θ is integrated using the CrystalClear program. The distances are obtained using Bragg's law: $d = \lambda/(2 \sin\Theta)$, with $\lambda = 1.5418$.
4. Data files are imported into MOSFLM and saved as TIFF files.

4. Notes

1. All aqueous solutions are based on filtered (0.22 μ*M*) ultrapure water with a resistivity of 18.2 MΩ.
2. Syntheses are carried out either manually or using a synthesizer. Various synthesizers are available commercially from Protein Technologies, CEM, Advanced Chemtech, and Applied Biosystems.
3. High-load resins (≥0.6 mmol/g resin) are preferred for synthesis scales greater than 0.1 mmol.
4. Cleavage cocktail components can be either mixed prior to adding to the resin or added separately. *Note*: Scavengers (TIS, EDT, water) are to be added before TFA. Arginine-containing peptides require prolonged or repeated cleavage when Fmoc-Arg(Pbf)-OH is used.
5. β-Alanines or ε-aminohexanoic acids can be used to extend the lysine ε-amino group prior to conjugations. These act as spacers and considerably improve yields.
6. Additional washings with DMF, DCM, or acetone may be required to remove unreacted fluorophores from the resin, particularly before cleavage.
7. Cys-based conjugation reactions are advised to be carried out in inert atmosphere (under nitrogen or argon) to avoid oxidation of thiol groups of cysteines.

8. Thiol groups of cysteine are nucleophilic enough to specifically react with maleimides in the presence of primary amines at neutral pH 7.0–7.4. Higher pH increases the probability of side reactions with the amines. If DTT is used, it should be removed by dialysis prior to conjugation.
9. Concentrations and degrees of labeling for the fluorophore-labeled peptides are determined using the following equations:

$$\text{Concentration (fluoresceuin)} = ((A_{280} - ((A_{494} \times 0.2)) \times \text{Dilution factor}) / 1280$$

$$\text{Concentration (rhodamine)} = ((A_{280} - ((A_{555} \times 0.3)) \times \text{Dilution factor}) / 1280$$

$$\text{Degree of labeling (fluoresceuin)} = (A_{494} \times \text{Dilution}) / 68{,}000 \times \text{Concentration}$$

$$\text{Degree of labeling (rhodamine)} = (A_{555} \times \text{Dilution}) / 65{,}000 \times \text{Concentration}$$

10. Both buffers and stock solutions have to be fresh to ensure reproducibility. The SAF peptides tend to aggregate in aqueous buffers within a week.
11. Decoration occurs within 30–60 min, and longer incubations do not improve fiber coatings.
12. Using this method may lead to an increased background, which complicates adequate analysis.
13. Binding to the biotinylated fibers is negligible in comparison to that for SGCs. Comparable background levels are observed for both naked and biotinylated fibers.
14. To dry, gently apply a filter paper against the edge of the grid without touching the drop.
15. Excessive amounts of free peptides and buffer (susceptible to interact with the stain contributive to a strong background) can be washed off the grid with water. Each wash is followed by drying.
16. When stain gathers around the edges of fibers, the observations should be referred to as negative staining. It is possible to remove this excess of stain by copious washings with water. Areas where the stain is still present would correspond to portions of the sample where the heavy atoms interact with the moieties on the fiber. This observation is referred to as positive staining.

References

1. Zhang S. (2003) Fabrication of novel biomaterials through molecular self-assembly. *Nat. Biotechnol.* **21**, 1171–1178.
2. Ryadnov MG, Woolfson DN. (2007) Self-assembling nanostructures from coiled coil peptides. In: Mirkin CA, Niemeyer CM, eds. *Nanobiotechnology II*. Weinheim: Wiley-VCH; pp. 17–38.
3. Woolfson DN, Ryadnov MG. (2006) Peptide-based fibrous biomaterials: some things old, new and borrowed. *Curr. Opin. Chem. Biol.* **10**, 559–567.
4. Whitesides GM, Boncheva M. (2002) Supramolecular chemistry and self-assembly special feature: beyond molecules: self-assembly of mesoscopic and macroscopic components. *Proc. Natl. Acad. Sci. U. S. A.* 99, 4769–4774.

5. Fairman R, Akerfeldt KS. (2005) Peptides as novel smart materials. *Curr. Opin. Struct. Biol.* 15, 453–463.
6. Lupas AN, Gruber M. (2005) The structure of alpha-helical coiled coils. In: Parry DAD, Squire JM, eds. *Advances in Protein Chemistry*, Vol. 70. New York: Academic Press; pp. 37–38.
7. Burkhard P, Stetefeld J, Strelkov SV. (2001) Coiled coils: a highly versatile protein folding motif. *Trends Cell Biol.* **11**, 82–88.
8. Woolfson DN. (2005) The design of coiled-coil structures and assemblies. In: Parry DAD, Squire JM, eds. *Advances in Protein Chemistry*, Vol. 70. New York: Academic Press; pp. 79–112.
9. Raman S, Machaidze G, Lustig A, Aebi U, Burkhard P. (2006) Structure-based design of peptides that self-assemble into regular polyhedral nanoparticles. *Nanomed. Nanotechnol. Biol. Med.* **2**, 95–102.
10. Ryadnov MG. (2007) A self-assembling peptide polynanoreactor. *Angew. Chem. Int. Ed.* **46**, 969–972.
11. Ryadnov MG, Ceyhan B, Niemeyer CM, Woolfson D. N. (2003) "Belt and braces": a peptide-based linker system of de novo design. *J. Am. Chem. Soc.* **125**, 9388–9394.
12. Wagner DE, Phillips CL, Ali WM, et al. (2005) Toward the development of peptide nanofilaments and nanoropes as smart materials. *Proc. Natl. Acad. Sci. U. S. A.* **102**, 12656–12661.
13. Ryadnov MG, Woolfson DN. (2003) Engineering the morphology of a self-assembling protein fibre. *Nat. Mater.* **2**, 329–332.
14. Ryadnov MG, Woolfson DN. (2003) Introducing branches into a self-assembling peptide fiber. *Angew. Chem. Int. Ed.* **42**, 3021–3023.
15. Ryadnov MG, Woolfson DN. (2005) MaP peptides: programming the self-assembly of peptide-based mesoscopic matrices. *J. Am. Chem. Soc.* **127**, 12407–12415.
16. Ryadnov MG, Woolfson DN. (2004) Fiber recruiting peptides: noncovalent decoration of an engineered protein scaffold. *J. Am. Chem. Soc.* **126**, 7454–7455.
17. Smith AM, Acquah SFA, Bone N, et al. (2005) Polar assembly in a designed protein fiber. *Angew. Chem. Int. Ed.* **44**, 325–328.
18. Zimenkov Y, Dublin SN, Ni R, et al. (2006) Rational design of a reversible pH-responsive switch for peptide self-assembly. *J. Am. Chem. Soc.* **128**, 6770–6771.
19. Pandya MJ, Spooner GM, Sunde M, Thorpe JR, Rodger A, Woolfson DN. (2000) Sticky-end assembly of a designed peptide fiber provides insight into protein fibrillogenesis. *Biochemistry* **39**, 8728–8734.
20. Zhang S. (2003) Fabrication of novel biomaterials through molecular self-assembly. *Nat. Biotechnol.* **21**, 1171–1178.
21. Whitesides GM. (2006) The origins and the future of microfluidics. *Nature* **442**, 368–373.
22. Cerasoli E, Sharpe BK, Woolfson DN. (2005) ZiCo: a peptide designed to switch folded state upon binding zinc. *J. Am. Chem. Soc.* **127**, 15008–15009.
23. Wilken J, Kent SBH. (1998) Chemical protein synthesis. *Curr. Opin. Biotechnol.* **9**, 412–426.

24. Nilsson BL, Soellner MB, Raines RT. (2005) Chemical synthesis of proteins. *Annu. Rev. Biophys. Biomol. Struct.* **34**, 91–118.
25. Zhou M, Bentley D, Ghosh I. (2004) Helical supramolecules and fibers utilizing leucine zipper-displaying dendrimers. *J. Am. Chem. Soc.* **126**, 734–735.
26. Severin K, Lee DH, Kennan AJ, Ghadiri MR. (1997) A synthetic peptide ligase. *Nature* **389**, 706–709.
27. Herrmann H, Aebi U. (2004) Intermediate filaments: molecular structure, assembly mechanism, and integration into functionally distinct intracellular scaffolds. *Annu. Rev. Biochem.* **73**, 749–789.
28. Potekhin SA, Melnik TN, Popov V, et al. (2001) De novo design of fibrils made of short alpha-helical coiled coil peptides. *Chem. Biol.* **8**, 1025–1032.
29. Smith AM, Banwell EF, Edwards WR, Pandya MJ, Woolfson DN. (2006) Engineering increased stability into self-assembled protein fibers. *Adv. Funct. Mater.* **16**, 1022–1030.
30. Zimenkov Y, Conticello VP, Guo L, Thiyagarajan P. (2004) Rational design of a nanoscale helical scaffold derived from self-assembly of a dimeric coiled coil motif. *Tetrahedron* **60**, 7237–7246.
31. Aletras A, Barlos K, Gatos D, Koutsogianni S, Mamos P. (1995) Preparation of the very acid-sensitive Fmoc-Lys(Mtt)-OH. Application in the synthesis of side-chain to side-chain cyclic peptides and oligolysine cores suitable for the solid-phase assembly of MAPs and TASPs. *Int. J. Pept. Protein Res.* **45**, 488–496.
32. Kates SA, Daniels SB. (1993) Automated allyl cleavage for continuous-flow synthesis of cyclic and branched peptides. *Anal. Biochem.* **212**, 303–310.
33. Papapostolou D, Smith AM, Atkins EDT, et al. (2007) Engineering nanoscale order into a designed protein fiber. *Proc. Natl. Acad. Sci. U. S. A.* **104**, 10853–10858.

4

Biomimetic Synthesis of Bimorphic Nanostructures

Joseph M. Slocik and Rajesh R. Naik

Summary

The widespread interest in the use of biomimetic approaches for inorganic nanomaterial synthesis have led to the development of biomolecules (peptides, nucleic acids) as key components in material synthesis. Using biomolecules as building blocks, additional functionalities can be introduced by engineering multifunctional peptides that are capable of binding, nucleating, and assembling multiple materials at the nanoscale. We describe methodologies that exploit peptides for the synthesis of bimorphic nanostructures.

Key Words: Bimetallic; bionanotechnology; hybrid; nanoparticles; peptides; phage display.

1. Introduction

Biology represents the ultimate paradigm for materials processing, synthesis, and assembly of complex functional nanostructures. In nature, structures range from single nanoparticles of iron oxide *(1,2)*, silica *(3)*, or silver *(4,5)* to exquisitely assembled multidimensional architectures such as diffracting arrays and large extended nanoparticle networks, with perhaps the most compelling example the intricate structure of the diatom *(3,6–8)*. For this reason, materials found in nature have inspired the synthesis of many new nanomaterials by exploiting biochemical processes, the diverse collection of biomolecule templates, and the extraordinary control biological systems offer. While it is not yet possible to replicate the complexity of the diatom structure in vitro, advances in biomimetic synthesis have resulted in nanoparticles produced within protein cages *(9)*, silica-encapsulated materials derived from biosilification reactions *(10,11)*, and more recently peptide-directed bimetallic *(12)* and metal-insulator structures *(13)*.

We describe the synthesis of bimetallic nanoparticles and extend the synthesis to include other compositions (i.e., metal sulfide/metal structures using

From: *Methods in Molecular Biology, vol. 474: Nanostructure Design: Methods and Protocols*
Edited by: E. Gazit and R. Nussinov © Humana Press, Totowa, NJ

peptides derived from phage display) *(12)*. These materials offer enhanced properties from their individual nanoparticle counterparts, such as improvements in catalytic activity, electrical conductivity, and optical properties. In addition, we address the physical characterization and assess nanoparticle catalytic activity as a means to evaluate nanoparticle structure, size, and composition.

2. Materials

1. Peptides: Gold-binding peptides are identified using a combinatorial phage display peptide library (phage display kit from New England Biolabs). Briefly, gold nanoparticles are incubated with the library of phage-displayed peptides (10^9 random peptide sequences), washed to remove unbound nonspecific phages over several rounds of panning against stringent conditions (buffers and detergents), eluted to yield gold-binding phages, amplified, subjected to additional rounds of washings, and sequenced by PCR (polymerase chain reaction). This procedure ensures the selection of peptide sequences that exhibit the highest affinity for gold ***(14)***. This process is then repeated for the selection of peptides that exhibit an affinity for palladium nanoparticle substrates (*see* **Note 1**). After a selection-and-identification process (**Fig. 1**), a multibinding peptide sequence is designed from both gold and palladium phage sequences (*see* **Note 2**). The final peptide design identified from phage display is then synthesized using a peptide synthesizer and standard Fmoc (9-fluorenylmethoxycarbonyl) protocols by New England Peptides. Peptides are obtained crude at about 90% purity and at a yield of 20 mg. A stock solution of peptide is prepared by weighing 1 mg of lyophilized peptide in a 1.5-mL microfuge tube and dissolving with 100 µL doubly deionized water to yield a peptide concentration of 10 mg/mL.
2. Buffers: 0.1*M* HEPES buffer, pH 7.4: Dilute 1 mL of sterile 1*M*, pH 7.4 *N*-2-hydroxyethylpiperazine-*N′*-2-ethanesulfonic acid (HEPES) buffer (Amresco) with 9 mL of doubly deionized water. 0.25*M* Tris buffer, pH 9.2: Dissolve 0.3 g of trishydroxymethylaminomethane hydrochloride (Aldrich) with 10 mL of deionized water. Adjust pH of buffer to 9.2 by adding 0–50 µL of dilute NaOH. Degass buffer for 30 min with N_2 (*see* **Note 3**).

Parent Peptides:

(Flg)	-Asp-Tyr-Lys-Asp-Asp-Asp-Asp-Lys-	Pd binding domain
(A3)	- Ala-Tyr-Ser-Ser-Gly-Ala-Pro-Pro-Met-Pro-Pro-Phe	Gold binding domain

Multi-Functional Peptide:

(Flg-A3) -Asp-Tyr-Lys-Asp-Asp-Asp-Asp-Lys-Pro- Ala-Tyr-Ser-Ser-Gly-Ala-Pro-Pro-Met-Pro-Pro-Phe

Modified Peptide:

(Flg-A3-Cys) -Asp-Tyr-Lys-Asp-Asp-Asp-Asp-Lys-Pro- Ala-Tyr-Ser-Ser-Gly-Ala-Pro-Pro-Met-Pro-Pro-Phe-Cys

Fig. 1. Design of multifunctional peptide template for gold and gold-palladium nanoparticle synthesis.

3. 0.1*M* stock solutions of metal ions: Dissolve 17.0 mg of $HAuCl_4$ and 19.9 mg of K_2PdCl_6 (Aldrich) in 500 μL of doubly deionized water. Store solutions at 4°C and covered in foil.
4. 0.1*M* stock solution of sodium borohydride reductant: Dissolve 1.9 mg $NaBH_4$ in 500 μL of double-deionized water in a microfuge tube. Prepare right before synthesis as $NaBH_4$ loses reducing strength over time.
5. 35 m*M* stock solution of Cd^{2+}: Weigh 7.5 mg of $CdCl_2$ in a 3-mL glass vial, seal with a rubber septum, purge with N_2 or Ar gas at about 2 psi using an inlet needle and exit needle for 10 min to remove all oxygen, and dissolve with 1.2 mL of N_2 degassed 0.01*M* HCl using a syringe. It is important to maintain an anaerobic environment during all steps of CdS synthesis (*see* **Note 4**).
6. 35 m*M* stock solution of sodium sulfide: Weigh 5 mg of Na_2S in a 3-mL glass vial, purge with N_2 for 10 min, and dissolve in 2 mL N_2 degassed deionized water. Again, maintain anaerobic environment and keep refrigerated at 4°C.

3. Methods

3.1. Gold Nanoparticle Synthesis

1. Add 10 μL of peptide from 10 mg/mL stock solution to 500 μL of HEPES buffer (0.1*M*, pH 7.4) in a 1.5-mL microfuge tube on benchtop in open air.
2. To peptide solution, add 2.5 μL of 0.1*M* $HAuCl_4$ and incubate for 4 h on benchtop (*see* **Note 5**). Over time, solution will change to dark red, indicating formation of gold nanoparticles (**Fig. 2**).
3. After 4-h incubation period, peptide-coated gold nanoparticles are purified from excess peptide and salts by centrifuging particles at 15,000 × g for 10 min (*see* **Note 6**). Remove excess peptide from the gold nanoparticle pellet by pipeting the supernatant off and discard. Redissolve gold nanoparticle pellet in 500 μL of doubly deionized water, repeat centrifugation, and repeat washing twice.
4. On final washing, dissolve pellet in 500 μL of water (*see* **Note 7**).

3.2. Gold-Palladium Nanoparticle Synthesis

1. Add 2.5 μL of K_2PdCl_6 (0.1*M*) to purified particles from above and incubate for 10 min (*see* **Note 8**).
2. To Pd^{4+}-Au(FlgA3) nanoparticles, add 10 μL of 0.1*M* $NaBH_4$ and incubate on benchtop for 2 h (**Fig. 2**).
3. Purify the Au-(FlgA3)-Pd nanoparticle product carefully by sedimentation on a sucrose gradient, fixed-density sedimentation, or column chromatography (*see* **Note 9**). For fixed-density sedimentation of particles, prepare 20 mL of a 1*M* sucrose by dissolving 6.95 g sucrose in 20 mL of water. Place 7 mL of 1*M* sucrose in a 15-mL centrifuge tube and carefully add 100 μL of crude Au-Pd nanoparticles to sucrose so that a colored band appears at the top of the sucrose. Centrifuge mixture at 300 × g for 30 min, observe placement of colored band, increase centrifugation speed to 500 × g for 30 min, and again observe migration of band. Continue increasing speed by 20 × g intervals at 30 min until multiple

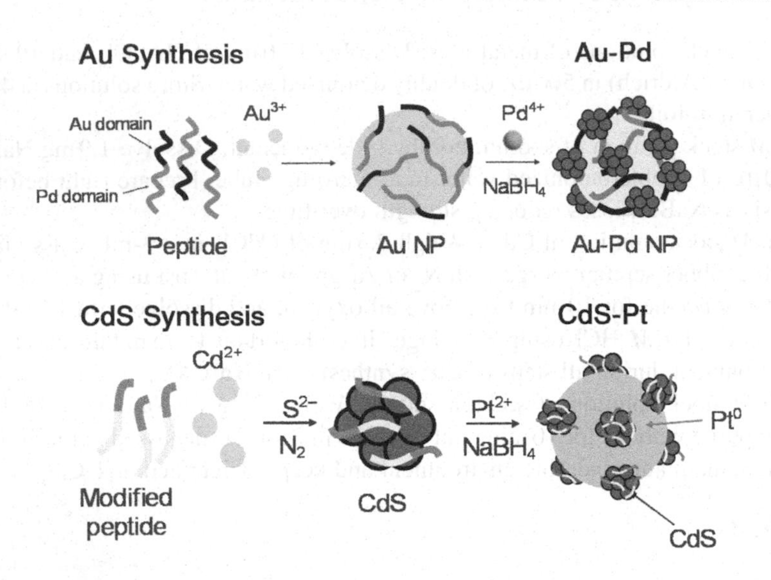

Fig. 2. Peptide-mediated synthesis of Au-Pd and CdS-Pt hybrid nanostructures.

bands appear. Individual Pd nanoparticles sediment differently from Au-Pd and will produce two colored bands. Pipet each band from the tube and analyze by transmission electron microscopy (TEM) or other physical analysis technique.

3.3. Synthesis of CdS Nanoparticles

1. Weigh 2 mg of lyophilized peptide-modified cysteine (*see* **Note 10**) into a 3 mL glass vial (8.66×10^{-7} mol peptide), add a micromagnetic stir bar, and seal with septum.
2. Purge peptide with N_2 using an inlet and exit needle for 10 min.
3. Dissolve peptide in 100 µL of degassed deionized water using a syringe.
4. To dissolved peptide, add 1 mL of degassed 0.25*M* Tris buffer, pH 9.2, by syringe.
5. Transfer 25 µL of Cd^{2+} from 35 m*M* stock solution to peptide in Tris buffer via microliter syringe.
6. Incubate Cd^{2+} with peptide for 30 min.
7. Add 25 µL of Na_2S from 35 m*M* stock solution to Cd^{2+}-peptide complex via syringe.
8. Cover in foil and stir on magnetic stir plate for 18 h. Over time, solution will slowly turn yellow.
9. After 18 h, particles can be exposed to air (*see* **Note 11**).
10. Purify by repeated ethanol precipitation. Add 30 mL of cold absolute ethanol to crude CdS-peptide reaction. Refrigerate at 4°C for 18 h to promote precipitation of CdS-peptide. Centrifuge precipitate at 15,000 × g for 10 min, redissolve pellet in 500 µL of 0.25*M* Tris buffer at pH 9.2, and add 30 mL of cold absolute ethanol. Repeat process two more times. After third round, dissolve pellet in 200 µL of deionized water. Store CdS-peptide at 4°C covered in foil.

3.4. Synthesis of CdS-Pt Nanohybrids

1. Add 3 µL of purified CdS-peptide to 100 µL of deionized water in a 1.5-mL microfuge tube.
2. Add 1.5 µL of 0.1*M* K_2PtCl_4 to CdS-peptide in water.
3. Incubate for 30 min.
4. Reduce CdS-peptide/Pt^{2+} with 10 µL of 0.1*M* $NaBH_4$.
5. Incubate for 3 h, at which time solution will turn light brown.

3.5. Characterization of Peptide-Coated Nanoparticles

3.5.1. Physical Characterization

1. Perform UV-visible spectroscopy on both particles in a 750-µL quartz cuvette scanning from 200 to 750 nm. Dilute 100 µL of nanoparticles from synthesis with 400 µL of doubly deionized water (*see* **Note 12**). Observe plasmon resonance peak of Au (~520–550 nm) or bandgap peak of CdS (~280–310 nm) for shape and wavelength.
2. Examine peptide-coated gold nanoparticles and peptide-coated Au-Pd bimetallic particles by TEM for structural details. Prepare TEM grids (200-mesh copper grids with carbon type A substrate, Ted Pella Inc.) by pipeting 10 µL of nanoparticle sample onto grid for each nanoparticle. Observe micrographs for structural features, geometry, sizes, and coverage of palladium nanoparticles on gold (**Fig. 3**).
3. Confirm elemental composition by energy dispersive X-ray spectroscopy. Most TEM microscopes are equipped with integrated EDAX systems (Energy dispersive X-ray spectroscopy).
4. Analyze particles by sedimentation in a sucrose gradient using a CPS particle size analyzer (CPS Instruments) or dynamic light scattering (DLS) for size distributions (*see* **Notes 13** and **14**). Inject 100 µL of nanoparticles onto sucrose gradient and collect sedimentation profile. For DLS, use a disposable cuvette, dilute sample to 3 mL with water, and collect size plot.

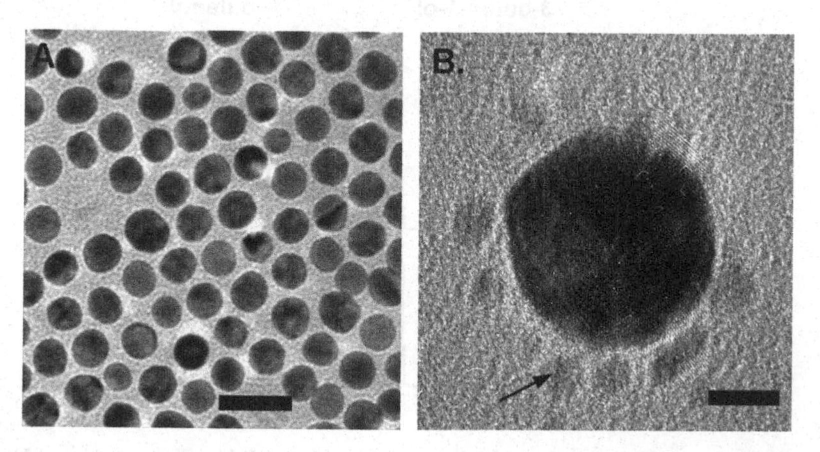

Fig. 3. TEM micrographs of (**A**) Au(FlgA3) nanoparticles (scale bar 40 nm) and (**B**) bimetallic Au-Pd nanoparticles (scale bar 7 nm). Arrow indicates palladium nanoparticle.

3.5.2. Assessment of Catalytic Activity

1. Add 100 μL of unsaturated alcohol (3-buten-1-ol; Aldrich), 500 μL of deuterium oxide (D_2O; Aldrich), Au-(FlgA3)-Pd particles prepared in Subheading 3.2., and a stir bar in a 50-mL conical flask.
2. Seal flask with a rubber septum and purge with H_2 gas for 15 min using an exit needle at about 2 psi. After 15 min, a slight positive pressure of hydrogen is applied.
3. Stir reaction contents on a magnetic stir plate for 4 h.
4. After 4 h, transfer contents from flask to microfuge tube and centrifuge at 15,000 × g for 10 min to remove Au-Pd catalyst particles (*see* **Note 15**).
5. Remove supernatant with products and place in a 5-mm bore nuclear magnetic resonance (NMR) tube (Wilmad/Labglass). Save nanoparticles for additional reactions by dissolving solid pellet in 50 μL of D_2O.
6. Analyze products by ^{1}H NMR spectroscopy using a 300-MHz NMR spectrometer. Collect ^{1}H NMR spectrum using proton pulse sequence over about 20,000 scans (16 h collection time) and default settings (0.2-Hz line broadening, 8000 spectral width). Assign resonances of proton spectrum to original 3-buten-1-ol substrate and to hydrogenation product of butanol. Resonances for the olefinic protons are located downfield at 5–6 ppm for 3-buten-1-ol; after hydrogenation, the pair of resonances shift upfield to 0.5–1.5 ppm, indicative of alkane protons (**Fig. 4**). Using NMR processing software, integrate these peaks and obtain a conversion ratio for butanol/3-buten-1-ol. With this ratio, calculate a turnover frequency for Au-Pd nanoparticle catalyst. The turnover frequency is defined as (moles of product)/(moles Pd·time).

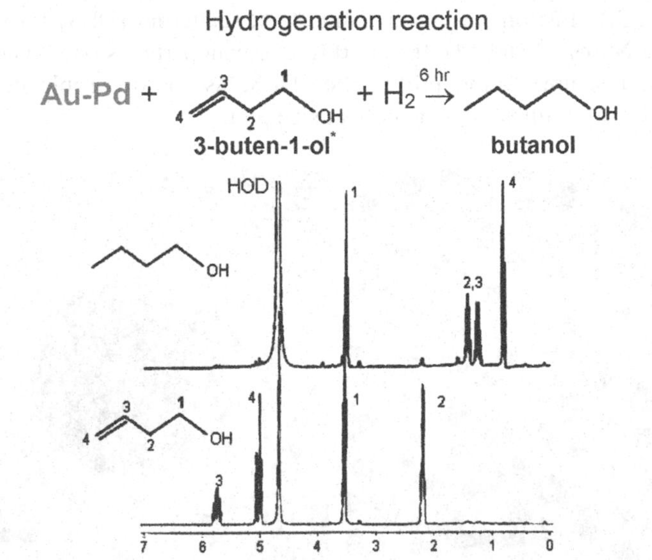

Fig. 4. Hydrogenation reaction of 3-buten-1-ol with Au-Pd in the presence of H_2. The 300-MHz ^{1}H nuclear magnetic resonance (NMR) spectra of 3-buten-1-ol reactant (bottom spectrum) and butanol product (top spectrum) after hydrogenation with Au-Pd in D_2O.

4. Notes

1. Peptides selected by phage display for a particular nanoparticle do not necessarily translate into an effective template for nanoparticle synthesis. Therefore, all peptide sequences should be tested for ability to synthesize and stabilize a nanoparticle.
2. The design of multibinding peptides should address the different permutations of possible sequences. For two specific sequences, two peptides should be examined; the Flg sequence is placed orthogonal to the N-terminus of the A3 sequence and then the C-terminus of A3.
3. Degass buffer with N_2 using a high-performance liquid chromatographic (HPLC) or schlenk line. With the schlenk line and double-manifold setup, "freeze-pump-thaw" represents the best method to completely deoxygenate solvents. Briefly, solvent is frozen, a vacuum is applied, the solvent is allowed to melt; this is repeated numerous times.
4. To obtain a N_2 or Ar purged environment, use of a glove box, schlenk techniques, or exit needle and gas inlet needle as described is required.
5. Sulfonated buffer like HEPES, 2-(*N*-morpholino)ethanesulfonic acid (MES), 3-(-*N*-morpholino)propanesulfonic acid (MOPS), in the absence of peptide, has been shown to reduce gold (see **ref.** ***18***) which quickly precipitates from solution.
6. After 10 min, supernatant should be clear but could require additional centrifugation. Peptide-coated gold particles should be washed with deionized water and centrifuged at least three times total to remove excess unreacted peptide and buffer salts.
7. Gold nanoparticles can be stored in solution for 9 mo at room temperature with no aggregation or degradation of particles.
8. Addition of metal ions like Pd^{4+}, Pt^{2+}, Ag^{+}, Cu^{2+}, Ni^{2+}, Zn^{2+}, and even Au^{3+} will promote aggregation of peptide-coated gold particles and a distinct color change instantly. Incubation time should not exceed 15 min; otherwise, particles will begin to crash out of solution. The amount of Pd^{4+} can be varied to yield different palladium nanoparticle coverage on gold.
9. Do not purify product by centrifugation or dialysis as this results in agglomeration of bare palladium particles.
10. Synthesis of CdS requires a cysteine residue for binding and passivating CdS surface. Cysteine ***(15)***, glutathione ***(16)***, and phytochelatin peptides ***(17)*** have been reported for synthesis of CdS.
11. Peptide protects CdS against oxidizing conditions.
12. Absorbance of nanoparticles from synthesis is high, >1 in corresponding spectrum, requiring dilution.
13. Size distribution obtained from CPS particle size analysis will most likely differ from TEM. CPS analysis accounts for total size population, while TEM provides sizes representing collections from several fields of vision.
14. CPS offers high resolution, sensitivity, and reproducibility for small-size particles, in contrast to DLS, which is biased toward larger particles.
15. Au-Pd particles could interfere with NMR analysis as well as contain additional resonances from peptide interface. Isolation of bimetallic catalyst particles from hydrogenation products ensures accurate integration of peaks.

References

1. Blakemore R. (1975) Magnetoctactic bacteria. *Science* **190**, 377.
2. Round FE, Crawford RM, Mann DG. (1985) *Magnetite Biomineralization and Magnetoreception in Organisms*. Plenum Press, New York.
3. Kroger N, Deutzmann R, Sumper M. (1999) Polycationic peptides from diatom biosilica that direct silica nanosphere formation. *Science* **286**, 6386.
4. Klaus T, Joerger R, Olsson E, Granqvist C-G. (1999) Silver-based crystalline nanoparticles, microbially fabricated. *Proc. Natl. Acad. Sci. U. S. A.* **96**, 13611.
5. Nair B, Pradeep T. (2002) Coalescence of nanoclusters formation of submicron crystallites by assisted by *Lactobacillus* strains. *Cryst. Growth Design* **2**, 293.
6. Kroger N, Deutzmann R, Bergsdor C, Sumper M. (2000) Species-specific polyamines from diatoms control silica morphology. *Proc. Natl. Acad. Sci. U. S. A.* **97**, 14133.
7. Round FE, Crawford RM, Mann RM. (1990) *The Diatoms: Biology and Morphology of the Genera*. Cambridge University Press, New York.
8. Kroger N, Sumper M. (2000) *Biomineralization: From Biology to Biotechnology and Medical Applications*. Wiley-VCH, Weinheim.
9. Douglas T, Stark VT. (2000) Nanophase cobalt oxyhydroxide mineral synthesized within the protein cage of ferritin. *Inorg. Chem.* **39**, 1828.
10. Naik RR, Tomczak MM, Luckarift HR, Spain JC, Stone MO. (2004) Entrapment of enzymes and nanoparticles using biomimetically synthesized silica. *Chem. Commun.* 1684.
11. Knecht MR, Wright DW. (2004) Dendrimer-mediated formation of multicomponent nanospheres. *Chem. Mater.* **16**, 4890.
12. Slocik JM, Naik RR. (2006) Biologically programmed synthesis of bimetallic nanostructures. *Adv. Mater.* **18**, 1988.
13. Banerjee IA, Regan MR. (2006) Preparation of gold nanoparticle templated germania nanoshells. *Mater. Lett.* **60**, 915–918.
14. Naik RR, Jones SE, Murray CJ, McAuliffe JC, Vaia RA, Stone MO. (2004) Peptide templates for nanoparticle synthesis derived from polymerase chain reaction-driven phage display. *Adv. Func. Mater.* **14**, 25.
15. Bae W, Abdullah R, Mehra RK. (1998) Cysteine-mediated synthesis of CdS bionanocrystallites. *Chemosphere* **37**, 363.
16. Nguyen L, Kho R, Bae W, Mehra RK. (1999) Glutathione as a matrix for the synthesis of CdS nanocrystallites. *Chemosphere* **38**, 155.
17. Dameron CT, Winge DR. (1990) Characterization of peptide-coated cadmium-sulfide crystallites. *Inorg. Chem.* **29**, 1343.
18. Xie J, Lee JY, Wang DIC. (2007) Seedless, surfactantless, high-yield synthesis of branched gold nanocrystals in HEPES buffer solution. *Chem. Matter.* **19**, 2823.

5

Synthesis and Primary Characterization of Self-Assembled Peptide-Based Hydrogels

Radhika P. Nagarkar and Joel P. Schneider

Summary

Hydrogels based on peptide self-assembly form an important class of biomaterials that find application in tissue engineering and drug delivery. It is essential to prepare peptides with high purity to achieve batch-to-batch consistency affording hydrogels with reproducible properties. Automated solid-phase peptide synthesis coupled with optimized Fmoc (9-fluorenylmethoxy-carbonyl) chemistry to obtain peptides in high yield and purity is discussed. Details of isolating a desired peptide from crude synthetic mixtures and assessment of the peptide's final purity by high-performance liquid chromatography and mass spectrometry are provided. Beyond the practical importance of synthesis and primary characterization, techniques used to investigate the properties of hydrogels are briefly discussed.

Key Words: Biomaterial; HPLC; hydrogel; peptide self-assembly; solid-phase peptide synthesis.

1. Introduction

Self-assembly forms an important process in the bottom-up approach to the design of nanostructural architectures *(1,2)*. For example, peptide self-assembly has been extensively utilized to design intricate well-ordered structures such as nanotubes *(3–12)* and ribbons *(13–18)*. Peptide self-assembly has also been employed in the development of hydrogels, heavily hydrated materials composed of dilute networks of assembled peptide. These materials are finding use in a variety of biomedical applications *(19–21)*. In addition, peptides can be designed to undergo triggered self-assembly, leading to the formation of hydrogel material in response to physiologically relevant changes in their external environment; this allows material formation to take place with temporal resolution *(22)*. The unique ability of peptide sequences to fold into specific secondary, tertiary, and quaternary structures has been exploited, and hydrogels based

From: *Methods in Molecular Biology, vol. 474: Nanostructure Design: Methods and Protocols*
Edited by: E. Gazit and R. Nussinov © Humana Press, Totowa, NJ

on the self-assembly of α-helices ***(23)***, β-sheets ***(24–29)***, coiled coils ***(30–32)***, and collagen mimetic peptides ***(33,34)*** have been reported in the literature. Moreover, biological function has been incorporated into hydrogels prepared from traditional polymers by ligating short peptide sequences to synthetic scaffolds ***(35–37)***.

For the reliable use of materials derived from peptides, it is essential to synthesize and purify the building blocks with high fidelity. With respect to hydrogels, batch-to-batch consistency can be achieved using peptides of high purity. In this chapter, solid-phase peptide synthesis is discussed in the context of the β-hairpin peptide hydrogels studied in our lab ***(25,38–42)***. Guidelines for the purification and primary characterization of the peptides are provided along with a brief discussion of select techniques utilized in the biophysical, structural, and mechanical characterization of the hydrogels. An example of the ramifications of purity on the material properties is also provided.

2. Materials

2.1. Reagents

2.1.1. Solid-Phase Peptide Synthesis

1. *N*-Methyl pyrrolidone (NMP) (EMD Biosciences).
2. Rink amide resin (PL-Rink Resin, loading = 0.64 mmol/g, 75–150 μm or 100–200 mesh) (Polymer Laboratories, Amherst, MA) (*see* **Note 1**).
3. Appropriately side-chain-protected 9-fluorenylmethoxycarbonyl (Fmoc) amino acids (Novabiochem).
4. 1H-Benzotriazolium1-[bis(dimethylamino)methylene]-5-chloro hexafluorophosphate (1-),3-oxide (HCTU) (Peptides International): 0.45*M* solution prepared in *N,N*-dimethylformamide (DMF) (Fisher).
5. Diisopropylethylamine (DIPEA) (Acros Organics): 2*M* solution prepared in NMP.
6. 20% piperidine (Sigma) in NMP or 19% piperidine in NMP containing 1% 1,8-diazabicyclo[5.4.0]-undec-7-ene (DBU) (Sigma).
7. Capping solution: 5% acetic anhydride (Ac_2O) (Acros Organics) in NMP.
8. Methylene chloride or dichloromethane (DCM) (Fisher).
9. Methanol (MeOH) (high-performance liquid chromatographic [HPLC] grade; Fisher).

2.1.2. Peptide Resin Cleavage and Side-Chain Deprotection

1. Trifluoroacetic acid (TFA) (Acros Organics).
2. Thioanisole (Acros Organics).
3. Ethanedithiol (Acros Organics).
4. Anisole (Acros Organics).
5. Diethyl ether (Fisher).
6. Nylon filter paper (MAGNA, nylon, supported, plain, 0.45 μm, 47 mm) (GE Water and Process Technologies).

2.1.3. Reverse-Phase High-Performance Liquid Chromatography

1. Solvent A: 0.1% TFA in water (*see* **Note 2**).
2. Solvent B: 90% acetonitrile (HPLC grade; Fisher), 10% water, and 0.1% TFA.
3. Protein or peptide C18 column (Vydac): For analytical purposes, the use of a 250-mm length, 4.6-mm internal diameter (id) column packed with 5 μm particles is used; for semipreparative scale, a column 250 mm long, 22-mm id packed with 10-μm particles is used.

2.1.4. Lyophilization

For lyophilization, use liquid N_2.

2.2. Instrumentation

1. A 433A peptide synthesizer from Applied Biosystems with SynthAssist software was used to carry out the synthesis of the peptides described here.
2. HP 1100 series HPLC equipment from Agilent Technologies equipped with a Vydac C18 peptide/protein column was utilized to perform analytical HPLC.
3. Bulk purification was carried out on Waters 600 series modular semipreparative-scale HPLC equipment with a Vydac C18 peptide/protein column.
4. Flexi-Dry freeze dryers from FTS, New York, were used for lyophilization.
5. Electrospray ionization mass spectrometry (ESI-MS) was carried out on a Thermo Finnigan LCQ mass spectrometer to characterize peptide mass.

3. Methods

3.1. Solid-Phase Peptide Synthesis

Although the following procedures are specific to the ABI 433A peptide synthesizer, they can be easily adapted for other automated synthesizers ***(43)***.

The ABI 433A is an automated batch peptide synthesizer that can perform syntheses from 0.1 to 1 mmol scale; however, the procedure described here is specific to a 0.25-mmol scale synthesis. In general, for the ABI 433A, the solid support resin is placed inside a reaction vessel. Filtered solvents or reagents are delivered to or drained from the reaction vessel by the application of N_2 pressure. NMP is used as the universal solvent ***(44)***. During synthesis, vortexing of the reaction vessel or bubbling N_2 through the reaction mixture facilitates mixing.

Although the instrumental software supports Fmoc-based synthesis using HBTU activation, we employ HCTU activation, which provides improved synthetic outcome. For any given synthesis, 1 mmol of dry powdered Fmoc-protected amino acid is packed in each cartridge, and the cartridges are sequentially arranged on a guideway. When the instrument couples an Fmoc amino acid to a resin-based free amine, first a pneumatic injector ruptures the cartridge septum to deliver NMP required for dissolving the amino acid. HCTU and the base (i.e., DIPEA) (stored in separate reservoirs) are also mixed with the amino acid to prepare active esters before delivery to the resin. A fourfold excess of the amino acid ensures that each coupling reaction reaches more than 99% completion. This is crucial since an accumulated decrease in the coupling efficacy can negatively affect the final yield and purity of even small peptides, such as the 20-residue sequences prepared here.

Typically, coupling is allowed to proceed for 15 min, after which the resin is washed several times with NMP. Neat piperidine from the reagent bottle is diluted with NMP in 1:4 ratio, resulting in a 20% piperidine solution, and is delivered to the reaction vessel for Fmoc deprotection. Fmoc deprotection is actively monitored by measuring the UV absorbance (301 nm) of the dibenzofulvene-piperidine adduct released from the resin-bound peptide. Based on the real-time deprotection data, successive rounds of deprotection are implemented automatically via the instrumental software. In addition, these extended rounds of deprotection automatically implement the application of a capping cycle (acetic anhydride) after the coupling of the next amino acid in the sequence is complete to minimize the formation of deletion sequences.

3.1.1. General Considerations

1. Scale of synthesis: Due to the time and expense of synthesizing peptides, we routinely synthesize new sequences initially in small scale (0.1 mmol) to map out difficult sequential couplings (*see* **Subheading 3.1.2.**).
2. Selection of the correct reaction vessel size: For efficient swelling, mixing, and washing of the resin, an appropriate size reaction vessel for the desired scale of synthesis should be used. A size that affords maximal resin swelling and mixing while minimizing the dead volume is desired.
3. Instrumental calibration: Routine instrumental calibration ensures optimal delivery of reagents, minimizing unsuccessful syntheses due to instrumental error.
4. Preparation of reagents: In our experience, reagents such as HCTU and piperidine do not store well in solution at room temperature; therefore, we recommend using fresh reagents for each new synthetic procedure (*see* **Note 1**).

3.1.2. Synthesis Optimization

Peptide sequences that have not been prepared previously in the lab are initially synthesized using a standard *non*optimized protocol. This allows problematic sequential positions to be identified where Fmoc amino acids may need to be double coupled to the growing resin-bound sequence. In this nonoptimized protocol, each residue of the sequence is single coupled to the growing chain, and based on the Fmoc deprotection profile, the instrument will conditionally cap the growing chain at problematic sequential positions. HCTU activation and 20% piperidine in NMP are used for the coupling and Fmoc deprotection steps, respectively.

Figure 1A shows the Fmoc deprotection profile for peptide **A** (VKVKV^{D}PPTKVKVKVKVKV-NH_2), which was prepared using this

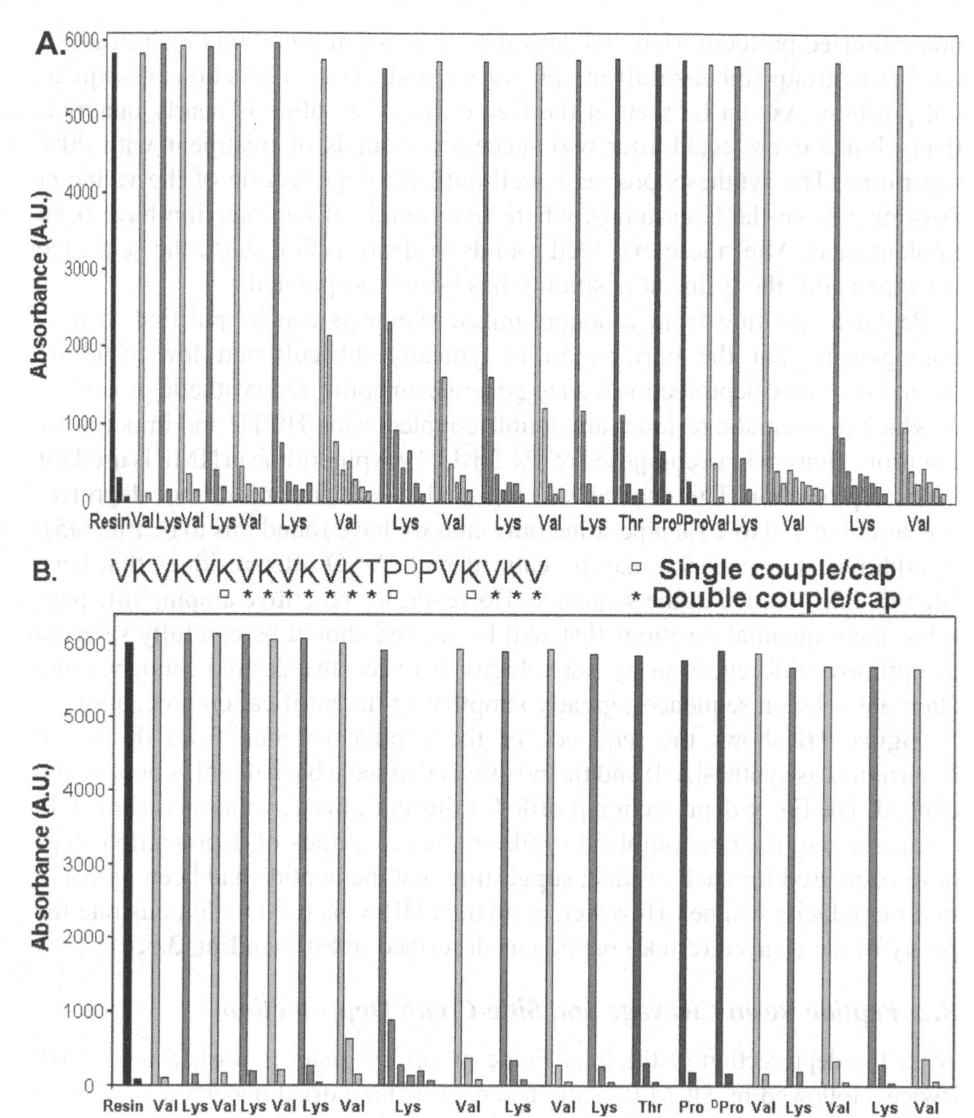

Fig. 1. Profiles monitoring UV absorbance of the dibenzofulvene-piperidine adduct at 301 nm for each residue in the synthesis of the peptide VKVKV^{D}PPTKVKVKVKVKV-KV-NH_2. (**A**) Nonoptimized synthesis in which each residue is single coupled, and 20% piperidine in *N*-methylpyrrolidone (NMP) is used for deprotection. (**B**) Optimized synthesis in which amino acid double coupling and N-terminal capping with acetic anhydride are employed at the positions indicated in the sequence (written from C- to N-terminus as synthesized). In addition, 1% 1,8-diazabicyclo[5.4.0]-undec-7-ene (DBU) and 19 % piperidine in NMP is employed for 9-fluorenylmethoxycarbonyl (Fmoc) deprotection in panel **B**.

nonoptimized protocol. Here, the absorbance at 301 nm monitors the release of the Fmoc group (dibenzofulvene-piperidine adduct) as a function of sequential position. As can be seen in the figure, the first valine is nearly quantitatively Fmoc deprotected after two successive rounds of treatment with 20% piperidine. The synthesis proceeds well until the deprotection of the valine at position 5 from the C-terminus, where five rounds of deprotection have been implemented. After these extended rounds of deprotection steps, the sequence is capped after the lysine at position 6 has been incorporated.

Peptides resulting from a nonoptimized synthesis can be purified to near homogeneity, but the purification is typically difficult and low yielding. We use the Fmoc deprotection data to generate an optimized synthetic procedure in which problematic residues are double coupled using HCTU and importantly a deprotection cocktail composed of 1% DBU, 19% piperidine in NMP is used for Fmoc deprotection. This cocktail was reported to be superior for Fmoc deprotection as compared to 20% piperidine only, and we have found this to be true ***(45)***. In addition, the sequence may be capped with Ac_2O after residues that have been double coupled in the sequence. However, for repetitive amphiphilic peptides, the sequential positions that will be capped should be carefully selected to optimize differences in hydrophobicity between the desired sequence and alternate deletion sequences, greatly simplifying the purification procedure.

Figure 1B shows the sequence of the peptide (written from the C- to N-terminus as synthesized) and the positions that have been double coupled and capped. The Fmoc deprotection profile for this optimized synthesis is shown. In comparison to the nonoptimized synthesis, fewer rounds of deprotection steps have been used for each residue, suggesting that the peptide had been prepared in a more facile manner. However, analytical HPLC is used to demonstrate the purity of the cleaved (crude) peptide as described in **Subheading 3.3.1**.

3.2. Peptide Resin Cleavage and Side-Chain Deprotection

After the deprotection of the final Fmoc group, the resin is washed with NMP (twice), followed by DCM (twice). The resin is then dried under vacuum for at least 1 h. It is advisable to carry out a test cleavage on a small quantity (30–40 mg) of resin to ensure selection of the correct cleavage reagent mixture and reaction time. With this said, we have found that the particular cleavage cocktail discussed next is extremely versatile in effecting resin cleavage and side-chain deprotection of almost all the sequences that have been prepared in our lab.

3.2.1. Cleavage Protocol

1. Prepare 10 mL of the cleavage reagent by mixing TFA/thioanisole/ethandedithiol/anisole in a 90:5:3:2 volume ratio ***(46)***.
2. Place the dry resin in a round-bottom flask containing a magnetic stir bar and slowly add enough cleavage reagent such that it completely covers the resin. Stir the resin

slowly under a N_2 atmosphere for 2h. (If the sequence contains 4-methoxy-2,3,6-trimethylbenzenesulfonyl [MTR]-protected or 2,2,5,7,8-pentamethylchroman-6-sulfonyl [PMC]-protected arginine, then increase the reaction time to 4h.)

3. Remove the resin via filtration through a sintered glass funnel using positive N_2 pressure to aid the filtration process. Wash the resin two or three times with a small amount of neat TFA to ensure that all the cleaved peptide is removed from the resin. Avoid using vacuum to aid the filtration process. This limits possible oxidation of the peptide that could result from pulling air through the apparatus.
4. Reduce the volume of the filtrate to 1/5 its original volume by flowing a stream of N_2 across the liquid.
5. Precipitate the peptide by adding ice-cold diethyl ether in small portions.
6. Immediately collect the peptide precipitate via filtration using a nylon filter. Wash the precipitate with copious amounts of cold ether. Again, filtration may be aided by using positive N_2 pressure instead of vacuum to limit possible oxidation of the peptide. The crude peptide is then dried under vacuum.

3.3. Purification and Primary Characterization

3.3.1. Initial Assessment of Peptide Purity and Establishing a Preparatory Reversed-Phase HPLC Gradient for Purification

The crude peptide obtained after resin cleavage and side chain deprotection is purified using reversed-phase HPLC (RP-HPLC). To assess the retention time of the desired peptide on a C18 peptide/protein column as well as to determine the impurity profile of the peptide synthesized, an analytical chromatogram of the crude material is collected. Typically, we analyze peptide solutions at a concentration of 1 mg/mL of solvent A (injection volume = 100 μL, eluent flow rate = 1 mL/min, column temperature = 20°C) on the analytical RP-HPLC. Before performing any HPLC experiment, the column is cleaned with 100% solvent B to eliminate any existing peptides adsorbed on the column. This is followed by equilibration with 100% solvent A prior to sample injection.

Figure 2A,B depicts an analytical HPLC trace and ESI-MS of the crude material obtained from the optimized synthetic procedure (**Fig. 1B**). In **Fig. 2A**, a linear gradient from 0% to 100% solvent B over 100 min is employed. The UV absorbance at 220 nm is monitored with respect to the retention time of the eluting species from the column. We typically monitor 220, 254, and 280 nm to detect peptide as well as aromatic species derived from the resin cleavage reaction. Here, we show only the data at 220 nm for clarity. Each eluted peak is manually collected as a separate fraction. Mass spectrometry of all the collected fractions indicates the retention time of the desired peptide. In the example discussed, the desired sequence elutes as the largest peak at 30 min, as shown in **Fig. 2A**. This indicates that an eluent mixture of 30% solvent B and 70% solvent A is necessary to elute the peptide from the C18 column since a linear gradient of 0% to 100% solvent B over 100 min was employed.

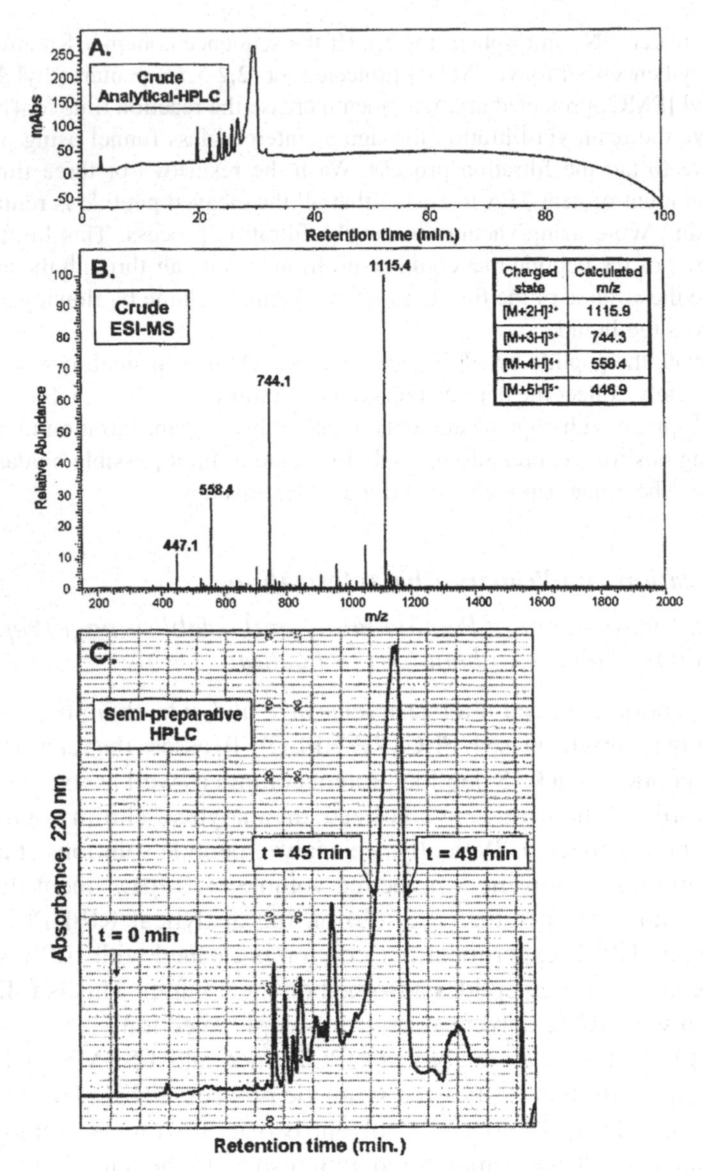

Fig. 2. (**A**) Reversed-phase high-performance liquid chromatographic (RP-HPLC) trace of the crude material isolated from the resin cleavage reaction. Absorbance at 220 nm is monitored versus retention time for peptide **A** (VKVKV^{D}PPTKVKVKVKVKV-NH_2) on a C18 column employing a linear gradient of 0% to 100% solvent B in 100 min. Desired peptide elutes at 30 min. (**B**) ESI-MS (electrospray ionization mass spectrometry) of crude material isolated from resin cleavage reaction. The molecular ions of the peptide are labeled and defined (inset). Unlabeled peaks in the spectrum are indicative of the impurities in the crude material. (**C**) Representative semipreparative RP-HPLC chromatogram. The desired peptide fraction is collected from 45 to 49 min.

It should be noted that despite the amphiphilic nature of the peptide, it is synthesized in high yield via the optimized synthetic protocol. The purification stage involves separation of the desired fraction from impurities on the semipreparative scale. For semipreparative-scale purification, we initially dissolve the crude material in solvent A (1–4 mg/mL) and inject 5-mL portions of this solution onto the column. A distinct semipreparative gradient is calculated and used for the HPLC purification of the peptide.

Typically, we will decrease the gradient steepness from 1% solvent B per minute, which was used for analytical HPLC, to either 0.5% solvent B per minute or 0.25% solvent B per minute depending on how similar the retention times of any impurities are to that of the desired peptide. In this example, we employ a gradient of 0.25% solvent B per minute. We have found through experience that small peptides should have a retention time greater than 30 min on the Vydac C18 columns used in our lab to maximize elution resolution. Therefore, a semipreparative gradient for peptide **A**, which eluted at 30% solvent B, can be calculated as follows:

Percentage solvent B that must be traversed during purification = 30 min on column × 0.25% solvent B per min ≈ 8% solvent B (1)

Initial column condition at start of gradient = 30% solvent B − 8% solvent B = 22% solvent B (2)

However, injecting crude material onto a column at high percentages of solvent B (in this example, 22% solvent B) decreases resolution; we typically introduce solutions of crude material to the column at 0% solvent B and subsequently rapidly approach the initial conditions using a steep gradient. The final gradient employed for peptide **A** is shown in **Table 1**.

The semipreparative chromatogram shown in **Fig. 2C** resulted from a protocol nearly identical to that discussed here. The peptide began to elute after about 40 min (**Fig. 2C**). A fraction was manually collected from 45 to 49 min. After this time, the gradient was aborted and the column immediately

Table 1
Semipreparative High-Performance Liquid Chromatographic (HPLC) Gradient Calculated for Peptide A

Time (min)	Solvent B (%)	Gradient
0	0	—
22	22	1% solvent B per min
312	100	0.25% solvent B per min

washed with 100% solvent B followed by reequilibration with 100% solvent A; the process was repeated for the remaining crude material. The isolated fractions from HPLC purification were combined and lyophilized, affording a white powder.

3.4. Assessment of Purity

Following lyophilization, the purity of the peptide is determined by analytical RP-HPLC and ESI-MS. A 1 mg/mL solution of the peptide is prepared in solvent A. Of this solution 100 µL are injected onto the analytical C18 column, and an analytical HPLC experiment employing a linear gradient of 0–100% solvent B in 100 min is carried out. **Figure 3 A** depicts the analytical HPLC chromatogram of peptide **A** purified with the aforementioned semipreparative HPLC

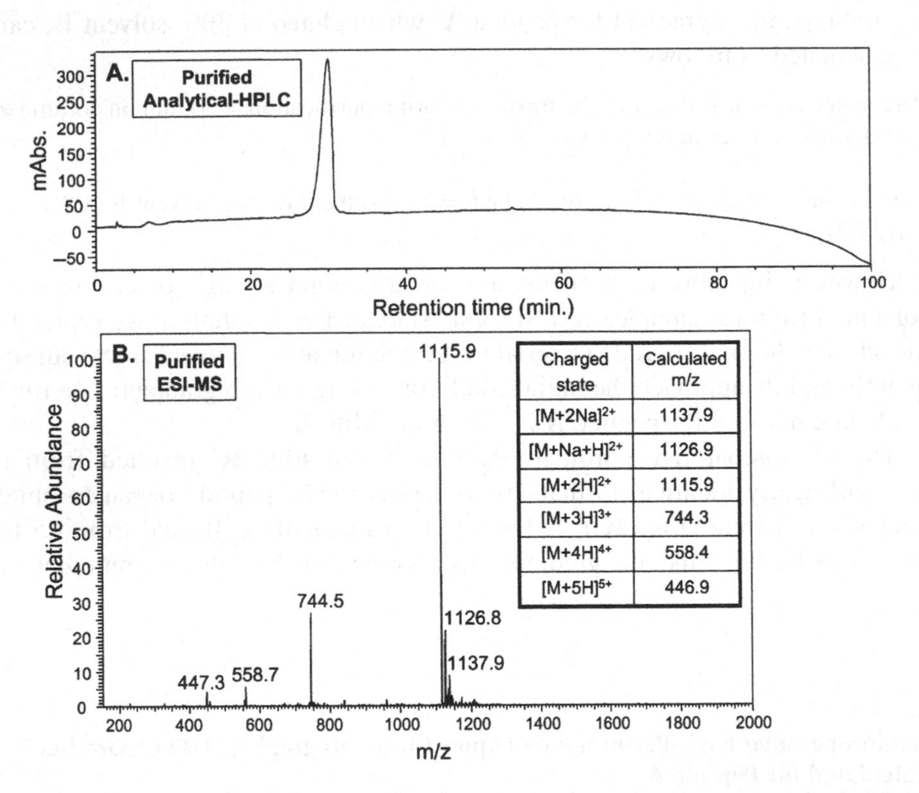

Charged state	Calculated m/z
$[M+2Na]^{2+}$	1137.9
$[M+Na+H]^{2+}$	1126.9
$[M+2H]^{2+}$	1115.9
$[M+3H]^{3+}$	744.3
$[M+4H]^{4+}$	558.4
$[M+5H]^{5+}$	446.9

Fig. 3. Assessment of the peptide purity following semipreparative-scale high-performance liquid chromatographic (HPLC) separation. (**A**) Analytical reversed-phase HPLC of the lyophilized peptide carried out using a linear gradient of 0% to 100% solvent B in 100 min on a C18 column. (**B**) ESI-MS (electrospray ionization mass spectrometry) of the purified peptide and appropriate calculated masses.

gradient. Peptide **A** was effectively separated from the prepeak and postpeak impurities present in the crude after resin cleavage and side-chain deprotection (**Fig. 2A**). The purity of this peptide was further assessed by mass spectrometry, as shown in **Fig. 3B**. The observed molecular mass ions at 1115.9, 744.5, 558.7, and 447.3 correspond to the +2, +3, +4, and +5 charged states of the peptide, respectively. The peptide mass determined from ESI-MS is in agreement with the calculated masses established from the sequence.

Both HPLC and mass spectral analysis suggest that peptide **A** has been purified to near homogeneity. Typically, hydrogels prepared from β-hairpin peptides of this level of purity afford consistent batch-to-batch material properties. There is always the possibility that an impurity may coelute with the purified peptide that is not observed by mass spectrometry. This uncommon scenario is usually realized after observing inconsistent properties from a given batch of peptide. This problem can usually be remedied by repurifying the peptide by RP-HPLC using a shallower gradient or isocratic conditions or warming or cooling the column. As long as the impurities have different temperature-dependent retention times, performing the purification at two different temperatures provides an excellent means of producing extremely pure samples. Of course, different column types may also be employed to maximize differences in retention times.

3.5. Importance of Purity

The importance of purifying peptides to the highest possible level with respect to achieving reproducible physical and biophysical properties is exemplified in **Fig. 4**.

MAX3 (VKVKVKTKV^{D}PPTKVKTKVKV-NH_2) is a β-hairpin peptide that was designed to undergo a thermally triggered intramolecular folding and self-assembly event, which affords hydrogel material *(38)*. **Figure 4A** shows the analytical chromatograms of two different batches of MAX3 that had been purified on separate occasions. The seemingly insignificant impurity (seen as a postpeak) in the "impure" batch grossly influences the temperature at which folding and consequent self-assembly occurs. The circular dichroism (CD) data in **Fig. 4B** show the mean residue ellipticity at 216 nm, an indicator of β-sheet structure, as a function of temperature. At low temperatures, peptide from both batches exists in random-coil conformations. As the temperature is increased, MAX3 folds and self-assembles into a β-sheet-rich hydrogel. It is clear from the data that the temperature at which this folding/assembly transition takes place is batch dependent; a small amount of impurity increased the temperature necessary to initiate peptide folding and self-assembly. Repurifying this batch to remove the impurity restored the peptide's normal temperature-dependent behavior. When possible, our lab routinely publishes an analytical HPLC chromatogram and the mass spectrum of each peptide discussed in a given

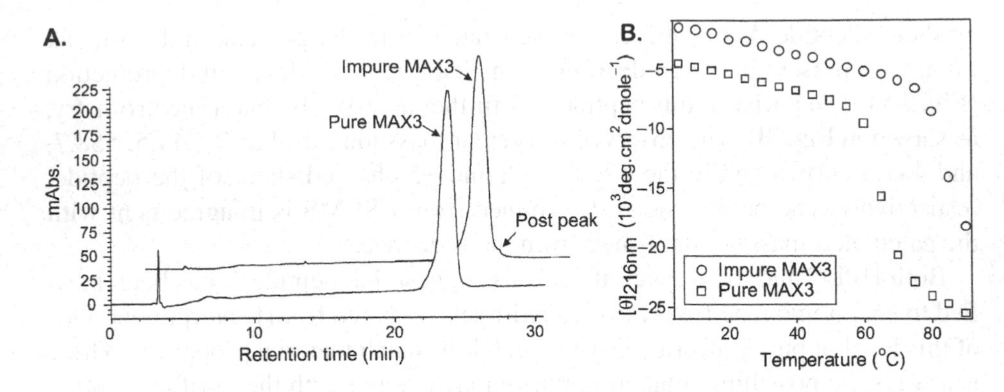

Fig. 4. (**A**) Analytical high-performance liquid chromatographic (HPLC) chromatograms of distinct batches of peptide MAX3. Impure MAX3 has a small postpeak not seen in the pure trace. (**B**) Temperature-dependant circular dichroism spectra of 150 μ*M* peptide at pH 9.0, 125 m*M* borate, 10 m*M* NaCl with purity corresponding to the chromatograms in (**A**). Figure depicts the secondary structure transition from random coil to β-sheet by monitoring the mean residue ellipticity at 216 nm as a function of temperature. Impure MAX3 folds and assembles at a higher temperature.

manuscript; these data are usually contained in the supporting information. This is important in that it establishes the level of purity needed to realize the observed biophysical/material properties reported.

3.6. Beyond Primary Characterization

Although this chapter is mainly concerned with the synthesis and primary characterization of peptides used in self-assembly, a brief introduction to several techniques that are common to the study of self-assembled peptide-based hydrogels is provided next. These techniques offer insight into the secondary structure of the peptide in the self-assembled state, the nanoscale morphology of the assembled structures that constitute the hydrogel, as well as the bulk mechanical properties of the hydrogel itself. These brief introductions are meant to acquaint those new to the field; comprehensive descriptions of each technique can be found in the literature as indicated.

Circular dichroism (CD) can be used to determine the secondary structure of peptides in the self-assembled state of optically clear hydrogels *(47)*. Characteristic dichroic signatures for α-helical, β-sheet, and β-turn secondary structures as well as random-coil conformations are easily detected. Importantly, CD spectroscopy provides an excellent means of monitoring changes in the secondary structure of peptides in response to changes in solution conditions (e.g., pH, temperature, ionic strength, chaotropes, etc.). However, obtaining spectra of hydrogel samples can sometimes be challenging due to the small path length cells that must be employed if the concentration of the peptide constituting the gel is high. If using small path length cells proves to be problematic, dilute preparations of assembled peptide can be studied employing larger path length cells as long as light scattering is minimized; this is the case in data that are shown in **Fig. 4B**.

Fourier transform infrared (FTIR) spectroscopy is another convenient technique to study the secondary structure of peptides in the self-assembled state *(48)*. One advantage of FTIR is that, unlike CD, it is less sensitive to light scattering; as a result, greater concentrations of peptides can be studied. Well-characterized absorptions are known for helical and β-sheet structures as well as random-coil conformations. Possible limitations of this technique are that TFA salts of peptides cannot be used since TFA absorbs strongly in the amide I′ region. In addition, H_2O cannot be used as a solvent for the same reason. Therefore, the TFA counterions of peptides are typically exchanged by dissolving the peptide in 0.1*M* HCl followed by lyophilization. The resulting HCl-peptide salt is subsequently dissolved in D_2O and lyophilized several times to exchange the water. Hydrogels can then be prepared using D_2O and studied.

Oscillatory rheology can be used to study the mechanical properties of peptide-based hydrogels *(49)*. Commonly, the mechanical rigidity of the hydrogel is assessed by measuring the storage and loss modulus of the gel as a function of time, frequency, or strain. In addition, detailed insight into the physical nature (crosslink type and density, response to shear strain, etc.) of the gels can be gleaned by performing rheological measurements.

Transmission electron microscopy (TEM) can be used to characterize the local nanostructure of self-assembled peptides *(50)*. Typically, dilute suspensions of assembled peptide are placed on grids and allowed to dry. Contrast-enhancing agents are often used to study the fine details of the nanostructure. Importantly, drawing appropriate conclusions from TEM necessitates that enough observations are recorded to provide meaningful statistics. For hydrogels, one possible limitation in employing conventional TEM is that samples are dehydrated; thus, inferences must be made to relate the self-assembled structure observed on the grid to that which actually exists in the hydrated state. In the limiting case, the observed structure may be different from that in the hydrated state. To overcome this possible limitation, the *in situ* structure of gels can be studied by cryogenic TEM. Here, the water in the hydrogel is vitrified to preserve the *in situ* nanostructure. However, this technique is difficult and lies in the hands of experts.

Complementary to TEM, atomic force microscopy (AFM) can be an important tool in studying the local nanostructure of self-assembled materials *(51)*. AFM is particularly well suited to define the height of assemblies deposited on a surface, a dimension not amenable to TEM analysis. Conversely, AFM is limited in its capacity to accurately measure in the *XY* dimension, which defines the width of a given assembly. For soft materials, taking measurements in the tapping mode provides a means of minimally invasive interrogation.

Small-angle neutron scattering (SANS) is an extremely powerful tool for globally defining the self-assembled structure present in a given hydrogel at both the local and network scales *(52)*. Scattering intensity is measured as a function of the reciprocal space, and the resulting data can be fit using different form factors that reveal information about the morphology and network properties of the self-assembly. In addition, SANS can be potentially used to track, in real time, the developing assembled structure; this affords information regarding the assembly mechanism. An important point worth mentioning is the use of ultra-small-angle neutron scattering (USANS) to characterize hydrogel morphology on the microscale. Such information can be critical for characterizing hydrogels used in biological applications.

These techniques, when used together, offer a powerful suite of analysis that allows relationships to be drawn among peptide sequence, secondary structure, and self-assembly mechanisms that ultimately dictate the assembly morphology and bulk material properties. These techniques can help establish the rules that govern the assembly of appropriately purified peptides so that custom materials can be fabricated for targeted applications.

4. Notes

1. Handling of synthesizer reagents: The resins, Fmoc amino acids, as well as HCTU should be stored under refrigeration to prevent degradation over time. All reagents are weighed and handled in the hood to minimize personal exposure.
2. Ultrapure water having a resistivity of 18.2 MΩ-cm obtained from a MilliQ (Millipore) purification system should be used to prepare all solutions and is referred to as "water" in this chapter.

Acknowledgments

We acknowledge the National Institutes of Health grant R01 DE016386-01. We also thank Lisa A. Haines-Butterick for optimization of the synthesizer chemistry and her helpful discussions for this chapter as well as Karthikan Rajagopal for performing the MAX3 studies.

References

1. Rajagopal K, Schneider JP. (2004) Self-assembling peptides and proteins for nanotechnological applications. *Curr. Opin. Struct. Biol.* **14**(4), 480–486.
2. Whitesides GM, Mathias JP, Seto CT. (1991) Molecular self-assembly and nanochemistry — a chemical strategy for the synthesis of nanostructures. *Science* **254**(5036), 1312–1319.

3. Carny O, Shalev DE, Gazit E. (2006) Fabrication of coaxial metal nanocables using a self-assembled peptide nanotube scaffold. *Nano Lett.* **6**(8), 1594–1597.
4. Ray S, Drew MGB, Das AK, Banerjee A. (2006) The role of terminal tyrosine residues in the formation of tripeptide nanotubes: a crystallographic insight. *Tetrahedron* **62**(31), 7274–7283.
5. Crisma M, Toniolo C, Royo S, Jimenez AI, Cativiela C. (2006) A helical, aromatic, peptide nanotube. *Org. Lett.* **8**(26), 6091–6094.
6. Leclair S, Baillargeon P, Skouta R, Gauthier D, Zhao Y, Dory YL. (2004) Micrometer-sized hexagonal tubes self-assembled by a cyclic peptide in a liquid crystal. *Angew. Chem. Int. Ed.* **43**(3), 349–353.
7. Horne WS, Stout CD, Ghadiri MR. (2003) A heterocyclic peptide nanotube. *J. Am. Chem. Soc.* **125**(31), 9372–9376.
8. Amorin M, Castedo L, Granja JR. (2005) Self-assembled peptide tubelets with 7 angstrom pores. *Chemistry* **11**(22), 6543–6551.
9. Block MAB, Hecht S. (2005) Wrapping peptide tubes: merging biological self-assembly and polymer synthesis. *Angew. Chem. Int. Ed.* **44**(43), 6986–6989.
10. Lu K, Jacob J, Thiyagarajan P, Conticello VP, Lynn DG. (2003) Exploiting amyloid fibril lamination for nanotube self-assembly. *J. Am. Chem. Soc.* **125**(21), 6391–6393.
11. Gao XY, Matsui H. (2005) Peptide-based nanotubes and their applications in bionanotechnology. *Adv. Mater.* **17**(17), 2037–2050.
12. Woolfson DN, Ryadnov MG. (2006) Peptide-based fibrous biomaterials: some things old, new and borrowed. *Curr. Opin. Chem. Biol.* **10**(6), 559–567.
13. Aggeli A, Nyrkova IA, Bell M, et al. (2001) Hierarchical self-assembly of chiral rod-like molecules as a model for peptide beta-sheet tapes, ribbons, fibrils, and fibers. *Proc. Natl. Acad. Sci. U. S. A.* **98**(21), 11857–11862.
14. Bitton R, Schmidt J, Biesalski M, Tu R, Tirrell M, Bianco-Peled H. (2005) Self-assembly of model DNA-binding peptide amphiphiles. *Langmuir* **21**(25), 11888–11895.
15. Deechongkit S, Powers ET, You SL, Kelly JW. (2005) Controlling the morphology of cross beta-sheet assemblies by rational design. *J. Am. Chem. Soc.* **127**(23), 8562–8570.
16. Elgersma RC, Meijneke T, Posthuma G, Rijkers DTS, Liskamp RMJ. (2006) Self-assembly of amylin(20–29) amide-bond derivatives into helical ribbons and peptide nanotubes rather than fibrils. *Chemistry* **12**(14), 3714–3725.
17. Lowik D, Garcia-Hartjes J, Meijer JT, van Hest JCM. Tuning secondary structure and self-assembly of amphiphilic peptides. *Langmuir* 2005;21(2):524–526.
18. Matsumura S, Uemura S, Mihara H. (2004) Fabrication of nanofibers with uniform morphology by self-assembly of designed peptides. *Chemistry* **10**(11), 2789–2794.
19. Zhang SG. (2003) Fabrication of novel biomaterials through molecular self-assembly. *Nat. Biotechnol.* **21**(10), 1171–1178.
20. Fairman R, Akerfeldt KS. (2005) Peptides as novel smart materials. *Curr. Opin. Struct. Biol.* **15**(4), 453–463.

21. Bonzani IC, George JH, Stevens MM. (2006) Novel materials for bone and cartilage regeneration. *Curr. Opin. Chem. Biol.* **10**(6), 568–575.
22. Mart RJ, Osborne RD, Stevens MM, Ulijn RV. (2006) Peptide-based stimuli-responsive biomaterials. *Soft Matter* **2**(10), 822–835.
23. Nowak AP, Breedveld V, Pakstis L, et al. (2002) Rapidly recovering hydrogel scaffolds from self-assembling diblock copolypeptide amphiphiles. *Nature* **417**(6887), 424–428.
24. Stendahl JC, Rao MS, Guler MO, Stupp SI. (2006) Intermolecular forces in the self-assembly of peptide amphiphile nanofibers. *Adv. Funct. Mater.* **16**(4), 499–508.
25. Schneider JP, Pochan DJ, Ozbas B, Rajagopal K, Pakstis L, Kretsinger J. (2002) Responsive hydrogels from the intramolecular folding and self-assembly of a designed peptide. *J. Am. Chem. Soc.* **124**(50), 15030–15037.
26. Caplan MR, Schwartzfarb EM, Zhang SG, Kamm RD, Lauffenburger DA. (2002) Control of self-assembling oligopeptide matrix formation through systematic variation of amino acid sequence. *Biomaterials* **23**(1), 219–227.
27. Collier JH, Messersmith PB. (2004) Self-assembling polymer-peptide conjugates: nanostructural tailoring. *Adv. Mater.* **16**(11), 907–910.
28. Ramachandran S, Trewhella J, Tseng Y, Yu YB. (2006) Coassembling peptide-based biomaterials: effects of pairing equal and unequal chain length oligopeptides. *Chem. Mater.* **18**(26), 6157–6162.
29. Zhang SG. (2002) Emerging biological materials through molecular self-assembly. *Biotechnol. Adv.* **20**(5–6), 321–339.
30. Yang JY, Xu CY, Wang C, Kopecek J. (2006) Refolding hydrogels self-assembled from *N*-(2-hydroxypropyl)methacrylamide graft copolymers by antiparallel coiled-coil formation. *Biomacromolecules* **7**(4), 1187–1195.
31. Shen W, Zhang KC, Kornfield JA, Tirrell DA. (2006) Tuning the erosion rate of artificial protein hydrogels through control of network topology. *Nat. Mater.* **5**(2), 153–158.
32. Ciani B, Hutchinson EG, Sessions RB, Woolfson DN. (2002) A designed system for assessing how sequence affects alpha to beta conformational transitions in proteins. *J. Biol. Chem.* **277**(12), 10150–10155.
33. Lee HJ, Lee J-S, Chansakul T, Yu C, Elisseeff JH, Yu SM. (2006) Collagen mimetic peptide-conjugated photopolymerizable PEG hydrogel. *Biomaterials* **27**(30), 5268–5276.
34. Kotch FW, Raines RT. (2006) Self-assembly of synthetic collagen triple helices. *Proc. Natl. Acad. Sci. U. S. A.* **103**(9), 3028–3033.
35. Yang ZM, Liang GL, Wang L, Bing X. (2006) Using a kinase/phosphatase switch to regulate a supramolecular hydrogel and forming the supramoleclar hydrogel in vivo. *J. Am. Chem. Soc.* **128**(9), 3038–3043.
36. Jun HW, Yuwono V, Paramonov SE, Hartgerink JD. (2005) Enzyme-mediated degradation of peptide-amphiphile nanofiber networks. *Adv. Mater.* **17**(21), 2612–2617.

37. Lutolf MP, Hubbell JA. (2005) Synthetic biomaterials as instructive extracellular microenvironments for morphogenesis in tissue engineering. *Nat. Biotechnol.* **23**(1), 47–55.
38. Pochan DJ, Schneider JP, Kretsinger J, Ozbas B, Rajagopal K, Haines L. (2003) Thermally reversible hydrogels via intramolecular folding and consequent self-assembly of a de Novo designed peptide. *J. Am. Chem. Soc.* **125**(39), 11802–11803.
39. Ozbas B, Kretsinger J, Rajagopal K, Schneider JP, Pochan DJ. (2004) Salt-triggered peptide folding and consequent self-assembly into hydrogels with tunable modulus. *Macromolecules* **37**(19), 7331–7337.
40. Kretsinger JK, Haines LA, Ozbas B, Pochan DJ, Schneider JP. (2005) Cytocompatibility of self-assembled ss-hairpin peptide hydrogel surfaces. *Biomaterials* **26**(25), 5177–5186.
41. Haines LA, Rajagopal K, Ozbas B, Salick DA, Pochan DJ, Schneider JP. (2005) Light-activated hydrogel formation via the triggered folding and self-assembly of a designed peptide. *J. Am. Chem. Soc.* **127**(48), 17025–17029.
42. Rajagopal K, Ozbas B, Pochan DJ, Schneider JP. (2006) Probing the importance of lateral hydrophobic association in self-assembling peptide hydrogelators. *Eur. Biophys. J. Biophys. Lett.* **35**(2), 162–169.
43. Chan WC, White PD. (2000) *Fmoc Solid Phase Peptide Synthesis: A Practical Approach*. Oxford University Press, New York.
44. Fields GB, Fields CG. (1991) Solvation effects in solid-phase peptide-synthesis. *J. Am. Chem. Soc.* **113**(11), 4202–4207.
45. Kates SA, Sole NA, Beyermann M, Barany G, Albericio F. (1996) Optimized preparation of deca(L-alanyl)-L-valinamide by 9-fluorenylmethyloxycarbonyl (Fmoc) solid-phase synthesis on polyethylene glycol-polystyrene (PEG-PS) graft supports, with 1,8-diazobicyclo[5.4.0]-undec-7-ene (DBU) deprotection. *Peptide Res.* **9**(3), 106–113.
46. Angell YM, Alsina J, Albericio F, Barany G. (2002) Practical protocols for stepwise solid-phase synthesis of cysteine-containing peptides. *J. Peptide Res.* **60**(5), 292–299.
47. Fasman GD. (1996) *Circular Dichroism and the Conformational Analysis of Biomolecules*. Plenum Press, New York.
48. Cantor CR, Schimmel PR. (1980) *Biophysical Chemistry*. Freeman, New York.
49. Larson RG. (1999) *The Structure and Rheology of Complex Fluids*. Oxford University Press, New York.
50. Williams DB, Carter CB. (1996) *Transmission Electron Microscopy: A Textbook for Materials Science*. Springer, New York.
51. Cohen SH, Lightbody ML. (1997) *Atomic Force Microscopy/Scanning Tunneling Microscopy 2*. Springer, New York.
52. Higgins JS, Benoît HC. (1997) *Polymers and Neutron Scattering*. Oxford University Press, New York.

6

Periodic Assembly of Nanospecies on Repetitive DNA Sequences Generated on Gold Nanoparticles by Rolling Circle Amplification

Weian Zhao, Michael A. Brook, and Yingfu Li

Summary

Periodical assembly of nanospecies is desirable for the construction of nanodevices. We provide a protocol for the preparation of a gold nanoparticle (AuNP)/DNA scaffold on which nanospecies can be assembled in a periodical manner. AuNP/DNA scaffold is prepared by growing long single-stranded DNA (ssDNA) molecules (typically hundreds of nanometers to a few microns in length) on AuNPs via rolling circle amplification (RCA). Since these long ssDNA molecules contain many repetitive sequence units, complementary DNA-attached nanospecies can be assembled through specific hybridization in a controllable and periodical manner.

Key Words: DNA; DNA amplification; enzymatic manipulation; gold; nanotechnology; rolling circle amplification; self-assembly.

1. Introduction

The precise assembly of nanostructures is a key step toward fabricating nanodevices such as nanoelectronics, nanophotonics, and biosensors *(1,2)*. In many cases, the nanospecies are required to be assembled in a well-controllable, periodical fashion, which yet still faces great challenges *(1,2)*. While the "top-down" lithographic methods approach their resolution limit, the "bottom-up" self-assembly strategies have recently gained tremendous attention in the construction of nanostructures *(1,2)*. In particular, DNA has proven to be desirable as building blocks for programmable assemblies in the material world due to the high specificity of DNA basepairing and the fact that DNA molecules can be manipulated by a number of enzymatic reactions (e.g., polymerization, digestion and ligation, etc.) *(3,4)*. We provide a protocol for the preparation of a gold nanoparticle (AuNP)/DNA scaffold and its use for further periodical assembly

From: *Methods in Molecular Biology, vol. 474: Nanostructure Design: Methods and Protocols*
Edited by: E. Gazit and R. Nussinov © Humana Press, Totowa, NJ

of nanospecies *(5)*. The AuNP/DNA scaffold is constructed by performing rolling circle amplification (RCA) reaction on a 13-nm AuNP core. RCA is a biochemical method that can generate long ssDNA with a repeating sequence unit on which the complementary DNA-attached nanospecies can be assembled periodically *(6–9)*. Moreover, the distance between two assembled nanospecies can be readily tuned by adjusting the length of circular DNA template used in the RCA reaction given that the length of each repeating unit in the RCA product is identical to that of circular template *(5)*. Furthermore, multiple assemblies in which two or more different types of nanospecies are assembled simultaneously can be potentially achieved by designing the sequence of complementary DNA or by performing different RCA reactions on each single AuNP *(5)*.

The present protocol provides experimental details of the preparation of the AuNP/DNA scaffold and the periodical assembly of 5-nm AuNPs attached with complementary DNA molecules. The techniques (e.g., polyacrylamide gel electrophoresis [PAGE], transmission electron microscopy [TEM], and atomic force microscopy [AFM]) used to characterize the RCA product, AuNP/DNA scaffold, or assembled superstructures are also provided.

2. Materials

2.1. Preparation of 13-nm AuNPs

1. Concentrated hydrochloric acid (HCl) (Fisher, 36.5–38%). Store in acid storage cabinet (*see* **Note 1**). *Caution*: Concentrated HCl can cause burns.
2. Concentrated nitric acid (HNO_3) (BDH, Toronto, Canada; 68–70%). Store in acid storage cabinet. *Caution*: Concentrated HNO_3 can cause burns.
3. Sodium citrate dihydrate ($HOC(COONa)(CH_2COONa)_2 \cdot 2H_2O$) (Sigma; >99.0%).
4. Hydrogen tetrachloroaurate (III) trihydrate ($HAuCl_4 \cdot 3H_2O$) (Sigma; >99.0%). Wrap in aluminum foil and store in a desiccator.
5. Sodium bicarbonate ($NaHCO_3$) (Sigma; >99.5%).
6. Double-deionized water (ddH_2O): produced by a Millipore water purifier.

2.2. Preparation of DNA Primer-Functionalized AuNPs (AuNP-Primer)

1. 13-nm AuNPs (~13 n*M*): Synthesized according to **Subheading 3.1**. Store at 4°C.
2. DNA1 (13 μ*M*; *see* **Note 2**): A DNA oligonucleotide with the sequence 5′-thiol-TTTTTTTTTTTTTTTTTTTTGGCGAAGACAGGTGCTTAGTC (*see* **Note 3**). It can be purchased from Keck Biotechnology Resource Laboratory (Yale University, New Haven, CT). Purified using standard PAGE procedure (*see* **Subheading 3.3., step 8**). Store at −20°C.
3. Phosphate buffer: Prepare 0.5*M* sodium phosphate monobasic ($NaHPO_4$) (Merck; molecular biology grade) and 0.5*M* sodium phosphate dibasic (Na_2HPO_4) (EMD; molecular biology grade) separately; adjust to pH 7.0 by adding $NaHPO_4$ dropwise to Na_2HPO_4. Store at 4°C.

4. 1X RCA buffer: 50 m*M* Tris-HCl (Bioshop; BioUltraPure grade), 10 m*M* $MgCl_2$ (Bioshop, >99.0%), 10 m*M* $(NH_4)_2SO_4$ (Bioshop; >99.0%), pH 7.5 (*see* **Note 4**).

2.3. Preparation of Circular DNA Template

1. DNA2 (30.6 µ*M*): A DNA oligonucleotide with the sequence 5′-TGTCTTCGCCTTCTT GTTT CCTTTCCTTGAAACTTCTTCCTTTCTTTCT TTCGACTAAGCACC (*see* **Note 5**). It can be purchased from Integrated DNA Technologies (IDT). Purify using the standard PAGE procedure (*see* **Subheading 3.3., step 8**) and store at −20°C.
2. DNA3 (78.5 µ*M*): A DNA oligonucleotide with the sequence 5′-GGCGAAGACAGGTGCT TAGTC. It can be purchased from IDT. Store at −20°C.
3. T4 polynucleotide kinase (PNK; MBI Fermentas, Burlington, Canada). Store at −20°C.
4. 10X PNK buffer A (supplied by MBI Fermentas with PNK): 500 m*M* Tris-HCl (pH 7.6), 100 m*M* $MgCl_2$, 50 m*M* dithiothreitol (DTT), 1 m*M* spermidine, and 1 m*M* EDTA (ethylenediaminetetraacetic acid). Store at −20°C.
5. T4 DNA ligase (MBI Fermentas). Store at −20°C.
6. 10X T4 DNA ligase buffer (supplied by MBI Fermentas with T4 DNA ligase): 400 m*M* Tris-HCl, 100 m*M* $MgCl_2$, 100 m*M* DTT, 5 m*M* adenosine triphosphate (ATP), pH 7.8. Store at −20°C.
7. ATP (100 m*M*; MBI Fermentas). Store at −20°C.
8. 0.5 *M* EDTA (pH 8.0): Add 186.1 g EDTA (Bioshop; biotechnology grade) to 800 mL ddH_2O. Bring to pH 8.0 using NaOH pellets. Adjust the final volume to 1 L using double-deionized water. Autoclave and store at 4°C.
9. 10X TBE stock solution: Add 432 g Tris-HCl (Bioshop; BioUltraPure grade), 220 g boric acid (Bioshop; biotechnology grade), and 80 mL of 0.5 *M* EDTA (pH 8.0) to a 5-L plastic beaker; add double-deionized water to a final volume of 4 L. Autoclave and store at 4°C.
10. 2X gel loading buffer stock: Mix 8 g sucrose (Bioshop; ultrapure grade), 10 mg bromophenol blue (Bioshop; ACS grade meets or exceeds the stringent specification set forth in the American Chemistry Society's Reagent handbook), 10 mg xylene cyanol FF (Sigma; molecular biology grade), 400 µL 10% sodium dodecyl sulfate (Bioshop; electrophoresis grade), and 4 mL of 10X TBE with enough double-deionized water to bring the volume to 40 mL; dissolve with mild heat and stirring. To the mixture, add 44 g urea (Bioshop; molecular biology grade); dissolve with mild heat and stirring. Store at 4°C.
11. 10% acrylamide stock solution: Mix 1681.7 urea (Bioshop; molecular biology grade), 400 mL of 10X TBE, and 1 L of 29:1 acrylamide/bisacrylamide premix (Bioshop; molecular biology grade); add double-deionized water to 4 L; dissolve with mild heat and stirring. Store at 4°C.
12. TEMED (*N,N,N′,N′*-tetramethylethylenediamine) (Bioshop; >99.0%). Store at 4°C.
13. 10% ammonium persulfate (APS) (Bioshop; electrophoresis grade). Store at 4°C.
14. 3 *M* sodium acetate (Sigma; >99.0%). Adjust to pH 7.0 using acetic acid (Sigma; ACS grade).

15. Elution buffer stock: Mix 8 mL of 5*M* NaCl, 2 mL of 1*M* Tris-HCl (pH 7.5), 0.4 mL of 0.5*M* EDTA (pH 8.0); add double-deionized water to 200 mL. Autoclave and store at 4°C.
16. Polyacrylamide gel electrophoresis: Standard gel electrophoresis apparatus (EC3000-90, E-C Apparatus Corp.); glass plates; spacers; comb; bulldog clamps; aluminum plate; Kimwipes; plastic wrap; spatula; handheld UV lamp; razor blade; 100-mL plastic beaker.

2.4. Preparation of AuNP/DNA Scaffold

1. AuNP-primer (~13 n*M*): *See* **Subheading 3.2**. for preparation. Store at 4°C.
2. Circular DNA template (20 μ*M*): *See* **Subheading 3.3**. for preparation. Store at −20°C.
3. 10X RCA buffer: 500 m*M* Tris-HCl, 100 m*M* $MgCl_2$, 100 m*M* $(NH_4)_2SO_4$, pH 7.5; 1X RCA buffer.
4. dNTP (deoxyribonucleotide 5′-triphosphate) mixture (deoxyadenosine 5′-triphosphate [dATP], deoxythymidine 5′-triphosphate [dTTP], deoxyguanosine 5′-triphosphate [dGTP], deoxycytidine 5′-triphosphate [dCTP]; 10 m*M* each; Fermentas). Store at −20°C.
5. [α-^{32}P]-dGTP (GE Healthcare). Store in a special radioisotope storage cabinet.
6. ϕ29 (phi29) DNA polymerase (New England Biolabs; NEB). Store at −20°C.
7. Polyacrylamide gel electrophoresis (*see* **Subheading 2.3., item 16**).
8. 2-Mercaptoethanol (MCE) (Sigma; >99.0 %).

2.5. Preparation of 5-nm AuNPs Attached With a DNA Oligonucleotide

1. DNA4 (91.8 μ*M*): A 5′-biotinylated DNA oligonucleotide with the sequence 5′-biotin-CCTTGAAACTTCTTCCTTTCTTTCT (purchased from Keck Biotechnology Resource Laboratory). Purify using standard PAGE procedure (*see* **Subheading 3.3., step 8**). Store at −20°C.
2. Streptavidin-coated 5-nm AuNPs (~83 n*M*; Bbinternational). Store at 4°C.
3. 10X streptavidin-biotin coupling buffer: 200 m*M* Tris-HCl, 200 m*M* EDTA, 3*M* NaCl, pH 7.5; 1X streptavidin-biotin coupling buffer.
4. 1X RCA buffer: 50 m*M* Tris-HCl, 10 m*M* $MgCl_2$, 10 m*M* $(NH_4)_2SO_4$, pH 7.5.

2.6. Assembly of 5-nm AuNPs on AuNP/DNA Scaffold

1. AuNP/DNA scaffold solution (~2.6 n*M*): *See* **Subheading 3.4**. for preparation.
2. 5-nm AuNPs attached with DNA3 (~83 n*M*): *See* **Subheading 3.5**. for preparation.
3. 1X RCA buffer: 50 m*M* Tris-HCl, 10 m*M* $MgCl_2$, 10 m*M* $(NH_4)_2SO_4$, pH 7.5.

3. Methods

3.1. Preparation of 13-nm AuNPs (10,11)

1. Prepare 200 mL of aqua regia solution by slowly adding 50 mL of concentrated HNO_3 to 150 mL of concentrated HCl in a 1000-mL glass beaker under stirring. (*Caution*: Aqua regia is highly corrosive. Always wear appropriate protective equipment such as goggles, gloves, and laboratory coat. Work in a chemical fume

hood. Aqua regia should be freshly prepared and never be stored. After use, aqua regia should be disposed after dilution and neutralization.)

2. Clean the glassware and magnetic stir bar with aqua regia, then rinse them with copious double-deionized water.
3. Dissolve 0.098 g $HAuCl_4 \cdot 3H_2O$ in a 500-mL two-neck flask with 250 mL ddH_2O (the final Au^{3+} concentration is 1 m*M*). The solution has a pale-yellow color.
4. Assemble the flask with a condenser and a stopper. Heat the flask on a hot plate to vigorous reflux while stirring (generally, the solution starts refluxing with stirring at about 110°C or above).
5. While the gold solution is being heated, prepare 25 mL of 38.8 m*M* sodium citrate solution in a 50-mL glass beaker by adding 0.28 g sodium citrate dihydrate to 25 mL ddH_2O.
6. When the Au^{3+} solution is refluxing vigorously (crucial), remove the stopper, quickly (crucial) add the sodium citrate solution using a 60-mL syringe. A color change, first from yellow to black and then to deep red, should be observed in about 3 min.
7. Once the color change starts, the solution is refluxed for another 15 min. Then, remove the hot plate and allow the solution to cool to room temperature under stirring. Transfer the AuNP solution to an amber-color glass storage container and store at 4°C (do *not* freeze). Normally, this AuNP solution is stable for at least 6 mo without any precipitation and color change.
8. Clean all the glassware with aqua regia. Safely dispose aqua regia as follows: First dilute with water and then neutralize by sodium bicarbonate (*caution*: add sodium bicarbonate slowly to the diluted acid solution; CO_2 gas produced in the process may cause acid spill).
9. Characterize the AuNPs with TEM on a JEOL 1200 EX instrument. The TEM sample is prepared by dropping AuNP solution (4 µL) onto a carbon-coated copper grid. After 1 min, the solution is wicked from the edge of the grid with a piece of filter paper. The TEM images can then be measured with a standard transmission electron microscope. The as-prepared AuNPs are spherical and about 13 nm in diameter (*see* **Note 6**).
10. Characterize the AuNPs with a UV-visible spectrophotometer (Cary 300). A typical UV-visible spectrum of 13-nm AuNP solution has a characteristic surface plasmon band at about 520 nm. The extinction coefficient for 13-nm AuNPs is about 2.7×10^8 M^{-1} cm^{-1} *(12)*. The concentration of as-prepared AuNP solution is about 13 n*M*.

3.2. Preparation of DNA Primer-Functionalized AuNPs (10,11)

1. Place 30 µL of 13 µ*M* (*see* **Note 7**) (390 pmol) thiol-modified DNA primer (DNA1) in a 1.5-mL microcentrifuge tube. Add 50 µL of AuNPs (prepared in **Subheading 3.1.**) to the tube with gentle shaking by hand (*see* **Note 8**).
2. Let the solution stand for 8 h or longer (e.g., overnight) at room temperature. Add 2 µL of 0.5*M* phosphate buffer (pH 7.0) to obtain a final phosphate concentration of 10.8 m*M*. Add 10 µL of 1*M* NaCl to obtain a final NaCl concentration of 108.6 m*M*. Shake the tube gently by hand to mix. Leave the tube at room temperature for 12 h.

3. Add 1 μL of 0.5*M* phosphate buffer and 25 μL of 1*M* NaCl to the tube. The final concentrations of phosphate and NaCl are 12.5 m*M* and 296.7 mM, respectively. Mix the solution well by gently shaking. Let the tube stand at room temperature for another 12 h.
4. Centrifuge the tube at 18,800*g* at room temperature on a benchtop centrifuge for 15 min. After centrifugation, the reddish AuNP-primer precipitate should be at the bottom of the vial, and the supernatant should be clear (if there is still some AuNP left in the supernatant, centrifuge for another 5 min). Gently remove as much supernatant as possible using a pipet.
5. Wash AuNP-primer by adding 200 μL of 1X RCA buffer (well-functionalized AuNP should be easily redispersed) and gently vortexing for about 15 s. Centrifuge the tube at 18,800*g* at room temperature for 15 min. Gently pipet off the supernatant. Wash the pellet twice more using the same centrifugation/redispersion cycles. Collect all the supernatant solutions.
6. Redisperse AuNP-primer in 50 μL of 1X RCA buffer (the final concentration of AuNP-primer is about 13 n*M* assuming there are no AuNPs lost during the washing steps) and store at 4°C.
7. Measure the absorbance of the supernatant solutions at 260 nm using a UV-visible spectrophotometer (Cary 300). Calculate the number of picomoles x of uncoupled primer using Oligo Calculator (MCLAB). The number N of primers on each 13-nm AuNP can then be calculated: $N = (390\,\text{pmol} - x\,\text{pmol})/(50 \times 0.013\,\text{pmol})$. Generally, there are about 150–200 DNA primers on each 13-nm AuNP.

3.3. Preparation of Circular DNA Template (13)

1. Mix 16.3 μL of 30.6 μ*M* linear DNA (DNA2) (500 pmol), 26.7 μL ddH_2O, 5 μL of 10X PNK buffer A, and 1 μL of 100 m*M* ATP in a 1.5-mL microcentrifuge tube. Gently vortex. Add 1 μL of PNK and mix well by pipeting up and down. Place the tube in a thermomixer at 37°C for 30 min.
2. Place the tube in the dry heat block at 90°C for 5 min to quench the reaction. Remove the tube from the heat block and allow it to cool at room temperature for 10 min.
3. Add 7.6 μL of 78.5 μ*M* (600 pmol) template oligonucleotide (DNA3) and 209.4 μL ddH_2O. Vortex thoroughly. Heat the tube in a dry heat block at 90°C for 30 s and allow it to cool at room temperature for 10 min.
4. Add 30 μL of 10X T4 DNA ligase buffer and 3 μL of T4 DNA ligase. Mix well by pipeting. Let it stand at room temperature overnight (i.e., 12 h).
5. Heat the tube in the dry heat block at 90°C for 10 min to quench the reaction. After brief cooling, add 900 μL of 100% ethanol. Vortex thoroughly. Place the tube in a −20°C freezer overnight.
6. Centrifuge the tube at 18,800*g* for 30 min. The DNA (often pellets with salt) should be at the bottom of the tube. Gently pipet off the supernatant. Dry the tube using a Speedvac concentrator at room temperature (usually takes about 10 min).
7. Add 10 μL ddH_2O and 10 μL of 2X gel loading buffer. Vortex for at least 1 min to completely dissolve the circular DNA product. Purify by 10% denaturing PAGE (*see* below).

8. During the centrifugation and drying process, set up 10% denaturing PAGE ***(14)*** as follows: Rinse two glass plates with double-deionized water followed by 95% ethanol. Dry the plates with Kimwipes. Clean two spacers, one comb, and one 100-mL plastic beaker with double-deionized water. Place the spacers in between the glass plates. Clamp the glass plates with four bulldog clamps and lay the plates flat on a plastic stand. Prepare acrylamide solution (always wear goggles, gloves, and laboratory coat) by adding 40 mL of 10% acrylamide stock solution, 40 μL of TEMED solution, and 400 μL of APS solution to a 100-mL plastic beaker. Mix thoroughly by pipet tip. Press the bottom of the plates by hand to provide a shallow slope from top to bottom. Pour the solution across the top of the plates and allow it to run between the plates. Continue to add acrylamide solution (do not let the top run dry or bubbles will form) until the solution fills the space between the plates (do not overfill). Lay the plates flat and insert comb. Polymerize for about 30 min. After polymerization, remove comb and rinse the top of gel with double-deionized water right away. Mount the plates together with an aluminum plate (to allow heat dispersion) in apparatus. Fill the upper and lower reservoirs of the electrophoresis tank with 1X TBE. Wash the wells with 1X TBE using a syringe with a needle. Prerun the gel for about 30 min at 600 V.
9. Heat the ligation solution (prepared in **step 7**) at 90°C for 1 min. Load the sample into one well. Run the gel at 600 V until the tracking dyes indicate that the DNA has migrated one-half to three-fourths of the way through the gel.
10. After the DNA is sufficiently resolved, turn off the power supply and disassemble the plates from the electrophoresis tank. Gently lift the top plate using a spatula. Cover the gel with plastic wrap evenly and smoothly and then flip over the gel. Peel a corner of the gel away from the remaining plate using the spatula and cover the gel by another sheet of plastic wrap.
11. Place the wrapped gel on a fluorescent thin-layer chromatographic (TLC) plate (20 × 20 cm silica gel $60F_{254}$; EM Science). Visualize the gel using a handheld UV lamp. The circular DNA product band will appear as the darkest black shadow right above the top tracking dye (xylene cyanol FF migrates as a 55-nt DNA fragment in 10% denaturing PAGE; ***15***) (*see* **Note 9**).
12. Cut out the band with a razor blade. Chop the gel into small pieces and put them in 1.5-mL microcentrifuge tube. Add 400 μL of elution buffer to the tube and place the tube in a thermomixer at 37°C with slightly shaking (101 g) overnight.
13. Spin down the gel with a benchtop minicentrifuge. Gently pipet off the solution to another 1.5-mL microcentrifuge tube. Add 1 mL of 100% ethanol and 40 μL of 3*M* sodium acetate (pH 7.0) to the tube. Place it in a −20°C freezer overnight.
14. Centrifuge the tube at 18,800*g* for 30 min. Gently pipet off the supernatant. Carefully add 100 μL of 70% cold ethanol, close the lid, and gently (but thoroughly) rinse the inside surface of the tube. Centrifuge the tube at 18,800*g* for about 10 min and remove the supernatant. Wash one more time with 70% cold ethanol. Dry the tube using a Speedvac concentrator at room temperature. Dissolve the circular DNA product with 100 μL ddH_2O. Determine the DNA concentration based on the absorbance at 260 nm. Store at −20°C.

3.4. Preparation of AuNP/DNA Scaffold

1. Mix 4 μL (*see* **Note 10**) of about 13 n*M* AuNP-primer (prepared in **Subheading 3.2.**), 1 μL of 20 μ*M* circle DNA template (20 pmol; prepared in **Subheading 3.3.**), 2 μL of 10X RCA reaction buffer, and 13 μL ddH_2O in a 1.5-mL microcentrifuge tube. Place the vial in a heating block at 60°C (*see* **Note 11**) for 5 min. Remove the tube from the heating block and allow it to cool at room temperature for 30 min.
2. Centrifuge the tube at 18,800*g* for 15 min. Gently remove the supernatant by a pipet. Wash the pellet twice with 50 μL of 1X RCA buffer using a centrifugation/redispersion cycle. Finally, redisperse the pellet in 20 μL of 1X RCA buffer.
3. Add 1 μL of dNTP mixture (10 m*M*) (for the synthesis of radioactive RCA product, 1 μL of [α-^{32}P]-dGTP [10 μCi] is incorporated. *Caution*: Be extremely careful handling radioisotopes. Always wear gloves, goggles, and protective clothing. All the experiments are performed behind an appropriate shield). Add 2 μL of phi29 DNA polymerase. Mix thoroughly using the pipet tip. Place the tube in a thermomixer at 30°C for 30 min (*see* **Note 12**).
4. Wash the product twice using centrifugation/redispersion cycles using 50 μL of 1X RCA buffer. The final AuNP/DNA scaffold is redispersed in 20 μL of 1X RCA buffer (the concentration of AuNP/DNA scaffold is about 2.6 n*M* assuming no AuNPs are lost during the washing steps).
5. Characterize the AuNP/DNA scaffold by AFM as follows: Attach mica onto an AFM substrate using double-sided adhesive tape. Cleave the mica by peeling off external layers of mica using Scotch tape. Dilute the AuNP/DNA scaffold solution (prepared in **Subheading 3.4., step 4**) five times using 1X RCA buffer. Drop about 50 μL of diluted solution onto the mica substrate. Use a piece of filter paper to wick away most of the solution from the edge of the mica substrate and allow it to dry in air for about 10 min. Mount the sample on the microscope (Digital Instruments Nanoscope II) and take images. A typical AuNP/DNA scaffold contains a few to tens of DNA molecules (typically a few hundreds of nanometers) that are attached on the AuNP surface (**Fig. 1A**). Due to the drying process, in some cases two or more ssDNA molecules (gray arrow) may stick together to form a double-stranded (ds) DNA-like structure (white arrow) (**Fig. 1B**).
6. Characterize the RCA product using PAGE: Treat the AuNP/DNA (radiolabeled) scaffold solution (prepared in **Subheading 3.4., step 4**) with MCE solution (250 m*M* final concentration of MCE) at 37°C for 2 h to displace the RCA product from the AuNP surface. Centrifuge the tube at 18,800*g* for 15 min. Analyze the DNA in the supernatant by denaturing PAGE (*see* **Subheading 3.2., steps 8–10**); the radiolabeled RCA product in the PAGE gel is visualized using a phosphorimaging plate and a Typhoon 9200 scanner (GE Healthcare). An intensive DNA band can be observed at the top of 10% denaturing PAGE gel as RCA product migrates very slowly due to its large size (typically >10,000 nt) *(5)*.

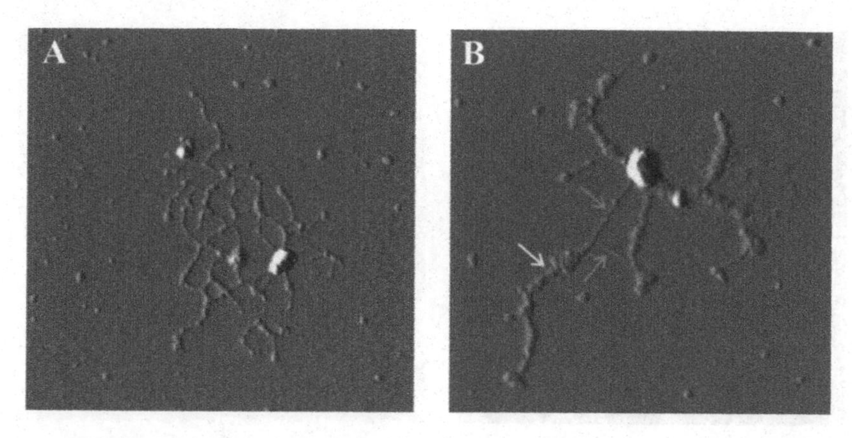

Fig. 1. Representative atomic force microscopy (AFM) images of gold nanoparticle (AuNP)/DNA scaffold *(5)*. This figure is adapted from **ref. 5**, with permission.

3.5. Preparation of 5-nm AuNPs Attached With a DNA Oligonucleotide Complementary to the RCA Product

1. Mix 2.5 µL of 91.8 µ*M* DNA4 (a 5′-biotinylated DNA oligonucleotide), 50 µL of streptavidin-coated 5-nm AuNPs (~83 n*M*), 10 µL of 10X streptavidin-biotin coupling buffer, and 37.5 µL of ddH_2O in a 1.5-mL microcentrifuge tube. Allow the mixture to stand at room temperature for 30 min.
2. Wash the DNA2-attached 5-nm AuNPs twice with a centrifugation/redispersion cycle using 100 µL of 1X RCA buffer. The sample is then redispersed in 50 µL of 1X RCA buffer (~83 n*M* assuming there are no AuNPs lost during the washing steps).

3.6. Assembly of 5-nm AuNPs on AuNP/DNA Scaffold

1. Mix the AuNP/RCA product solution (20 µL, ~2.6 n*M*, prepared in **Subheading 3.4.**) with DNA4-attached 5-nm AuNPs (50 µL, ~83 n*M*, prepared in **Subheading 3.5.**) in a 1.5-mL microcentrifuge tube. Bring the temperature of the mixture to 65°C using a thermomixer and allow it to cool very slowly to room temperature (~0.5°C/min).
2. Centrifuge the tube at 2400*g* for 10 min. Wash the assembled superstructure once with 100 µL of 1X RCA buffer using centrifugation. Redisperse the pellet in 50 µL of 1X RCA buffer.
3. Characterize the assembled superstructure by AFM (*see* **Subheading 3.4., step 5**) and TEM (*see* **Subheading 3.1., step 9**). A typical assembled superstructure shows that 5-nm AuNPs periodically assembled on the DNA molecules attached to the 13-nm AuNP core (**Fig. 2**) *(5)*. The distance of two assembled 5-nm AuNPs is about 21 nm (**Fig. 2B**), which is similar to the length of the repeating DNA unit in the RCA product *(5)* (*see* **Note 13**).

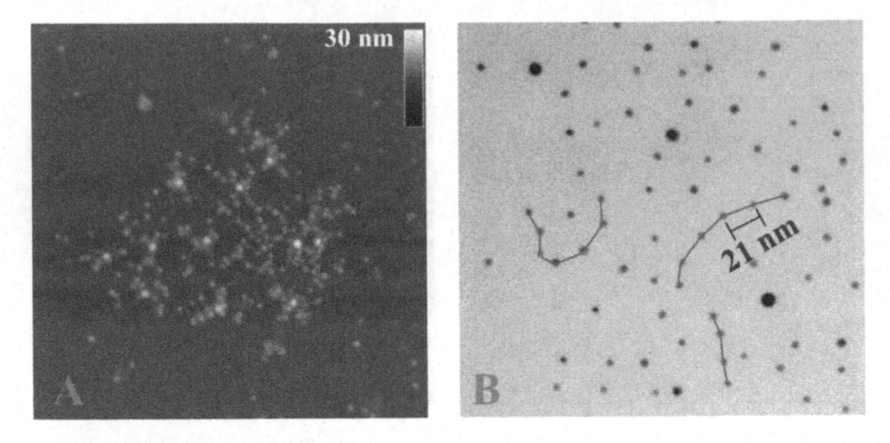

Fig. 2. Representative atomic force microscopy (AFM) image (**A**) and transmission electron microscopy (TEM) image (**B**) for the assembled superstructures. This figure is adapted from **ref. *5***, with permission.

4. Notes

1. All reagents and samples are stored at room temperature (~23°C) unless stated otherwise.
2. DNA concentrations are determined using a UV-visible spectrophotometer based on the absorbance at 260 nm and calculated according to Oligo Calculator (MCLAB).
3. The sequence of an RCA primer should contain at least 15 nucleotides for effective hybridization with a circular template. It was found that the sequence of primer and circular template had little effect on the performance of RCA reactions ***(16,17)***. When designing DNA sequences, it is important to check for potential hairpin or internal duplex formation using the Mfold program ***(18)***.
4. DTT should be absent in 1X RCA reaction buffer as this reagent may cause the dissociation of thiol-modified primer from the AuNP surface.
5. Generally, a linear ssDNA that is 50–100 nt long can be easily ligated to produce a circular DNA; however, a linear ssDNA of 30 nt or shorter may be difficult to convert into a circular DNA.
6. The 13-nm AuNPs are used in our studies due to their high quality and high reproducibility ***(10,11)***. However, AuNPs of other sizes (10–50 nm) could also be used.
7. Other DNA concentrations could also be used, but the reaction volume should be varied accordingly to obtain the same amount of product.
8. The reaction volume in the preparation of AuNP-primer, AuNP/DNA scaffold, 5-nm AuNPs attached with complementary DNA oligonucleotides, or assembled superstructures can be scaled up to prepare more product.
9. A circular DNA migrates slower than the linear DNA with the same sequence. When analyzing ligation reaction mixtures for the production of a circular DNA, lighter bands that migrate slower than the circular DNA may also be observed;

these bands correspond to linear dimmers, trimers, and so on produced by intermolecular ligation.

10. ϕ29 DNA polymerase becomes less effective when larger amounts of AuNP-primer are used *(5)*.
11. Heating at elevated temperature (e.g., 90°C) may cause the dissociation of thiol-modified primer from the AuNP surface *(19)*.
12. A longer reaction time can be used to produce more RCA product. It is important to note, however, that a longer reaction time (e.g., 2 h) may cause instability of AuNPs as DTT (from ϕ29 DNA polymerase storage buffer) present in the reaction solution can undergo ligand exchange with the thiol-modified primer on the AuNP surface.
13. In some cases, aggregated AuNPs are observed as each 5-nm AuNP may carry multiple complementary DNA that can cause the crosslink between the RCA product and 5-nm AuNPs. This problem could potentially be solved by assembling mono-DNA-functionalized nanospecies *(20)*.

Acknowledgments

This work was supported by the Natural Sciences and Engineering Research Council of Canada (NSERC); SENTINEL: Canadian Network for the Development and Use of Bioactive Paper; and the Canadian Institutes for Health Research (CIHR). We wish to thank Dr. Yan Gao for help with the AFM and Dr. Srinivas A. Kandadai for the helpful discussion. Y.L. holds a Canada Research Chair.

References

1. Katov NA. (2006) *Nanoparticle Assemblies and Superstructures*. CRC Press, Taylor and Francis, Boca Raton, FL.
2. Zhang J, Wang Z, Liu J, Chen S, Liu G. (2003) *Self-Assembled Nanostructures*. Kluwer Academic/Plenum, New York.
3. Seeman NC. (2003) DNA in a material world. *Nature* **429**, 427–431.
4. Rosi NL, Mirkin C. (2005) Nanostructures in biodiagnostics. *Chem. Rev.* **105**, 1547–1562.
5. Zhao W, Gao Y, Kandadai SA, Brook MA, Li Y. (2006) DNA polymerization on gold nanoparticles through rolling circle amplification: towards novel scaffolds for three-dimensional periodic nanoassemblies. *Angew. Chem. Int. Ed.* **45**, 2409–2413.
6. Fire A, Xu S. (1995) Rolling replication of short DNA circles. *Proc. Natl. Acad. Sci. U. S. A.* **92**, 4641–4645.
7. Liu D, Daubendiek SL, Zillman MA, Ryan K, Kool ET. (1996) Rolling circle DNA synthesis: small circular oligonucleotides as efficient templates for DNA polymerases. *J. Am. Chem. Soc.* **118**, 1587–1594.
8. Beyer S, Nickels P, Simmel FC. (2005) Periodic DNA nanotemplates synthesized by rolling circle amplification. *Nano Lett.* **5**, 719–722.

9. Deng Z, Tian Y, Lee S, Ribbe AE, Mao C. (2005) DNA-encoded self-assembly of gold nanoparticles into one-dimensional arrays. *Angew. Chem. Int. Ed.* **44**, 3582–3585.
10. Hill HD, Mirkin CA. (2006) The bio-barcode assay for the detection of protein and nucleic acid targets using DTT-induced ligand exchange. *Nat. Protocol.* **1**, 324–336.
11. Liu J, Lu Y. (2006) Preparation of aptamer-linked gold nanoparticle purple aggregates for colorimetric sensing of analytes. *Nat. Protocol.* **1**, 246–252.
12. Jin R, Wu G, Li Z, Mirkin CA, Schatz GC. (2003) What controls the melting properties of DNA-linked gold nanoparticle assemblies? *J. Am. Chem. Soc.* **125**, 1643–1654.
13. Billen LP, Li Y. (2004) Synthesis and characterization of topologically linked single-stranded DNA rings. *Bioorg. Chem.* **32**, 582–598.
14. Rickwood D, Hames BD. (1990) *Gel Electrophoresis of Nucleic Acids: A Practical Approach*. IRL Press, Oxford, NY.
15. Sambrook J, Russell DW. (2001) *Molecular Cloning: A Laboratory Manual*, 3rd ed. Cold Spring Harbor Laboratory Press, Cold Spring Harbor, NY.
16. Kuhn H, Demidov VV, Frank-Kamenetskii MD. (2002) Rolling-circle amplification under topological constraints. *Nucleic Acids Res.* **30**, 574–580.
17. Lin C, Xie M, Chen J, Liu Y, Yan H. (2006) Rolling-circle amplification of a DNA nanojunction. *Angew. Chem. Int. Ed.* **45**, 7537–7539.
18. Zuker M. (2003) Mfold web server for nucleic acid folding and hybridization prediction. *Nucleic Acids Res.* **31**, 3406–3415.
19. Li Z, Jin R, Mirkin CA, Letsinger RL. (2002) Multiple thiol-anchor capped DNA-gold nanoparticle conjugates. *Nucleic Acids Res.* **30**, 1558–1562.
20. Xu X, Rosi NL, Wang Y, Huo F, Mirkin CA. (2006) Asymmetric functionalization of gold nanoparticles with oligonucleotides. *J. Am. Chem. Soc.* **128**, 9286–9287.

II

Computational Approach

7

Protocols for the *In Silico* Design of RNA Nanostructures

Bruce A. Shapiro, Eckart Bindewald, Wojciech Kasprzak, and Yaroslava Yingling

Summary

Recent developments in the field of nanobiology have significantly expanded the possibilities for new modalities in the treatment of many diseases, including cancer. Ribonucleic acid (RNA) represents a relatively new molecular material for the development of these biologically oriented nanodevices. In addition, RNA nanobiology presents a relatively new approach for the development of RNA-based nanoparticles that can be used as crystallization substrates and scaffolds for RNA-based nanoarrays. Presented in this chapter are some methodological shaped-based protocols for the design of such RNA nanostructures. Included are descriptions and background materials describing protocols that use a database of three-dimensional RNA structure motifs; designed RNA secondary structure motifs; and a combination of the two approaches. An example is also given illustrating one of the protocols.

Key Words: Molecular dynamics; molecular mechanics; RNA building blocks; RNA 3D modeling; RNA databases; RNA motifs; RNA nanostructure; RNA secondary structure; RNA structure.

1. Introduction

The field of nanobiology holds great opportunities for the development of methodologies for treatment of various diseases, including cancer, as well as for the development of tools that can be used for the diagnosis and prognosis of these diseases. RNA (ribonucleic acid) represents a relatively new molecular modality for the development of these biologically oriented nanodevices. Without associated proteins, RNA-based therapeutics have a relatively low immune response and therefore represent a safe and effective means for the delivery of multiple therapeutic agents. These include siRNAs (small interfering RNAs) for downregulation of gene expression of multiple genes in multiple cell types; RNA aptamers (RNA sequences and structure that target specified binding sites);

From: *Methods in Molecular Biology, vol. 474: Nanostructure Design: Methods and Protocols*
Edited by: E. Gazit and R. Nussinov © Humana Press, Totowa, NJ

ribozymes (RNAs with catalytic properties); antisense RNA (for binding specific sites for gene regulation); and molecular beacons and sensors. In addition, RNA nanoscale entities can be used for the development of self-assembling nanoarrays, which may contain functional molecular entities such as biosensors or act as substrates for crystallization.

2. Structural and Functional Capabilities of RNA for Nanotechnology

RNA is a biopolymer that is composed of four nucleotides, two purines (adenine and guanine), and two pyrimidines (uracil and cytosine). These nucleotides are usually abbreviated as A, G, U, and C, respectively. RNA is synthesized as a single-stranded molecule that can fold on itself via complementary basepairing interactions, usually between purines and pyrimidines. The strongest basepair is G-C, followed by A-U and G-U and other possible combinations. These basepairs tend to stack and form helices, which are energetically favorable. The helical region of RNA has a well-known nanometer-scale structural geometry, approximately 2.86 nm per helical turn with 11 basepairs and an approximate 2.3-nm diameter. The unpaired nucleotides can form bulges, symmetric and asymmetric internal loops and hairpin loops (*see* **Fig. 1** for some typical examples), single-stranded overhangs or sticky tails, and other unpaired motifs. Unpaired nucleotides usually aid in the formation of helical junctions and bends. Moreover, the unpaired nucleotides can facilitate further structural assembly into more complex structures by forming complementary basepairing or inter- and intramolecular interactions of the different single-stranded regions in the RNA. One of the most important examples of such assembly would be loop–loop contacts, such as kissing loops (*see* **Fig. 1**). Thus, even though RNA is chemically simple, involving only four nucleotides, it can form complex structures that are directly related to its complex function. However, artificial RNA structures can also be created in which the basepairing can be relatively simply controlled and predicted due to the natural tendency of RNA to form basepairs.

RNAs have very versatile functionalities, and the importance and capabilities of these are still being discovered. RNA functionalities include a carrier of genetic information (messenger RNA), catalytic properties, RNA editing, gene silencing, and transcriptional and translational control, to name a few.

The versatility of RNA's function and structure makes it a very good molecule for use in nanobiology. Moreover, a protein-free RNA does not induce significant immune response and thereby makes RNA nanoparticles attractive for medicinal use. Such all-RNA nanoparticles can thus limit antibody production and cellular immune reaction. Thus, nanoparticles constructed entirely from RNA can potentially be used in long-term treatment of chronic diseases such as cancer, hepatitis B, or AIDS. However, a limiting factor for RNA nanoparticles is their stability in the bloodstream and nonspecific cellular uptake, posing a

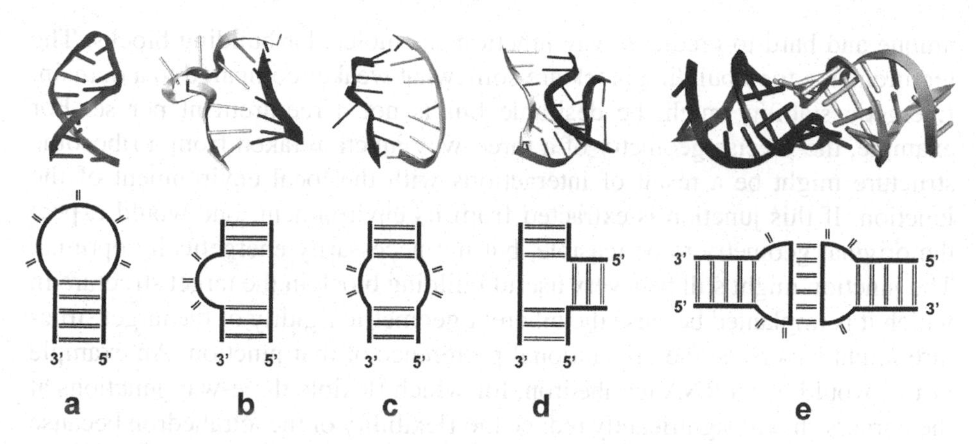

Fig. 1. Typical RNA motifs. Illustrated are the secondary and the associated three-dimensional (3D) structural motifs that are frequently found in RNA structures. These are representative of some of the common building blocks that can be used to construct RNA nanoparticles. Seen are (**a**) a hairpin loop, (**b**) a triple base bulge loop, (**c**) a symmetric internal loop, (**d**) a multibranch three-way junction, and (**e**) a kissing loop structure construct with two complementary hairpin loops.

requirement for high dosage. The relative stability can be controlled by the chemical modification of the backbone; addition of proteins, lipids *(1)*, or polymeric chains; as well as the formation of relatively compact structures *(2,3)*. Targeting can be controlled by the use of appropriate RNA aptamers.

3. Building Block Approach for the Design of RNA Nanoparticles

The prediction and design of large macromolecular structures is drastically simplified if one can adopt a modular approach *(4)*. This means designing different regions of the target structure independently. This approach works best if the target structure is separable into clearly defined domains. Ideal domains are portions of the macromolecular assembly that fold virtually independently into energetically stable units.

RNA can, in many cases, be regarded as a set of hierarchical modular objects *(4,5)*. This is exemplified by the fact that pseudoknot-free RNA secondary structures of single sequences can be represented as tree data structures *(6,7)*. The more interactions the RNA nanostructure possesses beyond a simple tree representation (like pseudoknots, or non-canonical tertiary interactions), the more the modularity of the RNA structure will be weakened.

Even if semi-independent domains cannot be identified, it is still possible to use fragments of RNA molecules as building blocks. Pseudoknot-free RNA structures can be separated into *n*-way junctions, hairpin structures, single-stranded regions, and helices. One approach is to extract from known structures unique and hard-to-predict *n*-way junctions as molecular building blocks. The requirements for a building block are somewhat weaker compared to a domain. Energetic stability might be desirable but is not a requirement per se. For example, the precise geometry of a three-way junction taken from a ribosome structure might be a result of interactions with the local environment of the junction. If this junction is extracted from its environment, one would expect the original geometry to be feasible, but not necessarily energetically optimal. The junction might still be a very useful building block in the target structure in which it is implanted because the inherent geometric rigidity of the target structure might supersede the orientational preference of that junction. An example of this would be an RNA tetrahedron, for which flexible three-way junctions at the corners should significantly reduce the flexibility of the tetrahedron because of the inherent overall rigidity of the structure.

Several experimental groups have shown that RNA can be used as efficient nanoparticles and scaffolds that are built from specific building blocks. Small structural fragments found in the ribosome and HIV have been used in the design of artificial RNA building blocks, called *tectoRNAs* ***(8)***. Each tectoRNA contains four right-angle motifs. These right-angle motifs have a structural element that forms a 90° angle between two adjacent helices, which are capped with a hairpin loop. The hairpin loops are programmed to interact with each other in a precise manner via formation of specific noncovalent loop–loop interactions, which are based on HIV kissing loops. Each right-angle motif also has an interacting single-stranded 3′ end, called a *sticky tail*. The sticky tails further allow the assembly of tetramers into complex nanoarrays. Thus, these tectoRNAs can be programmed to self-assemble into novel nano- and mesoscale biofabrics with controllable directionality, topology, and geometry ***(8,9)***. Moreover, such RNA–RNA interactions can guide the precise deposition of gold nanoparticles ***(10)***. For example, self-assembling tectoRNA ladders have been shown to induce a precise linear arrangement of cationic gold nanoparticles, demonstrating that RNA can control regular spacing of gold nanoparticles and can act as a nanocrown scaffold ***(11)***.

Another example is derived from the use of the bacteriophage phi29-encoded RNA (pRNA), which has been reengineered to form dimers, trimers, rods, hexamers, and three-dimensional (3D) arrays several microns in size through interactions of interlocking loops ***(12,13)***. Most prominently, a nanoparticle containing a pRNA trimer as a delivery vehicle was used to carry siRNAs to a cell. The cell targeting was accomplished with an RNA receptor-binding aptamer. This RNA nanoparticle was able to block cancer development in cell culture and living mice ***(14,15)***.

Also, H-shaped RNA molecular units were built from a portion of the group I intron domain ***(5,16,17)***. They were shown to be able to form oriented filaments ***(9,17)***. Specific RNA nanoarrangements based on HIV dimerization initiation site stem loops ***(18,19)*** were also shown to be capable of thermal isomerization to alternative structures ***(20)***.

4. RNA Fabrication Techniques

Two main fabrication techniques can be used for programmable self-assembly of nucleic acid nanostructures *(21)*. Assembly may be accomplished through a single-step process, as exemplified by the building of DNA nanostructures *(22,23)*, or in a stepwise fashion, as was used in the building of RNA nanostructures *(8)*. In the single-step approach, all the component molecules are placed together. This mixed set of molecules is then slowly cooled. The methodology works if the building block structure that is desired is more stable than any other possible structures. Thus, through the annealing process, the building block structures should form first at higher temperatures. As the temperature is lowered, the weaker interactions associated with the nanostructure self-assembly will occur next. Sometimes it is difficult to design sequences that, when folded, have the desired structure well separated energetically from non-desirable alternatives. If this is the case, it may be necessary first to form the building blocks, then at a later stage these building blocks are mixed together to allow for their self-assembly. In addition, it may be necessary to add Mg^{2+} ions to stabilize tertiary interactions. The multistep approach is more time consuming, and it is also important that the melting temperatures of the populations be well separated.

5. Computational Design

Computer modeling is well suited for assisting the experimental community in the design of novel nanostructures. The major advantage of computational nanodesign is that it provides a relatively inexpensive and fast way to explore many structural designs and assess their properties. One can view the design process as proceeding in at least two ways: using an RNA 3D motif approach or using a RNA secondary structure motif approach. These two methodologies have their individual strengths and weaknesses. It is also possible, where appropriate, to iterate back and forth between some of the steps in either protocol at particular points. **Figure 2** illustrates the basic flow associated with the two methodologies. We discuss the individual computational components concerned with computer-driven RNA nanodesign, describe some of the specifics of the protocols, and finally show how a protocol can be used in a particular example.

5.1. RNA Databases

RNA databases are a valuable resource for finding structural motifs and characterizations that can be used in building block design. Atomic structures, derived from the application of X-ray crystallography and nuclear magnetic resonance (NMR), containing RNA can be downloaded from the Protein Data Bank (PDB) *(24)*. The PDB home page now contains various annotations of nucleic acid structures that can be very useful, including backbone torsion angles, basepair geometry parameters, and hydrogen-bonding classification. Several databases

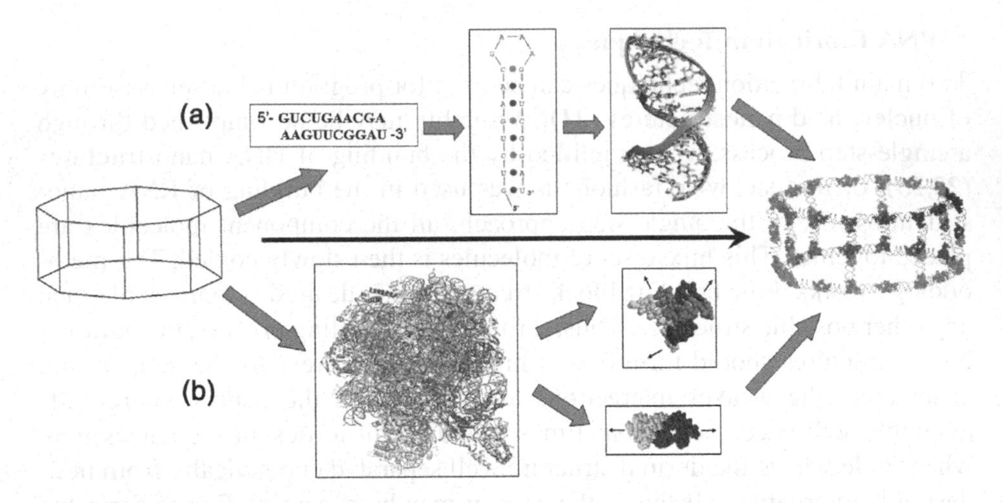

Fig. 2. Illustration of the two RNA nanostructure building protocols. From a user-defined shape, two basic paths can be followed. In path 'a', an RNA sequence and its associated secondary structure are determined. A three-dimensional (3D) model of the secondary structure is then built and later used as one of the RNA building block motifs. In path 'b', RNA 3D motifs are extracted from a database for the construction of the RNA nanoparticle. One can mix either of the two protocols described in the text to produce the desired structure.

contain RNA structures derived from the PDB, offering annotations, classifications, and substructures going beyond the information that the PDB offers.

The Nucleic Acid Database (NDB) is a repository of structural RNA and DNA data ***(25)***. It contains nucleic acid structures classified by type (for example, transfer RNAs, ribozyme structures, RNA helices) and experimental method (X-ray crystallography or NMR).

The Structural Classification of RNA (SCOR) database contains a classification of RNA internal loop and hairpin loop structures ***(26,27)***. It allows the user to search for motifs by sequence, key word, or PDB/NDB identifier. A novel motif, the extruded helical single strand has been identified with the help of the SCOR database ***(28)***.

The NCIR is a database containing structural information about noncanonical RNA basepairs ***(29,30)***. It provides access to all RNA structures in which a specified basepair has been found. This is useful for obtaining the original data corresponding to a rare (noncanonical) basepair.

For designing novel RNA structures, it is useful to have a database that contains curated and extracted fragments of RNA structures that can be used as building blocks. This prompted us to develop the RNAJunction database (30a). It provides structural and sequence information of RNA junctions that are extracted from RNA coordinate data files. As mentioned in the section introducing the RNA building block approach, it is the RNA junctions and the relative orientation of its connector helices that are most interesting for the design of RNA structures. The assumption here is that the designer knows the outline of the RNA structure to be built (for example, in the form of a 3D graph). In other words, one typical case is that one knows the number of branches in a junction (for example, a three-way junction) as well as the angles between the direction vectors of the junction's connector helices. Thus, the motivation for RNAJunction is to provide search capabilities with respect to junction order, interhelix angles, sequence, and PDB identifier. **Figure 3** shows a screen shot of the database describing an example RNA junction.

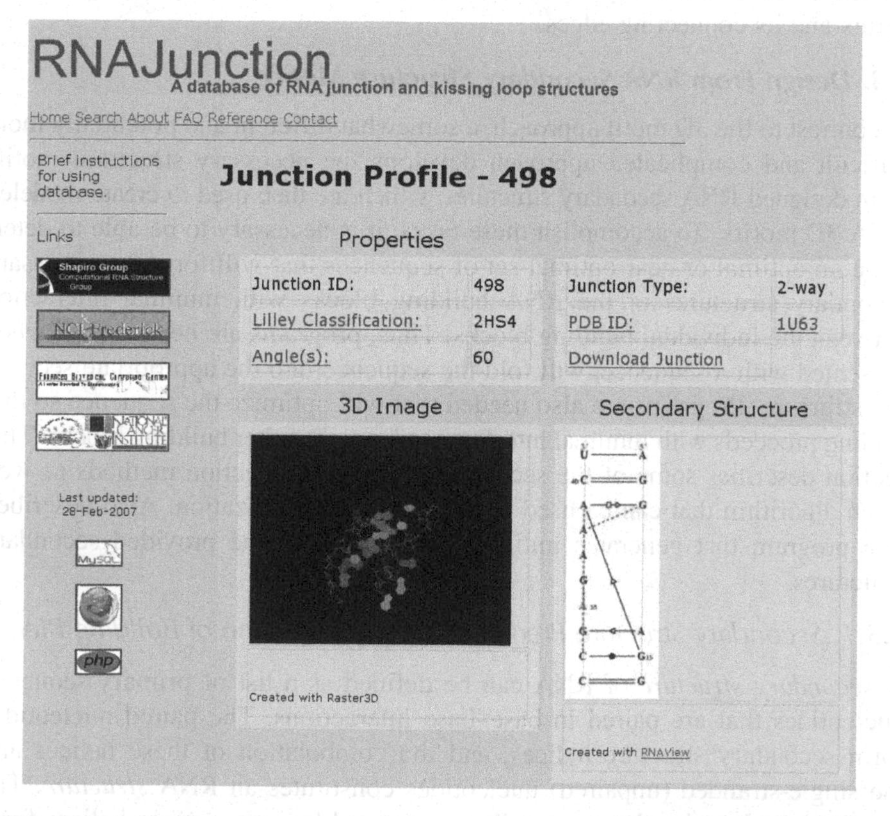

Fig. 3. Screen shot illustrating the results of a three-dimensional (3D) RNA motif search. In this particular case, information associated with a two-way junction is portrayed. Seen are the Protein Data Bank (PDB) identifier from which the junction was derived, the approximate angle that the junction attachment points make with each other (angle 60°), the NC=IUBMB classification (indicating that there are two helices separated by a single strand consisting of four bases), an RNAView ***(102)*** depiction of the motif, and a 3D rendering of the actual motif.

5.2. Design From Known RNA Three-Dimensional Motifs

The main approach in 3D motif design of RNA structures starts from an abstract model of the target structure (a 3D graph). One then identifies suitable RNA n-way junctions corresponding to the vertices of order n in that graph. These junction structures are then superimposed onto the graph vertices such that their connector helices are parallel to the edges connecting the vertices. When two junctions of two adjacent vertices have been placed, one then attempts to attach the connector helices corresponding to the connecting edge with an ideal RNA double helix. While the precise order of placing junctions and connecting them with helices depends on the algorithm used, one can see that in this way, in principle, an arbitrarily complex 3D graph can be traced with RNA motifs, provided one can identify suitable junction structures corresponding to each vertex and its connecting edges.

5.3. Design From RNA Secondary Structure Motifs

In contrast to the 3D motif approach, a somewhat different and potentially more difficult and complicated approach develops the necessary structural motifs from designed RNA secondary structures, which are then used to create modeled RNA 3D motifs. To accomplish these tasks, it is necessary to be able to determine an optimal or near-optimal set of sequences that will form the necessary secondary structures of the RNA building blocks with minimal interaction between the individual building blocks. Thus, programs are needed that, when presented with a sequence, will fold the sequence into the appropriate secondary structure. Programs are also needed that will optimize the sequence so that folding proceeds with minimal interference from the other building blocks. This section describes some of the secondary structure prediction methods as well as an algorithm that can be used to do sequence optimization. Also described is a program that generates initial 3D models from the provided secondary structures.

5.3.1. Secondary Structure Prediction for Optimization of Building Blocks

A *secondary structure* of RNA can be defined as a list of primary sequence nucleotides that are paired in base–base interactions. The paired nucleotides form secondary structure helices, and the combination of these helices and the single-stranded (unpaired) nucleotides constitutes an RNA *structure*. The unpaired nucleotides that are totally constrained by one or more helices form the so-called loops (hairpin, bulge, internal, and multibranch). Considering helices and loops as the key motifs, a secondary structure can be represented by a tree topology in which loops form nodes and helices form edges between them. Such a topology representation is also suitable for free-energy calculations since they are based on the assumption that the total free energy of a secondary structure (a tree) is a sum of the free energies of its independent elements (branches). In most general terms, loops tend to destabilize a secondary structure (raise the free energy), while helices and coaxial stacks tend to stabilize it (lower the free energy).

Computational prediction of a secondary structure is a nontrivial task, considering that a sequence of n nucleotides can theoretically form on the order of 1.8^n secondary structures. Numerous approaches to solving the problem have been used, aiming to combine the strengths of both computational and experimental methods. Computational secondary structure prediction algorithms can be divided into two general categories. In the first category, structures of individual sequences are predicted using free-energy minimization algorithms. In the second category, information provided by multiple sequence alignments is used to predict structures. Some programs combine these approaches. A comprehensive review and evaluation of RNA secondary structure prediction programs was presented by Gardner and Giegerich in 2004 ***(31)***. Mathews and Turner presented a review of free-energy minimization methods, with an emphasis on dynamic programming-based algorithms (DPAs) ***(32)***. For nanoscale structure designs in which the inclusion of pseudoknot interactions could be beneficial, there are several programs capable of pseudoknot prediction based on empirical energy parameters ***(33–38)***. Shapiro et al. ***(39)*** reviewed RNA structure prediction programs with an emphasis on RNA 3D modeling from secondary structure data.

5.3.1.1. RNA Secondary Structure Prediction for a Single Sequence

When only a single sequence is considered as input, that is, there is no phylogenetically related sequence data to help (see next subheading), the method most often used for secondary structure prediction is free-energy minimization. While less accurate than the multisequence-based methods, it is a more practical approach in the majority of real-life problems.

The most widely used secondary structure prediction programs, such as Mfold ***(40)***, RNAfold ***(41)***, and RNAstructure ***(40,42)***, use the concepts associated with DPAs. These programs produce a minimum free-energy structure and a sample of energetically suboptimal structures within a requested suboptimal energy range. When it comes to the prediction of wild-type secondary structures, the functional conformation or multiple conformations of a given RNA sequence do not always correspond to a minimum free-energy structure. Therefore, the programs Sfold ***(43,44)*** and RNAshapes ***(45–47)*** narrow the search for the most likely solutions to a relatively few representative structures. Some secondary structure prediction algorithms use statistical sampling of known RNA secondary structures to create a model that can then be used to predict the secondary structure of a given new sequence ***(48)***.

In general, the DPA-based programs and their derivatives offer a fast and reasonably accurate way of testing what the dominant secondary structure of a given RNA sequence may be. In a somewhat limited way, the solution space sampling programs can indicate if more than one stable conformation for the same sequence is possible. Thus, they make good evaluators of the building blocks nanodesign.

An RNA molecule may pass through several active or inactive conformations over its lifetime. These states may be a function of the kinetics of full-sequence folding, cotranscriptional folding (i.e., folding during sequence elongation), or environmental factors, such as proteins locking a molecule in one of two or more folding states.

Programs approximating folding kinetics implement stochastic simulations ***(49,50)*** or use genetic algorithms (GAs). GAs are based on the concepts of biological evolution and the survival of the fittest individuals ***(51)***. They include intermediate folding states and are thus capable of capturing unfolding and refolding of domains in transitional structures. GA implementations include our massively parallel GA MPGAfold ***(52–56)*** and the smaller-scale program STAR ***(57,58)***. Because of their stochastic nature, they need to be run multiple times to produce consensus structures. We have shown that MPGAfold can capture, in a coarse-grained fashion, the dynamic folding process inherent in many RNA molecules ***(55,59–64)***. MPGAfold and some other algorithms can simulate cotranscriptional folding and thus are capable of capturing folding kinetics unique to RNA transcription. Since the amount of information produced by programs such as MPGAfold is enormous (multiple stochastic runs), we have developed a suite of visual analysis tools, also suitable for DPA results analysis, called StructureLab ***(61,65)***.

Some of the algorithms mentioned give you the advantage of exploring folding states, which is important when multistable structural motifs or structural switches are a desired feature of the designed nanostructure. Inversely, if one wants to make sure that a structural building block does not have a likely alternate conformation one may use these folding algorithms.

5.3.1.2. Prediction of RNA Secondary Structures and Pseudoknots Using Multiple Sequence Alignments

Programs predicting RNA secondary structures based on alignments of multiple sequences produce a consensus secondary structure from both the energetic and evolutionary information provided by the alignment. This is a potentially more accurate approach than that based on free-energy minimization of single-sequence structures, but it depends on a set of well-aligned sequences with enough diversity to indicate structure-preserving mutations. Programs in this category follow the same basic two-stage paradigm. First, a matrix of scores for each basepair is computed. These scores typically incorporate information based on the thermodynamics of basepairs and covariation for a given pair of positions in the sequences considered. Second, this score matrix is mapped into one unique secondary structure. In our review paper on bridging the gap between 2D and 3D RNA structure prediction ***(39)***, we included a detailed discussion of these programs, including our own KNetFold ***(36,66–71)***.

5.3.2. Optimization of Sequence Structure Design

The secondary structure prediction algorithms just described provide useful "structural surveys" of potential building blocks from various wild-type sequences. The next step in a rational structure design (although it may equally well be the first step) is determining the best sequence to fold into a desired (user-defined) secondary structure ***(72–75)***. In the case of a wild-type sequence folding naturally into or close to the desired structure, we may want to check if it could be modified to match the desired structure better or made more stable in its most likely conformation. If the only known constraint is a defined secondary structure, we may want to find a sequence "from scratch" that would fold into it.

The program RNAinverse ***(75)*** works by optimizing a sequence, given a secondary structure definition and a possibly partially defined sequence. If no starting sequence is provided, the program generates a random starting sequence. Whichever starting sequence and constraints on its modification are chosen, the program iterates over modified sequences folded by RNAfold ***(41)*** until a sequence folding into a minimum free energy (MFE) structure matching the specified structural constraints is produced. Since the MFE stopping criterion may result in only a marginally stable structure, RNAinverse allows for a better approach to achieving a stronger preference for the specified secondary structure by optimizing the frequency of the specified secondary structure in the thermodynamic ensemble (refer to the subheading on suboptimal solution space sampling).

When it comes to nanoscale structure design, the approach taken by RNAinverse for the optimization of one single-stranded sequence has to be extended to multiple strands to deal with multiple building blocks. In this case, the stability of the individual building blocks has to be optimized in parallel with the optimization of the desired building block. Interactions and minimization of unwanted interference between the building blocks is a desired goal. We have developed a program that utilizes RNAinverse to produce sequences for individual building blocks (individual strands), followed by minimization of the unspecified interstrand interactions and optimization of the user-specified interstrand interactions. This procedure is iterative, with RNAfold used for comparing the designed sequence folds to the desired target structures.

5.3.3. Three-Dimensional Modeling of the Structure of the Building Blocks and the RNA Nanostructure

The knowledge of 3D models of RNA nanoparticles is crucial for the optimization of design and the complete understanding of their function and properties. The use of 3D RNA building blocks can be obtained from known solution structures extracted from databases, as discussed, or can be generated using specific 3D modeling software, such as MANIP *(76)*, RNA2D3D *(77)*, or NAB *(78)*. The modeled fragments can then be put together via template-assisted assembly by using software such as RNA2D3D and NanoTiler.

The concept of modeling RNA building blocks by creating the starting coordinates of such an entity from a given secondary structure generated by algorithms such as those described can be done using the program RNA2D3D. This includes some special features for modeling RNA tectosquares. The use of RNA2D3D is exemplified by our modeling of the telomerase pseudoknot domain *(79)*. RNA2D3D can generate, view, and compare 3D RNA molecules. In this software, the 3D atomic coordinates of a nucleotide are initially embedded in a planar representation of an RNA secondary structure and are generated from a reference triad of atoms. The stems are created from the reference triad of its 5′ nucleotide using helical coordinates taken from the Biosym® database. The unpaired nucleotides, bulges, hairpin loops, branching loops, and other nonhelical motifs are generated using the coordinates of their reference triad relative to the 5′ neighboring nucleotide. As a result, a first-order approximation of the actual 3D molecule is established. Structure refinement uses molecular modeling or interactive editing. The interactive editing involves a rotation and translation of a nucleotide or a group of nucleotides and is used for the removal of structural clashes, enforcing tertiary interactions, and modification of mutual stacking.

Sometimes, the structures of the building blocks can benefit from mutating one or several residues. Mutation of residues in the 3D RNA structures can be produced by replacing the original residue with the desired one in the structure using a combination of CHIMERA® and DSViewer® software.

5.4. Mixing of the Three-Dimensional Motif Protocol With the Secondary Structure Motif Protocol

Different approaches can combine the 3D motif and secondary structure motif protocols. For example, one can start with 3D information that is derived from a database of motifs. Using the assumption that similar secondary structures possess similar tertiary interactions, an algorithm maps the RNA 3D structures found in PDB to their secondary structure equivalents. Other algorithms then measure secondary structure similarity and pick 3D structure motifs from a specialized database, thus providing fragments from which an entire secondary structure and

3D structure of a given sequence can be assembled ***(80)***. This approach can be combined with or used separately from the specification of RNA secondary structure templates that will determine the sequences that will fold into some of the RNA building blocks.

Thus, the elements that comprise the final nanostructure design can be derived from database motifs, modeled motifs that may be derived from optimized secondary structures, or motifs that merge sequence and structure from both protocols. The ultimate utility of the designed nanoparticles can only be determined by understanding the structure/function relationships that exist in the underlying RNA components and by having an understanding of the properties of RNA folding. The correct assembly of RNA-based nanoparticles requires both a fundamental understanding of the role that RNA plays in biological processes and an understanding of how these processes can be designed into functional nanoparticles.

5.5. Checking Stability and Dynamic Characteristics

Once the 3D coordinates of the individual building blocks and nanoparticles have been established, we can use molecular dynamics (MD), molecular mechanics (MM), geometric structural analysis, energetic analysis, and high-temperature simulations to refine the structural characteristics and elucidate the stability, flexibility, and effect of environmental factors on the constructed entities. Described next are some of the tools for refining and determining these 3D structural features.

5.5.1. Molecular Dynamics

The main purpose of molecular dynamics (MD) simulations is to describe molecular motions. MD is a technique in which the time evolution of the molecular system is followed by numerical integration of the equations of motion. The MD method can provide highly detailed information not accessible by other methods, including atomically resolved conformational changes and response to environmental and chemical changes, such as concentration of ions, high temperature, pressure, mutations, and chemical modifications. We have successfully used atomistic MD simulations to model, predict, and characterize structures and dynamic behaviors of various RNA molecules, such as the minimal telomerase RNA pseudoknot domain ***(79,81,82)*** and HIV kissing loops ***(83)***, for characterization of single-bulge motifs ***(84)***, to find the characteristics of the 16s ribosomal RNA S15 binding site ***(85)***, and to determine the unfolding and folding characteristics of a tetraloop ***(86)*** and RNA nanoparticles.

Molecular dynamics makes possible the dynamic characterization and an exploration of the conformational energy landscape of biomolecules and their surroundings. Recent reviews outlined the successful use of MD simulations to characterize a wide variety of nucleic acid structures ***(87–92)***. The limitations of MD simulations include the size of biomolecules and the relatively short simulation timescale, limited to several tens of nanoseconds. Most important, the reliability of MD depends on accurate force fields for both nucleic acids and solvent. Explicit solvent simulations are most accurate and can provide information on specific water and ion interactions; however, they are more computationally extensive. MD simulations can be significantly accelerated by treating the solvent with continuum dielectric methods. In this case, only the intrasolute electrostatics need to be evaluated ***(93,94)***. Even though implicit MD simulations are less accurate than simulations with explicit solvent, they not only permit much longer simulations and larger molecules but also provide a variety of sampled conformations. Considerable improvements in the force field have also been achieved, making simulations more reliable and accurate. The typical force field used for RNA molecules is ff99 ***(95)***. There are three major MD software packages generally used for RNA molecules: Amber ***(96)***, CHARMM ***(97)***, and NAMD (Nanoscale Molecular Dynamics) ***(98)***.

5.5.2. Molecular Mechanics

Molecular mechanics (MM) is a method to calculate the structure and energy of molecules based on nuclear motions. MM implements energy minimization methods to study the potential energy surfaces of different molecular systems. MM can also provide important energy-related information, such as the existence of energy barriers between different conformers or steepness of a potential energy surface around a local minimum. MD and MM are usually based on the same classical force fields and thus can be found in the same software packages. For example, the MM-PB(GB)SA module in Amber can be used to calculate the contributions of gas-phase and solvation free energies for snapshots of the MD trajectories. Total MM energies E_{gas}; internal energies E_{int} (i.e., bonds, angles, and dihedrals); and van der Waals E_{vdw} and electrostatic E_{elec} components can be determined. Moreover, the same module can help estimate the enthalpy for folding by calculating the energy difference between folded and coiled RNA structures with the same sequence. The enthalpies for folding can then estimated as $\Delta H = E_{tot}$ folded – E_{tot} coiled.

5.6. Structural Analysis

Groove widths, backbone torsion angles, and local basepair parameters (twist, tilt, roll, shift, slide, and rise) for each strand and stem can be analyzed by the program CURVES ***(99)*** and compared to standard A-RNA, B-DNA, and A-DNA triplex helical parameters ***(100)***. The standard A-RNA, B-DNA, and A-DNA triplex helices can be built using Insight II®.

Ptraj, a module in Amber, which can also be used as a standalone package, is useful for analyzing MD trajectories, including the calculation of bond angles;

dihedral angles; the root mean square differences between various structures; displacements, including atomic positional fluctuations; correlation functions; and many other features.

The evaluation of all of these features is extremely useful for determining the 3D structural characteristics of the designed nano-building blocks and nanostructures whether they are designed using the 3D motif, the secondary structure motif, or the mixed approach. However, one has to keep in mind that the nanoconstructs can ultimately be quite large. This may require significant computing power or the use of coarse-graining techniques to reduce the combinatorics of the calculations.

6. Summary of the Three-Dimensional and Secondary Structure Motif Protocols for RNA Nanostructure Design

6.1. Protocol for Three-Dimensional Motif RNA Nanostructure Design as Defined by NanoTiler

1. Determine the shape of the RNA nanostructure desired.
2. Specify the shape in the form of a 3D graph.
3. Input the 3D graph coordinate file.
4. Read the database of available building blocks (RNA junctions).
5. For each vertex in the input graph, identify the junction from the building block database that results in the smallest fitting error.
6. Generate double-stranded helices interpolating between the fragments.
7. For randomly chosen vertices, attempt to exchange the used junction building blocks to reduce the fitting errors between the junction and the double-stranded helices that were generated to connect the neighboring junctions.
8. Optimize sequences of the generated model strands (optional).
9. Apply MM and MD to further refine the individual building blocks and the structure as a whole.
10. Analyze the structural and functional characteristics of the nanostructure to ensure that they conform to the desired attributes.
11. Iterate the procedure as necessary to optimize the desired structural characteristics.

Note that the user can specify the acceptable fitting errors for placing junctions and generating interpolating stems. NanoTiler is used for **steps 3–8**.

6.2. Secondary Structure Motif Protocol

1. Determine the shape of the RNA nanostructure desired.
2. Specify the shape in the form of a 3D graph.
3. Input the 3D graph coordinate file.
4. Determine the secondary structure templates for the building blocks.
5. Run the sequence optimizer to obtain the optimized sequences for the building blocks.
6. Use RNA2D3D and other molecular modeling software to obtain the 3D coordinates for the building blocks.

7. Apply MM and MD to further refine the individual building blocks and the structure as a whole.
8. Analyze the structural and functional characteristics of the nanostructure to ensure that they conform to the desired attributes.
9. Iterate the procedure as necessary to optimize the desired structural characteristics.

NanoTiler is used for **steps 3–5**.

6.3. Mixed Protocol

The mixed protocol involves the use of database structures when obtainable, and the prediction, optimization, and modeling of structural components if necessary. Molecular dynamics and mechanics should also be applied, including analysis of the structural characteristics to ensure that they conform to the desired structural and functional attributes.

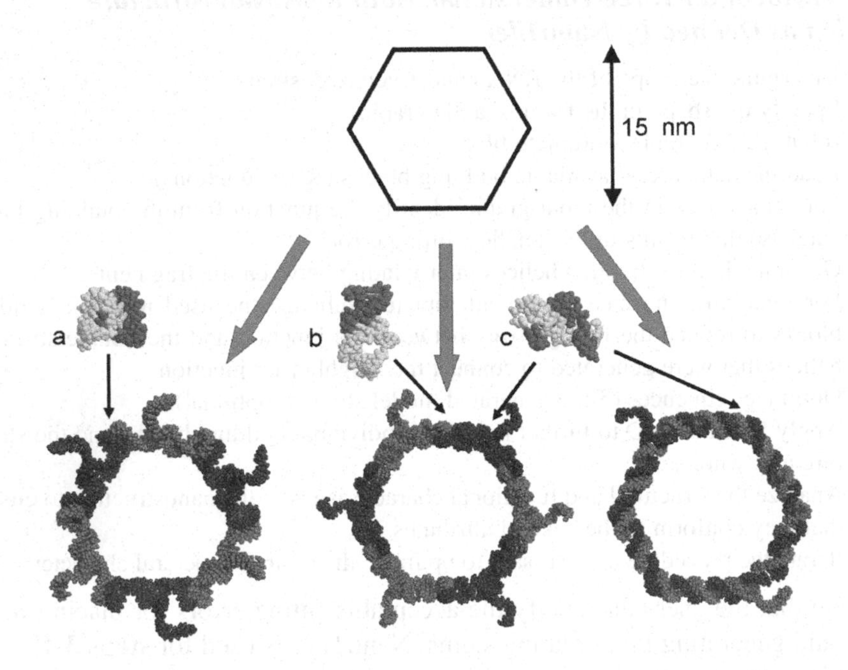

Fig. 4. An example of the workflow in the three-dimensional (3D) motif protocol for generating computer models of RNA nanostructures with NanoTiler and the RNAJunction database ***(30a)***. From an abstract 3D graph (a hexagon in this case), an algorithm scans for suitable building blocks in the RNAJunction database. Kissing-loop building blocks (**a**) or two-way junctions (**c**) have a 120° angle and can be placed at the graph vertices. The kissing-loop building blocks with a 180° angle (**b**) are placed in the middle of the graph edges. The three different hexagon RNA structures were built from a library consisting of only these three building blocks. Sticky tails have been added manually by editing the molecular structures.

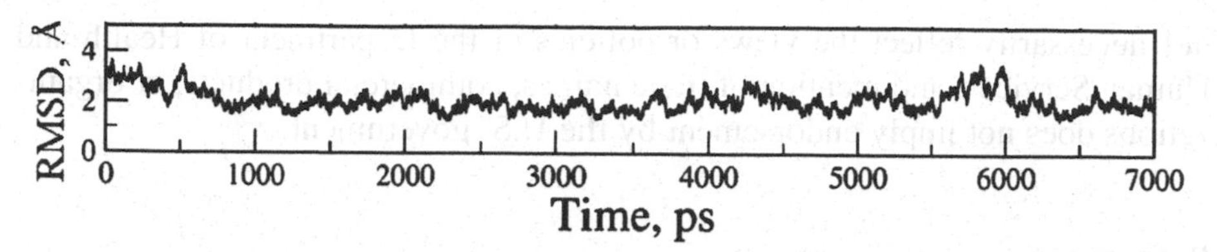

Fig. 5. Illustrated is a 7-ns molecular dynamics plot of the 120° kissing-loop motif shown in **Fig. 4a** *(102)*. The plot depicts the root mean square deviation (RMSD) of the motif relative to its average structure. The plot is relatively flat, indicating a stable structure.

6.4. Example of RNA Nanostructure Design Using the Three-Dimensional Motif Protocol With NanoTiler

Computer-assisted nanodesign can use a shape-based approach by which the desired shape guides the choice of the specific building blocks. As an example, we show how the computational methods described can be used for the design of an RNA hexagonal shape that is approximately 15 nm in diameter ***(101, 102)***. The building blocks shown here were derived from the RNAJunction database. Examples of this concept applied to the automatic model building of hexameric rings are shown in **Fig. 4a–c**. The figure shows that kissing loops can be placed either in the corners (graph vertices) or in the middle of stem regions (corresponding to graph edges). These models have been generated with our newly developed software NanoTiler. Input to the program is a simple representation of the graph as well as a library of possible building blocks. The input graph consists of a hexagonal wire frame structure; the building block library consists of three elements: one two-way junction with a 120° angle, a kissing loop with a 120° angle, and a kissing loop (originating from the structure of the HIV dimer initiation site) corresponding roughly to a 180° angle between the connector stems. The computation time to generate these models with NanoTiler was less than 5 min in each case. **Figure 5** illustrates the use of molecular dynamics on the 120° kissing-loop motif shown in **Fig. 4a**. The results show that the motif is quite stable.

Acknowledgments

We wish to thank Robert Hayes, Christine Viets, and Calvin Grunewald for their contributions to the development of our research tools. Computational support was provided in part by the National Cancer Institute's Advanced Biomedical Computing Center. This publication was supported by the Intramural Research Program of the National Institutes of Health, National Cancer Institute, Center for Cancer Research. This publication has been funded in whole or in part with federal funds from the National Cancer Institute, National Institutes of Health, under contract N01-CO-12400. The content of this publication does not necessarily reflect the views or policies of the Department of Health and Human Services, and mention of trade names, commercial products, or organizations does not imply endorsement by the U.S. government.

References

1. Yano J, Hirabayashi K, Nakagawa S, et al. (2004) Antitumor activity of small interfering RNA/cationic liposome complex in mouse models of cancer. *Clin. Cancer Res.* **10**(22), 7721–7726.
2. Schiffelers RM, Ansari A, Xu J, et al. (2004) Cancer siRNA therapy by tumor selective delivery with ligand-targeted sterically stabilized nanoparticle. *Nucleic Acids Res.* **32**(19), e149.
3. Howard KA, Rahbek UL, Liu X, et al. (2006) RNA interference in vitro and in vivo using a novel chitosan/siRNA nanoparticle system. *Mol. Ther.* **14**(4), 476–484.
4. Westhof E, Masquida B, Jaeger L. (1996) RNA tectonics: towards RNA design. Fold. Des. **1**(4), R78–R88.
5. Jaeger L, Westhof E, Leontis NB. (2001) TectoRNA: modular assembly units for the construction of RNA nano-objects. *Nucleic Acids Res.* **29**(2), 455–463.
6. Shapiro BA, Zhang KZ. (1990) Comparing multiple RNA secondary structures using tree comparisons. *Comput. Appl. Biosci.* **6**(4), 309–318.
7. Shapiro BA. (1988) An algorithm for comparing multiple secondary structures. *Comput. Appl. Biosci.* **4**(3), 387–93.
8. Chworos A, Severcan I, Koyfman AY, et al. (2004) Building programmable jigsaw puzzles with RNA. *Science* **306**(5704), 2068–2072.
9. Nasalean L, Baudrey S, Leontis NB, Jaeger L. (2006) Controlling RNA self-assembly to form filaments. *Nucleic Acids Res.* **34**(5), 1381–1392.
10. Bates AD, Callen BP, Cooper JM, et al. (2006) Construction and characterization of a gold nanoparticle wire assembled using Mg^{2+}-dependent RNA–RNA interactions. *Nano. Lett.* **6**(3), 445–448.
11. Koyfman AY, Braun G, Magonov S, Chworos A, Reich NO, Jaeger L. (2005) Controlled spacing of cationic gold nanoparticles by nanocrown RNA. *J. Am. Chem. Soc.* **127**(34), 11886–11887.
12. Shu D, Moll W-D, Deng Z, Mao C, Guo P. (2004) Bottom-up assembly of RNA arrays and superstructures as potential parts in nanotechnology. *Nano. Lett.* **4**(9), 1717–1723.
13. Guo P. (2005) RNA nanotechnology: engineering, assembly and applications in detection, gene delivery and therapy. *J. Nanosci. Nanotechnol.* **5**(12), 1964–1982.
14. Khaled A, Guo S, Li F, Guo P. (2005) Controllable self-assembly of nanoparticles for specific delivery of multiple therapeutic molecules to cancer cells using RNA nanotechnology. *Nano. Lett.* **5**(9), 1797–1808.
15. Guo S, Huang F, Guo P. (2006) Construction of folate-conjugated pRNA of bacteriophage phi29 DNA packaging motor for delivery of chimeric siRNA to nasopharyngeal carcinoma cells. *Gene Ther.* **13**(10), 814–820.

16. Jaeger L, Leontis NB. (2000) Tecto-RNA: one-dimensional self-assembly through tertiary interactions. *Angew. Chem. Int. Ed. Engl.* **39**(14), 2521–2524.
17. Hansma HG, Oroudjev E, Baudrey S, Jaeger L. (2003) TectoRNA and "kissing-loop" RNA: atomic force microscopy of self-assembling RNA structures. *J. Microsc.* **212**(Pt. 3), 273–279.
18. Horiya S, Li X, Kawai G, et al. (2002) RNA LEGO: magnesium-dependent assembly of RNA building blocks through loop–loop interactions. *Nucleic Acids Res.* **Suppl 2**, 41–42.
19. Horiya S, Li X, Kawai G, et al. (2003) RNA LEGO: magnesium-dependent formation of specific RNA assemblies through kissing interactions. *Chem. Biol.* **10**(7), 645–654.
20. Li X, Horiya S, Harada K. (2006) An efficient thermally induced RNA conformational switch as a framework for the functionalization of RNA nanostructures. *J. Am. Chem. Soc.* **128**(12), 4035–4040.
21. Jaeger L, Chworos A. (2006) The architectonics of programmable RNA and DNA nanostructures. *Curr. Opin. Struct. Biol.* **16**(4), 531–543.
22. Chelyapov N, Brun Y, Gopalkrishnan M, Reishus D, Shaw B, Adleman L. (2004) DNA triangles and self-assembled hexagonal tilings. *J. Am. Chem. Soc.* **126**(43), 13924–13925.
23. Mathieu F, Liao S, Kopatsch J, Wang T, Mao C, Seeman NC. (2005) Six-helix bundles designed from DNA. *Nano. Lett.* **5**(4), 661–665.
24. Berman HM, Westbrook J, Feng Z, et al. (2000) The Protein Data Bank. *Nucleic Acids Res.* **28**(1), 235–242.
25. Berman HM, Olson WK, Beveridge DL, et al. (1992) The nucleic acid database. A comprehensive relational database of three-dimensional structures of nucleic acids. *Biophys. J.* **63**(3), 751–759.
26. Tamura M, Hendrix DK, Klosterman PS, Schimmelman NR, Brenner SE, Holbrook SR. (2004) SCOR: Structural Classification of RNA, version 2.0. *Nucleic Acids Res.* **32**(Database issue), D182–D184.
27. Klosterman PS, Tamura M, Holbrook SR, Brenner SE. (2002) SCOR: a Structural Classification of RNA database. *Nucleic Acids Res.* **30**(1), 392–394.
28. Klosterman PS, Hendrix DK, Tamura M, Holbrook SR, Brenner SE. (2004) Three-dimensional motifs from the SCOR, Structural Classification of RNA database: extruded strands, base triples, tetraloops and U-turns. *Nucleic Acids Res.* **32**(8), 2342–2352.
29. Nagaswamy U, Larios-Sanz M, Hury J, et al. (2002) NCIR: a database of non-canonical interactions in known RNA structures. *Nucleic Acids Res.* **30**(1), 395–397.
30. Nagaswamy U, Voss N, Zhang Z, Fox GE. (2000) Database of non-canonical base pairs found in known RNA structures. *Nucleic Acids Res.* **28**(1), 375–376.
30a. Bindewald E, Hayes R, Yingling YG, Kasprzak W, Shapiro BA. (2008) RNAJunction: a database of RNA junctions and kissing loops for three-dimensional structural analysis and nanodesign. *Nucleic Acids Res.* **36**, D392–D397.
31. Gardner PP, Giegerich R. (2004) A comprehensive comparison of comparative RNA structure prediction approaches. *BMC Bioinform.* **5**, 140–157.

32. Mathews DH, Turner DH. (2006) Prediction of RNA secondary structure by free energy minimization. *Curr. Opin. Struct. Biol.* **16**(3), 270–278.
33. Rivas E, Eddy SR. (1999) A dynamic programming algorithm for RNA structure prediction including pseudoknots. *J. Mol. Biol.* **285**(5), 2053–2068.
34. Dirks RM, Pierce NA. (2003) A partition function algorithm for nucleic acid secondary structure including pseudoknots. *J. Comput. Chem.* **24**(13), 1664–1677.
35. Reeder J, Giegerich R. (2005) Consensus shapes: an alternative to the Sankoff algorithm for RNA consensus structure prediction. *Bioinformatics* **21**(17), 3516–3523.
36. Ruan J, Stormo GD, Zhang W. (2004) An iterated loop matching approach to the prediction of RNA secondary structures with pseudoknots. Bioinformatics **20**(1), 58–66.
37. Ruan J, Stormo GD, Zhang W. (2004) ILM: a Web server for predicting RNA secondary structures with pseudoknots. *Nucleic Acids Res.* **32**(Web Server issue), W146–W149.
38. Ren J, Rastegari B, Condon A, Hoos HH. (2005) HotKnots: heuristic prediction of RNA secondary structures including pseudoknots. *RNA* **11**(10), 1494–504.
39. Shapiro BA, Yingling YG, Kasprzak W, Bindewald E. (2007) Bridging the gap in RNA structure prediction. *Curr. Opin. Struct. Biol.* **17**(2), 157–165.
40. Mathews DH, Sabina J, Zuker M, Turner DH. (1999) Expanded sequence dependence of thermodynamic parameters improves prediction of RNA secondary structure. *J. Mol. Biol.* **288**(5), 911–940.
41. Hofacker IL, Fontana W, Stadler PF, Bonhoeffer S, Tacker M, Schuster P. (1994) Fast folding and comparison of RNA secondary structures. *Monatsh. Chem.* **125**, 167–188.
42. Mathews DH, Disney MD, Childs JL, Schroeder SJ, Zuker M, Turner DH. (2004) Incorporating chemical modification constraints into a dynamic programming algorithm for prediction of RNA secondary structure. *Proc. Natl. Acad. Sci. U. S. A.* **101**(19), 7287–92.
43. Ding Y, Chan CY, Lawrence CE. (2005) RNA secondary structure prediction by centroids in a Boltzmann weighted ensemble. *RNA* **11**(8), 1157–1166.
44. Chan CY, Lawrence CE, Ding Y. (2005) Structure clustering features on the Sfold Web server. *Bioinformatics* **21**(20), 3926–3928.
45. Giegerich R, Voss B, Rehmsmeier M. Abstract shapes of RNA. *Nucleic Acids Res.* **32**(16), 4843–4851.
46. Steffen P, Voss B, Rehmsmeier M, Reeder J, Giegerich R. (2006) RNAshapes: an integrated RNA analysis package based on abstract shapes. *Bioinformatics* **22**(4), 500–503.
47. Voss B, Giegerich R, Rehmsmeier M. (2006) Complete probabilistic analysis of RNA shapes. *BMC Biol.* **4**(1), 5–27.
48. Do CB, Woods DA, Batzoglou S. (2006) CONTRAfold: RNA secondary structure prediction without physics-based models. *Bioinformatics* **22**(14), e90–e98.
49. Xayaphoummine A, Bucher T, Isambert H. (2005) Kinefold Web server for RNA/DNA folding path and structure prediction including pseudoknots and knots. *Nucleic Acids Res.* **33**(Web Server issue), W605–W610.

50. Isambert H, Siggia ED. (2000) Modeling RNA folding paths with pseudoknots: application to hepatitis delta virus ribozyme. *Proc. Natl. Acad. Sci. U. S. A.* **97**(12), 6515–6520.
51. Holland JH. (1992) *Adaptation in Natural and Artificial Systems: An Introductory Analysis With Applications in Biology, Control, and Artificial Intelligence*. MIT Press, Cambridge, MA.
52. Shapiro BA, Navetta J. (1994) A massively parallel genetic algorithm for RNA secondary structure prediction. *J Supercomputing* **8**, 195–207.
53. Shapiro BA, Wu JC. (1996) An annealing mutation operator in the genetic algorithms for RNA folding. *Comput. Appl. Biosci.* **12**(3), 171–180.
54. Shapiro BA, Wu JC. (1997) Predicting RNA H-type pseudoknots with the massively parallel genetic algorithm. *Comput. Appl. Biosci.* **13**(4), 459–471.
55. Shapiro BA, Bengali D, Kasprzak W, Wu JC. (2001) RNA folding pathway functional intermediates: their prediction and analysis. *J. Mol. Biol.* **312**(1), 27–44.
56. Shapiro BA, Wu JC, Bengali D, Potts MJ. (2001) The massively parallel genetic algorithm for RNA folding: MIMD implementation and population variation. *Bioinformatics* **17**(2), 137–148.
57. van Batenburg FH, Gultyaev AP, Pleij CW. (1995) An APL-programmed genetic algorithm for the prediction of RNA secondary structure. *J. Theor. Biol.* **174**(3), 269–280.
58. Gultyaev AP, van Batenburg FH, Pleij CW. (1995) The computer simulation of RNA folding pathways using a genetic algorithm. *J. Mol. Biol.* **250**(1), 37–51.
59. Kasprzak W, Bindewald E, Shapiro BA. (2005) Structural polymorphism of the HIV-1 leader region explored by computational methods. *Nucleic Acids Res.* **33**(22), 7151–7163.
60. Linnstaedt SD, Kasprzak WK, Shapiro BA, Casey JL. (2006) The role of a metastable RNA secondary structure in hepatitis delta virus genotype III RNA editing. RNA **12**(8), 1521–1533.
61. Shapiro BA, Kasprzak W, Grunewald C, Aman J. (2006) Graphical exploratory data analysis of RNA secondary structure dynamics predicted by the massively parallel genetic algorithm. *J. Mol. Graph. Model* **25**, 514–531.
62. Gee AH, Kasprzak W, Shapiro BA. (2006) Structural differentiation of the HIV-1 polyA signals. *J. Biomol. Struct. Dyn.* **23**(4), 417–428.
63. Tortorici MA, Shapiro BA, Patton JT. (2006) A base-specific recognition signal in the 5′ consensus sequence of rotavirus plus-strand RNAs promotes replication of the double-stranded RNA genome segments. *RNA* **12**(1), 133–146.
64. Zhang J, Zhang G, Guo R, Shapiro BA, Simon AE. (2006) A pseudoknot in a preactive form of a viral RNA is part of a structural switch activating minus-strand synthesis. *J. Virol.* **80**, 9181–9191.
65. Kasprzak W, Shapiro B. (1999) Stem Trace: an interactive visual tool for comparative RNA structure analysis. *Bioinformatics* **15**(1), 16–31.
66. Hofacker IL, Fekete M, Stadler PF. (2002) Secondary structure prediction for aligned RNA sequences. *J. Mol. Biol.* **319**(5), 1059–1066.

67. Witwer C, Hofacker IL, Stadler PF. (2004) Prediction of consensus RNA secondary structures including pseudoknots. *IEEE/ACM Trans. Comput. Biol. Bioinform.* **1**(2), 66–77.
68. Freyhult E, Moulton V, Gardner P. (2005) Predicting RNA structure using mutual information. *Appl. Bioinformatics* **4**, 53–59.
69. Jossinet F, Westhof E. (2005) Sequence to Structure (S2S), display, manipulate and interconnect RNA data from sequence to structure. *Bioinformatics* **21**(15), 3320–3321.
70. Bindewald E, Shapiro BA. (2006) RNA secondary structure prediction from sequence alignments using a network of *k*-nearest neighbor classifiers. *RNA* 12, 342–352.
71. Bindewald E, Schneider TD, Shapiro BA. (2006) CorreLogo: an online server for 3D sequence logos of RNA and DNA alignments. *Nucleic Acids Res.* **34**(Web Server issue), W405–W411.
72. Aguirre-Hernandez R, Hoos HH, Condon A. (2007) Computational RNA secondary structure design: empirical complexity and improved methods. *BMC Bioinform.* **8**(1), 34.
73. Andronescu M, Fejes AP, Hutter F, Hoos HH, Condon A. (2004) A new algorithm for RNA secondary structure design. *J. Mol. Biol.* **336**(3), 607–624.
74. Busch A, Backofen R. (2006) INFO-RNA—a fast approach to inverse RNA folding. *Bioinformatics* **22**(15), 1823–1831.
75. Hofacker IL, Fontana W, Stadler PF, Bonhoeffer S, Tacker M, Schuster P. (1994) Fast folding and comparison of RNA secondary structures. *Monatsh. Chem.* **125**, 167–188.
76. Massire C, Westhof E. (1998) MANIP: an interactive tool for modelling RNA. *J. Mol. Graph. Model.* **16**(4–6), 197–205, 55–57.
77. Martinez HM, Maizel JJ, Shapiro BA. RNA2D3D. (2008) A program for generating, viewing, and comparing 3-dimensional models of RNA. JBSD, **25**(6), 669–683.
78. Macke T, Case D. (1998) *Modeling Unusual Nucleic Acid Structures.* American Chemical Society, Washington, DC.
79. Yingling YG, Shapiro BA. (2006) The prediction of the wild-type telomerase RNA pseudoknot structure and the pivotal role of the bulge in its formation. *J. Mol. Graph. Model.* **25**, 261–274.
80. Barreda DCJ, Shigenobu Y, Ichiishi E, Del Carpio MC. (2004) RNA 3D structure prediction: (1) assessing RNA 3D structure similarity from 2D structure similarity. *Genome Inform.* **15**(2), 112–120.
81. Yingling YG, Shapiro BA. (2005) Dynamic behavior of the telomerase RNA hairpin structure and its relationship to dyskeratosis congenita. *J. Mol. Biol.* **348**(1), 27–42.
82. Yingling YG, Shapiro BA. (2007) The impact of dyskeratosis congenita mutations on the structure and dynamics of the human telomerase RNA pseudoknot domain. *J. Biomol. Struct. Dyn.* **24**(4), 303–320.
83. Pattabiraman N, Martinez HM, Shapiro BA. (2002) Molecular modeling and dynamics studies of HIV-1 kissing loop structures. *J. Biomol. Struct. Dyn.* **20**(3), 397–412.
84. Hastings WA, Yingling YG, Chirikjian GS, Shapiro BA. (2006) Structural and dynamical classification of RNA single-base bulges for nanostructure design. *J. Comp. Theor. Nanosci.* **3**, 63–77.

85. Li W, Ma B, Shapiro BA. (2003) Binding interactions between the core central domain of 16S rRNA and the ribosomal protein S15 determined by molecular dynamics simulations. *Nucleic Acids Res.* **31**(2), 629–638.
86. Li W, Ma B, Shapiro B. (2001) Molecular dynamics simulations of the denaturation and refolding of an RNA tetraloop. *J. Biomol. Struct. Dyn.* **19**(3), 381–396.
87. Cheatham TE, 3rd, Kollman PA. (2000) Molecular dynamics simulation of nucleic acids. *Annu. Rev. Phys. Chem.* **51**, 435–471.
88. Cheatham TEI, Young MA. (2001) Molecular dynamics simulations of nucleic acids: successes, limitations, and promise. *Biopolymers* **56**, 232–256.
89. Zacharias M. (2000) Simulation of the structure and dynamics of nonhelical RNA motifs. *Curr. Opin. Struct. Biol.* **10**(3), 311–317.
90. Auffinger P, Vaiana AC. (2005) *Molecular Dynamics of RNA Systems*. Wiley-VCH Verlag, Weinheim.
91. Auffinger P, Westhof E. (1998) Simulations of the molecular dynamics of nucleic acids. *Curr. Opin. Struct. Biol.* **8**(2), 227–236.
92. Giudice E, Lavery R. (2002) Simulations of nucleic acids and their complexes. *Acc. Chem. Res.* **35**(6), 350–357.
93. Lee MS, Salsbury FR Jr, Brooks CL III. (2002) Novel generalized Born methods. *J. Chem. Phys.* **116**, 10606–10614.
94. Scarsi M, Apostolakis J, Caflisch A. (1998) Comparison of a GB model with explicit solvent simulations: potentials of mean force and conformational preferences of alanine dipeptide and 1,2-dichloroethane. *J. Phys. Chem. B* **102**, 3637–3641.
95. Wang J, Cieplak P, Kollman PA. (2000) How well does a restrained electrostatic potential (RESP) model perform in calculating of organic and biological molecules? *J. Comput. Chem.* **21**, 1049–1074.
96. Case DA, Cheatham TE 3rd, Darden T, et al. (2005) The Amber biomolecular simulation programs. *J. Comput. Chem.* **26**(16), 1668–1688.
97. Brooks BR, Bruccoleri RE, Olafson BD, States DJ, Swaminathan S, Karplus M. (1983) CHARMM: A program for macromolecular energy, minimization, and dynamics calculations. *J. Comput. Chem.* **4**, 187–217.
98. Phillips JC, Braun R, Wang W, et al. (2005) Scalable molecular dynamics with NAMD. *J. Comput. Chem.* **26**(16), 1781–1802.
99. Lavery R, Sklenar H. (1988) The definition of generalized helicoidal parameters and of axis curvature for irregular nucleic acids. *J. Biomol. Struct. Dyn.* **6**(1), 63–91.
100. Schlick T. (2006) Molecular modeling and simulation: An interdisciplinary Guide. Springer, New York, New York.
101. Shapiro BA, Yingling YG. (2006) RNA nanoparticles and nanotubes. Patent pending.
102. Yingling YG, Shapiro BA. (2007) Computational design of an RNA hexagonal nanoring and an RNA nanotube. *Nano Lett.* **7**(8), 2328–2334.
103. Yang H, Jossinet F, Leontis N, et al. (2003) Tools for the automatic identification and classification of RNA base pairs. *Nucleic Acids Res.* **31**(13), 3450–3460.

8

Self-Assembly of Fused Homo-Oligomers to Create Nanotubes

Idit Buch, Chung-Jung Tsai, Haim J. Wolfson, and Ruth Nussinov

Summary

The formation of a nanostructure by self-assembly of a peptide or protein building block depends on the ability of the building block to spontaneously assemble into an ordered structure. We first describe a protocol of fusing homo-oligomer proteins with a given three-dimensional (3D) structure to create new building blocks. According to this protocol, a single monomer *A* that self-assembles with identical copies to create an oligomer A_l is covalently linked, through a short linker *L*, to another monomer *B* that self-assembles with identical copies to create the oligomer B_j. The result is a fused monomer *A* - *L* - *B*, which has the ability to self-assemble into a nanostructure $(A - L - B)_k$. We control the self-assembly process of *A* - *L* - *B* by mapping the fused building block onto a planar sheet and wrapping the sheet around a cylinder with the target's dimensions. Finally, we validate the created nanotubes by an optimization procedure. We provide examples of two nanotubes in atomistic model details. One of these has experimental data. In principal, such a protocol should enable the creation of a wide variety of potentially useful protein-based nanotubes with control over their physical and chemical properties.

Key Words: Building block (BB); homo-oligomers; oligomerization domain; nanotube; self-assembly; symmetry; unit-cell.

1. Introduction

Bionanotechnology is a new discipline that aims at designing novel biomaterials and molecular devices, often via self-assembly. Its rich variety of applications include electronic devices, membrane channels, scaffolding tissues, and targeted drug delivery systems ***(1,2)***. The self-assembly process of biomacromolecules, such as protein domains, results in the increase of internal organization and the stability of the system. Consequently, large complexes may be formed that have a specific functionality or no functionality at all ***(3)***.

From: *Methods in Molecular Biology, vol. 474: Nanostructure Design: Methods and Protocols*
Edited by: E. Gazit and R. Nussinov © Humana Press, Totowa, NJ

The natural ability of protein domains to self-assemble can be exploited to predict, design, construct, and validate novel molecular structures ***(4–7)***. During the past few years, there have been numerous reports describing the use of natural and artificial peptides to create nanostructures ***(8,9)***. Due to rapid development of peptide synthesis and molecular engineering techniques ***(10,11)***, the self-assembly of peptide chains can become a favorable route to obtain nanostructures, particularly those consisting of single or associated tubes, vesicles, or fibers. Computations are increasingly becoming a major tool in bionanotechnology and nanostructure design to obtain accurate measurements and allow better comprehension. Advanced computational tools, such as fast simulation methods and efficient modeling algorithms, as well as a constant growth and enrichment of data that can be retrieved from protein databases can considerably accelerate the design process, ruling out unlikely models. The ultimate goal of computational design is that the top-ranking candidates passed to experiment will be valid with a high chance of occurring in the test tube. Although a range of methods have been devised for engineering nanostructures from naturally occurring peptides (or proteins), very few general strategies for such a construction toward different architectures and symmetries arose.

In 2001, Padilla et al. described a general approach for engineering self-assembling nanostructures by combining naturally symmetric protein components ***(12)***. The method is based on the fact that many natural proteins were formed as a result of a self-association of protein domains through noncovalent interactions. Proteins made of two monomers (dimers), three monomers (trimers), or even four monomers (tetramers) are relatively common. These natural proteins are referred to as *oligomerization domains*. Following specific geometric rules, two oligomerization domains are connected covalently, through a short linker, into a single larger molecule called a *fusion protein*. Hence, a fusion protein is made of three segments, with those segments at both ends having a strong tendency to associate with other copies of themselves (*see* **Fig. 1**). As a consequence of this design, the fusion protein self-assembles with many identical copies of itself into a symmetric object, referred to as a *protein nanohedron*. A vast range of nanohedral structures can be designed, such as cages, shells, unbounded layers, or filaments. Each may be referred to as an *architectural class*. Within each architectural class, several different symmetries may be possible. Every such symmetry has a unique construction rule.

Another general strategy for constructing a nanostructure computationally was published by Tsai et al. ***(13)***. The strategy involves optimal mapping of candidate protein building blocks onto the nanostructure shape. The result is a full atomic model of the protein nanostructure. The mapping technique describes the construction of the simplest architecture — the nanotube; the tube construction procedure involves wrapping a planar sheet onto a tube surface.

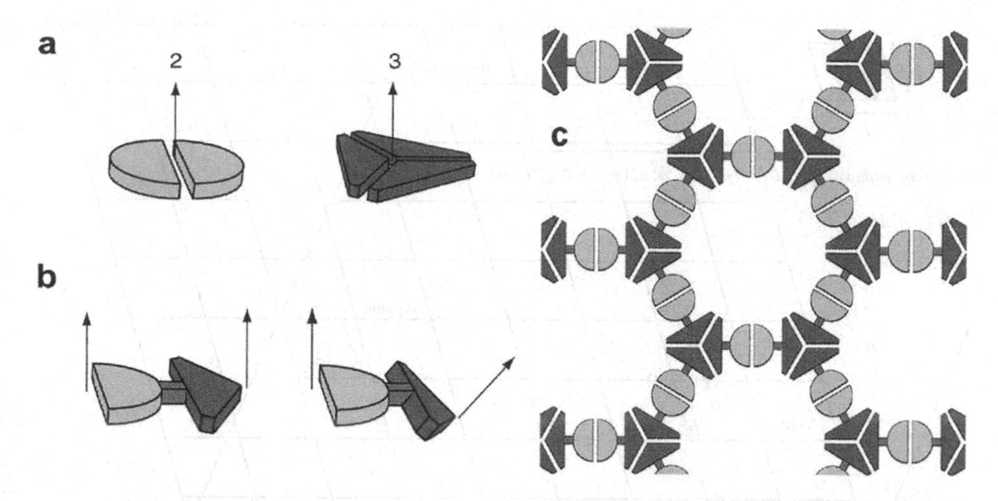

Fig. 1. The design strategy of fusion proteins that assemble into symmetric nanostructures. (**a**) The semicircles represent a natural dimeric protein, whereas the triangle represents a natural trimeric protein. The symmetry axes of the oligomers are shown. (**b**) The two natural monomers, referred to as *oligomerization domains*, are linked through a short linker to create a single fusion protein. The linkage of the oligomerization domains may produce two distinct geometries. (**c**) The designed fusion protein self-assembles into a nanostructure. The geometry of the nanostructure depends on the symmetry axes of its fusion proteins' oligomerization domains. As a result, a molecular layer is formed from the fusion protein like the one illustrated at the left in **b**. (Reproduced from **ref.** ***12*** with permission.)

If the arrangement of the building block has a two-dimensional (2D) repeating pattern, then a planar sheet can be shaped by a repeating 2D lattice that is described by three lattice constants $\{a, b, \gamma\}$. An additional two integers $\{n_1,n_2\}$ define the wrapping of the planar sheet onto the tube surface, where n_1 defines the number of cells needed for wrapping a full round of the tube, and n_2 defines the number of cells shifted after a complete round. **Figure 2** illustrates the five parameters. The tube structure is optimized using CHARMM 22 force field ***(14)*** and a local optimization method.

In this work, we combine the two strategies ***(12,13)*** to computationally design a prespecified shape, the nanotube; first, we define a set of symmetries of a 2D lattice cell. The lattice cell will eventually wrap around a cylinder surface. Then, for each lattice symmetry, we judiciously select homo-oligomers from the Protein Data Bank (PDB) ***(15)*** according to a defined set of criteria, as described in **Subheading 2**. To construct a fused protein as a building block, we first define a protocol for the selection of a specific linker that links two oligomerization domains and proceed to establish a method for connecting it to these domains. Next, we optimize the constructed cell by using CHARMM

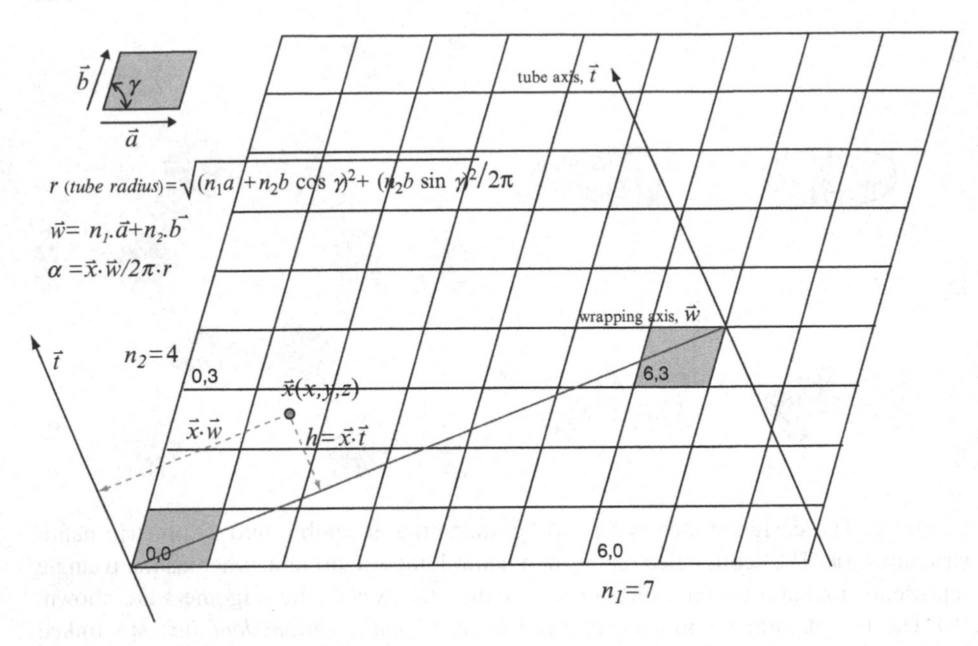

Fig. 2. The two-dimensional (2D) lattice wrapping system. The 2D lattice is highlighted with the angle **γ** between the two axes $\vec{a}$ and $\vec{b}$. This sketch is an example of wrapping with the parameters (n_1 = 7, n_2 = 4). (Adapted from **ref. *13*.**)

force field 22 and a local optimization method. The fused protein and its lattice symmetry define the exact values of the lattice constants $\{a, b, \gamma\}$. To complete the construction procedure, we wrap the cell as a planar sheet around a cylinder, optimizing the wrapping constants n_1 and n_2. The construction of the cylinder leads to an atomic model of the protein nanotube. To verify the model's stability, we optimize it using the same optimization procedure. Two examples of nanotubes are presented; one was first observed experimentally, and subsequently was constructed computationally by Tsai et al. in **ref. *13***.

2. Methods

To create a nanotube, we wrap a planar sheet around a cylinder, at least one full round. To create a planar sheet, we construct a single 2D unit cell from one or more building blocks. This unit cell has a specific symmetry. The cell parameters $\{a, b, \gamma\}$ are therefore set *a priori* by its symmetry and by the dimensions of its building block. The wrapping constants $\{n_1, n_2\}$ are optimized during the tube construction. Therefore, given the cell parameters and the wrapping constants, we can calculate the diameter of the tube. In other words, the tube construction procedure enables the calculation of the tube's diameter for any doublet {symmetry, building block}. Hence, to construct a nanotube with a given diameter, the appropriate doublet of {symmetry, building block} can be selected from a prepared database of such doublets. In **Subheading 2.1.**, we list representative lattice symmetries that were used for constructing a nanotube from fused oligomerization domain building blocks.

2.1. The Main Lattice Symmetries

Figure 3 illustrates symmetric schemes of a unit cell made of fused oligomerization domain building blocks. The oligomers that build a lattice cell include dimers, trimers, tetramers, and hexamers. The various illustrated schemes impose the following constraints on the composition of the fused building blocks:

1. Only homo-oligomers can be selected.
2. Only one binding linker must be selected for each nanotube system.
3. All the homo-oligomers consisting of the same number of monomers (dimers/trimers/etc.) that are selected for a specific nanotube system must be identical.

As a result of these constraints, we can draw an important conclusion: A single nanotube system that conforms to the illustrated schemes of lattice symmetries can be constructed from the same building block only

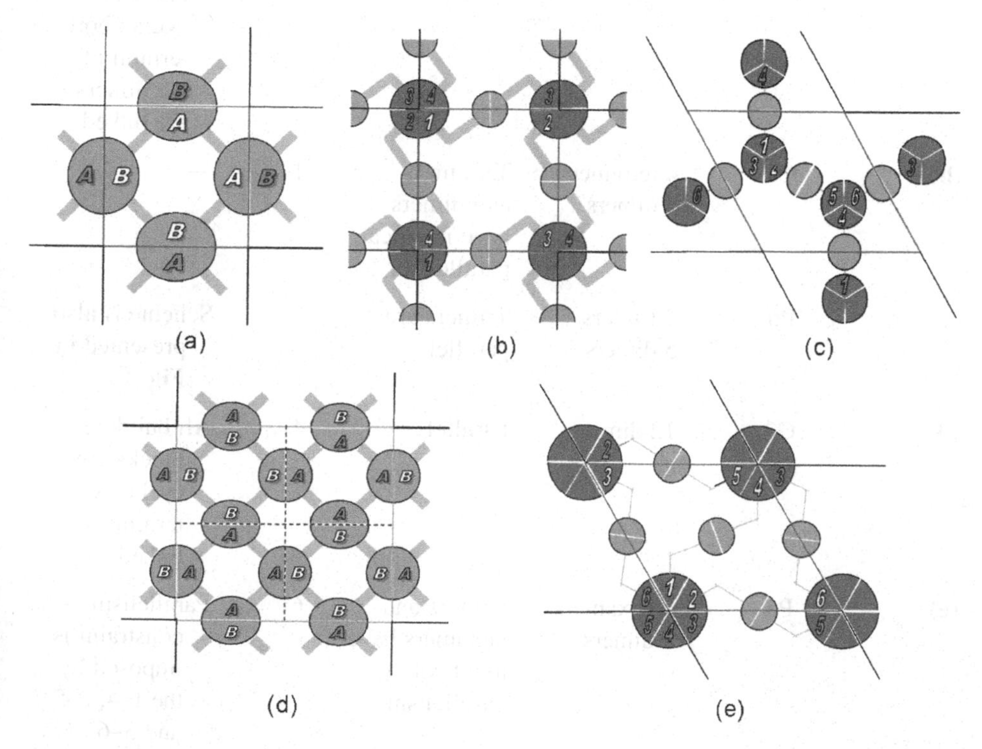

Fig. 3. The symmetries of a two-dimensional (2D) unit cell built from fused oligomerization domain building blocks. Every circle represents an oligomer, and a thick line within a cell represents a linker.

The illustrated schemes also impose constraints on the parallelism of the selected oligomers. Two identical monomers A_1 and A_2 are considered parallel if Distance (A_1 :N-terminus, A_2 :N-terminus) < Distance (A_1 :N-terminus, A_2 : C-terminus).

Table 1 describes the properties of the symmetry schemes in detail.

2.2. The Construction of a Fused Oligomerization Domain

As described by Padilla et al. *(12)*, to construct a fused oligomerization domain experimentally, the gene of one of the oligomers is amplified from a plasmid using polymerase chain reaction (PCR), and its DNA is extended using primers

Table 1
Properties of the Two-Dimensional Unit Cells Shown in Fig. 3

Subfigure	Symmetry of cell	Oligomers in cell	Parallelism	Building blocks	Comments
(a)	C4	4 dimers	Antiparallel	1	Building block is cyclic; both termini of the dimers are linked
(b)	C4	4 tetramers 4 dimers	Tetramers and dimers have identical parallelism	4	—
(c)	P6	2 trimers 5 dimers	Trimers are parallel	6	Scheme is also presented in **Fig. 1**
(d)	C2v	12 dimers	Parallel	4	All building blocks are cyclic; all termini are linked
(e)	P6	4 hexamers 5 dimers	Dimers and hexamers have identical parallelism	6	Parallelism constraint is imposed by the 1–4, 2–5, and 3–6 hexamer links

to include the linker's residues. This DNA is digested and ligated into a vector containing the gene of the second oligomer, and the resulting DNA is expressed in bacteria cells.

The above procedure can only be used to construct a noncircular protein. To construct a circular fused oligomerization domain experimentally (as shown schematically in **Fig. 3a**), the Fmoc (9-fluorenylmethoxycarbonyl) solid-phase peptide synthesis (SPPS) technique can follow the procedure to circulate the constructed fused oligomerization domain, as described in **ref. *9***.

Here, we define a protocol for the construction of a fused oligomerization domain computationally. The following subheadings describe our protocol's three main stages: selecting oligomers, selecting a linker, and covalently attaching the linker to the oligomers.

2.2.1. Selecting Oligomers

The following oligomer selection procedure was used:

1. All the oligomers were selected from the PDB ***(15)***. The selection criteria were:
   a. The oligomer must be a homo-oligomer.
   b. The parallelism is the one defined by **Table 1**.
   c. The chain size must be small, for example, fewer than 50 residues. The chain size has a major effect on the computational times and resources required for the validation procedures.
   d. Preferences: (1) The method of protein structure determination is X-ray diffraction to obtain the highest accuracy of the structure. (3) There are no missing atoms and residues in the structure.
   e. According to **ref. *12***, the secondary structure at the termini of the two monomers *A* and *B* must be α-helix. To link both termini by a rigid linker, it is necessary to choose a short α-helical linker so that all three α-helices will bind to create a single long, rigid α-helix. In our protocol, the secondary structure of the end residues at the termini of the monomers as well as the linker that links them together may have any secondary structure.
2. To obtain a complete atomic model of the selected oligomer, all the missing hydrogen atoms were added using the CHARMM 22 force field.
3. A local energy minimization method was run, in explicit water, using NAMD (Nanoscale Molecular Dynamics) ***(16)***. The objective was to stabilize the protein structure before we begin the actual fusion process.

2.2.1.1. Modeling Missing Residues

In case the selected oligomer has missing residues and the number of residues in its monomers is different, there is a need to equalize the number of residues in all the monomers. Since the selected proteins are homo-oligomers, the sequences of all the monomers are the same. Hence, we can model the missing residues of the short monomers by the same residues of the longest monomer.

To model the missing residues, we run MultiProt, an algorithm that detects multiple structural alignments of protein structures *(17)*.

2.2.2. Selecting a Linker

To link the C-terminus of one oligomer with the N-terminus of the second oligomer, an appropriate linker must be selected. The linker must:

1. Separate the oligomers and avoid steric clashes between them.
2. Be short (i.e., less than 12 residues long) to maintain the stability of the fused building block.
3. Bind to both termini of the oligomerization domains with favorable dihedral angles according to the predicted secondary structure of the residues at both termini and the linker itself.
4. Be flexible, as opposed to the rigidity required by **ref.** *12*. The flexibility of the linker ensures the ability to control the self-assembly process of the fused building blocks into a lattice cell with specific symmetry.

The following selection procedure was defined:

5. Get the sequence of five residues from the C-terminus of a monomer from oligomer $A(a_{n-4}...a_n)$ and the five residues from the N-terminus of a monomer from oligomer $B(b_1...b_5)$.
6. Use the FASTA algorithm for pairwise sequence alignment *(18)* against the National Center for Biotechnology Information (NCBI) PDB structures *(15)* with the following sequence (in FASTA format): $a_{n-4}...a_n$ $(*)^k$ $b_1...b_5$, where k is the number of residues allowed between the sequence of A and the sequence of B. From our experience, and according to the above-listed constraints, k must be in the range of [3,10]. Using a pairwise sequence alignment approach to select the linker ensures that there already exists a sequence of amino acids in nature that links a subset of the C-terminus residues from oligomer A with a subset of the N-terminus residues from oligomer B.
7. For all ks find the solution with the highest Smith-Waterman score *(19)* such that residues from A and residues from B are aligned.
8. Obtain the linker of $k + 2$ residues from the selected protein. The $k + 2$ residues include the linker and one additional residue on each terminus of the oligomerization domain. These additional residues will be used for the binding procedure.

2.2.3. Binding the Linker to the Oligomers

Since the linker was obtained using a pairwise sequence alignment algorithm, the two additional residues at the selected linker's termini are similar/identical to the residues a_n and b_1 of the corresponding oligomers A and B. Hence, the selected linker can be linked to both oligomers using the backbone orientation of its k residues with both terminal amino acids. Following is a general procedure for binding two oligomers A and B through one linker:

1. Prepare the C-terminus of monomer *A*, the N-terminus of monomer *B*, and both termini of the linker for a peptide bond formation. That is, remove two hydrogen atoms from the N-terminus and an oxygen atom from the C-terminus.
2. Transform the linker to the location of the C-terminus of monomer *A*. Delete the redundant residue from monomer *A* or the linker.
3. Transform monomer *B*'s N-terminus to the current location of the C-terminus of the linker. Delete the redundant residue from the linker or from monomer *B*.
4. Run a local minimization algorithm on the fused oligomerization domains, in explicit water, applying harmonic restraining force on the oligomers. The purpose is to solve any side-chain clashes that might have arisen as a result of the binding procedure and at the same time retain the conformation preferences of each fused oligomer.

The building blocks that comprise the lattice cell can be cyclic (as illustrated in **Fig. 3a**). A cyclic fused oligomerization domain is comprised of four oligomers and four linkers, whereas a noncyclic fused oligomerization domain is comprised of two oligomers and one linker only. To construct a cyclic peptide from monomers A_1, A_2, A_3, and A_4 and linkers l_1, l_2, l_3, and l_4, we must ensure that in chain $A_1l_1A_2l_2A_3l_3A_4l_4$ the C-terminus of l_4 will bind to the N-terminus of A_1. Therefore, the procedure for constructing a cyclic fused oligomerization domain is as follows:

1. Prepare the termini of monomers $A_1 \ldots A_4$ and the termini of the linkers $l_1 \ldots l_4$ for a peptide bond formation.
2. Place the monomers $A_1 \ldots A_4$ on the *x-y* plane such that the distance between consequent termini will fit the length of the linker. The center of mass of all oligomers should also reside on the *x-y* plane.
3. Place the linkers such that their termini will fit the termini of the consequent monomers. Delete the redundant residues from the monomers or the linkers.
4. Run a local minimization algorithm on the cyclic fused oligomerization domain, in explicit water, applying harmonic restraining force on each oligomer.

2.3. Constructing a Unit Cell and Building a Nanotube

To construct a unit cell, it is necessary to place all its building blocks in their relative position according to the chosen scheme (**Fig. 3**). Therefore, once the first building block is constructed, the other building blocks are duplicated and transformed according to the location of the unbound monomers of the oligomers.

Finally, given the lattice cell parameters $\{a, b, \gamma\}$ as derived from the cell symmetry, the set of wrapping parameters $\{n_1,n_2\}$, and the constructed unit cell from the fused oligomerization domains, the protein nanotube can be constructed as follows:

1. Register all the cells according to the portion of the tube that is specified to be built.
2. Iteratively generate the coordinates for each registered cell using the tube-wrapping transformation as described in **ref. 5**.

2.4. Validation Procedures

The process of self-assembly of fused oligomerization domains to create a nanotube takes place in a solvent, and so it is clearly important to consider how the solvent affects the behavior of the system. The solvent does not directly interact with the fused building blocks, but it provides an environment that strongly affects the behavior of the tube. Therefore, it is not necessary to model the solvent molecules. Currently, we optimize the tube by a conjugate gradient minimization, in vacuum, without applying any harmonic force restraints. In the near future, we plan to run MD simulations with implicit water to validate the stability of the nanotube.

2.5. Results

We present two examples of different lattice symmetries.

2.5.1. P6 Symmetry of Dimers and Trimers

We constructed a nanotube of a P6 symmetry lattice cell comprised of dimers and trimers. **Figure 3c** illustrates the corresponding 2D scheme. Each building block is a fusion of the GCN4-Pvsl coiled-coil trimer crystal structure *(20)*, PDB code 1ij3, with a designed p53 dimer nuclear magnetic resonance (NMR) model *(21)*, PDB code 1hs5. Each monomer of the trimer is a helix, and all the helices are parallel. Since in the atomic model chain A has 32 residues whereas the other two chains have 31 residues and the last residue of chain A is modeled inaccurately (the dihedral angles do not fit the Ramachandran plot), we picked the first 31 residues in all three monomers. For the dimer, we selected the first atomic model among the NMR models. Each monomer taken from the trimer was linked to the monomer from the dimer through a linker of four residues, HRQE, cut from the crystal structure of PDB code 1kj8 *(22)*. The linker was obtained using the FASTA pairwise sequence alignment *(18)*. The complete constructed cell was slightly manipulated so that each dimer will reside on the *x*-*y* plane and the trimer's N-termini will stick out of the *x*-*y* plane. This structure was then optimized using the NAMD local minimization algorithm while applying harmonic force restraints on all dimers and trimers *(16)*.

The constructed cell is shown in **Fig. 4a**. To construct a tube, we used lattice constants {120.0 Å, 120.0 Å, 120.0°} and tube wrapping constants {8, 2}. Hence, the tube radius is 137.72 Å. The constructed tube is shown in **Fig. 4b**. The geometry of the system implies that the N-termini of the trimer can be used as connectors to external molecules, such as ligands of specific receptors.

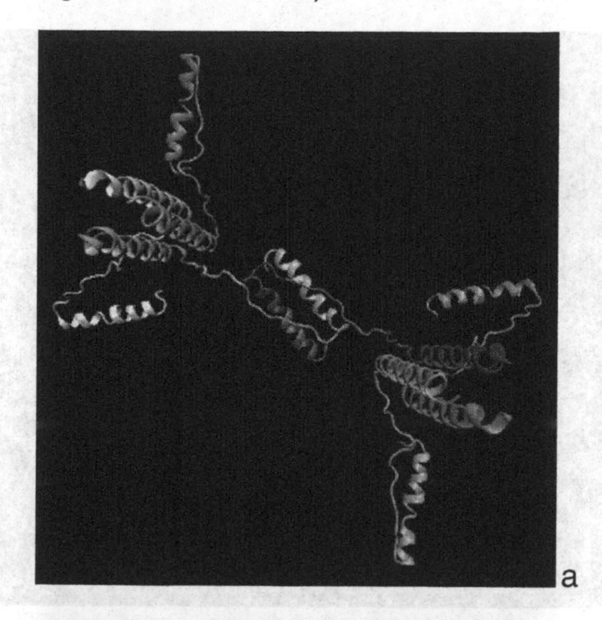

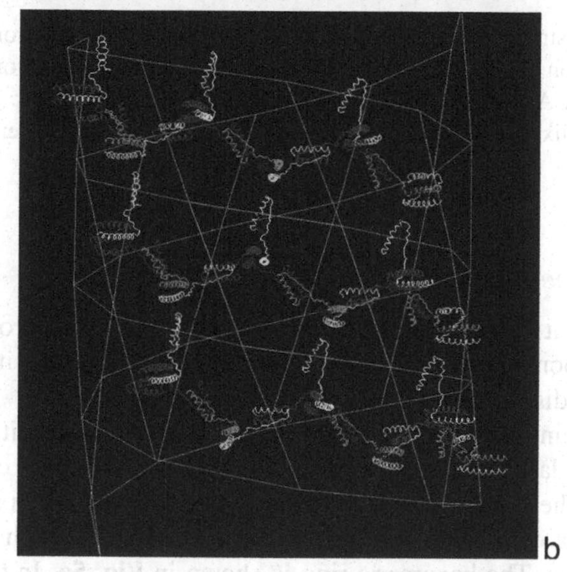

Fig. 4. (**a**) The unit cell of the Gcn4-Pvsl coiled-coil trimer fused with a designed P53 dimer. The symmetry of the lattice is P6, and the scheme is illustrated in **Fig. 3c**. Both trimer and dimer are parallel. The dimer's helices reside on the *x-y* surface, whereas the trimer's N-termini stick out. The geometry of the system implies that the trimer's helices can be used as connectors to external molecules, such as ligands of specific receptors. Each of the six chains is a different gray color. (**b**) Nine copies of the trimer-dimer cell, with wrapping parameters $n_1 = 8$, $n_2 = 2$.

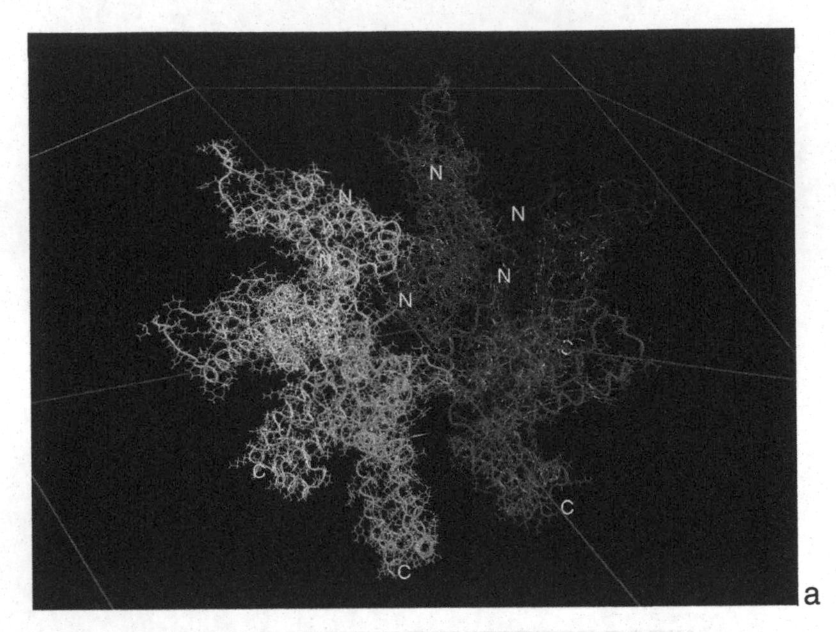

Fig. 5. (**a**) The structure of the HIV-1 CA protein tube. (Adapted from **ref.** ***13***.). The initial conformation of the hexameric ring with the N-termini sitting on the outer wall of the tube. Each CA monomer is highlighted in a different gray color. The ring is connected by a 12-helix bundle with 2 helices contributed from each N-terminus.

2.5.2. P6 Symmetry of Dimers and Hexamers

We briefly present the construction of the HIV-1 CA (Capsid) protein tube. This example corresponds to the scheme illustrated in **Fig. 3e**, and its construction was thoroughly discussed by Tsai et al. in **ref.** ***13***.

The CA protein from HIV-1 has an atomic model in which it is arranged in approximate P6 lattice symmetry ***(23)***. The structure is made of a hexameric ring, which is the outcome of the association between the six N-termini of the monomers; a dimerization of two C-termini connects each ring to its six neighboring rings. The hexameric ring is shown in **Fig. 5a**. In the CA protein nanotube construction, the initial structure of the CA protein was taken from the crystal structure ***(24)***, PDB code 1e6j, chain P. The tube was constructed according to the electron microscopic (EM) images with lattice constants {108.0 Å, 108.0 Å, 120.0°} and two wrapping constants of {12, 1}. Hence, its outer radius was 198.23 Å. The crystal coordinates were manipulated, translating and rotating the molecules to form a hexameric ring. The tube is shown in **Fig. 5b** with nine copies of the hexameric ring.

Fig. 5. (*continued*) (**b**) Nine copies of the hexameric ring sitting on the optimized protein tube. The wrapping parameters are $n_1 = 12$, $n_2 = 1$.

3. Note

1. If a selected protein structure was determined by NMR, the first model is selected.

Acknowledgments

We thank Nurit Haspel and Dan Fishelovitch for their help and support. The computation times were provided by the National Cancer Institute's Frederick Advanced Biomedical Supercomputing Center and by the high-performance computational capabilities of the Biowulf PC/Linux cluster at the National Institutes of Health (NIH), Bethesda, Maryland (http://biowulf.nih.gov). This project has been funded in whole or in part with federal funds from the National

Cancer Institute, National Institutes of Health, under contract number NO1-CO-12400. This research was supported in part by the Intramural Research Program of the NIH, National Cancer Institute, Center for Cancer Research. The content of this publication does not necessarily reflect the view or policies of the Department of Health and Human Services, and mention of trade names, commercial products, or organization does not imply endorsement by the U.S. government.

References

1. Ferrari M. (2005) Cancer nanotechnology: opportunities and challenges. *Nat. Rev. Cancer* **5**, 161–171.
2. Ferrari M. (2005) Nanovector therapeutics. *Curr. Opin. Chem. Biol.* **9**, 343–346.
3. Ma B, Nussinov R. (2002) Stabilities and conformations of Alzheimer's beta-amyloid peptide oligomers (Abeta 16–22, Abeta 16–35, and Abeta 10–35): sequence effects. *Proc. Natl. Acad. Sci. U. S. A.* **99**, 14126–14131.
4. Zhang S. (2003) Fabrication of novel biomaterials through molecular self-assembly. *Nat. Biotechnol.* **21**, 1171–1178.
5. Tsai CJ, Zheng J, Aleman C, Nussinov R. (2006) Structure by design: from single proteins and their building blocks to nanostructures. *Trends Biotechnol.* **24**, 449–454.
6. Haspel N, Zanuy D, Aleman C, Wolfson H, Nussinov R. (2006) De novo tubular nanostructure design based on self-assembly of beta-helical protein motifs. *Structure* **14**, 1137–1148.
7. Yemini M, Reches M, Rishpon J, Gazit E. (2005) Novel electrochemical biosensing platform using self-assembled peptide nanotubes. *Nano. Lett.* **5**, 183–186.
8. Beniash E, Hartgerink JD, Storrie H, Stendahl JC, Stupp SI. (2005) Self-assembling peptide amphiphile nanofiber matrices for cell entrapment. *Acta Biomater.* **1**, 387–397.
9. Horne WS, Stout CD, Ghadiri MR. (2003) (A heterocyclic peptide nanotube. *J. Am. Chem. Soc.* **125**, 9372–9376.
10. Kohli RM, Walsh CT, Burkart MD. (2002) Biomimetic synthesis and optimization of cyclic peptide antibiotics. *Nature* **418**, 658–661.
11. Frank R. (2002) The SPOT-synthesis technique. Synthetic peptide arrays on membrane supports—principles and applications. *J. Immunol. Methods* **267**, 13–26.
12. Padilla JE, Colovos C, Yeates TO. (2001) Nanohedra: using symmetry to design self assembling protein cages, layers, crystals, and filaments. *Proc. Natl. Acad. Sci. U. S. A.* **98**, 2217–2221.
13. Tsai CJ, Zheng J, Nussinov R. (2006) Designing a nanotube using naturally occurring protein building blocks. *PLoS Comput. Biol.* **2**, e42.
14. Brooks BR, Bruccoleri RE, Olafson BD, States DJ, Swaminathan S, Karplus M. (1983) CHARMM—a program for macromolecular energy, minimization, and dynamics calculations. *J. Comput. Chem.* **4**, 187–217.

15. Berman HM, Westbrook J, Feng Z, et al. (2000) The Protein Data Bank. *Nucleic Acids Res.* **28**, 235–242.
16. Phillips JC, Braun R, Wang W, et al. (2005) Scalable molecular dynamics with NAMD. *J. Comput. Chem.* **26**, 1781–1802.
17. Shatsky M, Nussinov R, Wolfson HJ. (2004) A method for simultaneous alignment of multiple protein structures. *Proteins* **56**, 143–156.
18. Pearson WR, Lipman DJ. (1988) Improved tools for biological sequence comparison. *Proc. Natl. Acad. Sci. U. S. A.* **85**, 2444–2448.
19. Smith TF, Waterman MS. (1981) Identification of common molecular subsequences. *J. Mol. Biol.* **147**, 195–197.
20. Akey DL, Malashkevich VN, Kim PS. (2001) Buried polar residues in coiled-coil interfaces. *Biochemistry* **40**, 6352–6360.
21. Davison TS, Nie X, Ma W, et al. (2001) Structure and functionality of a designed p53 dimer. *J. Mol. Biol.* **307**, 605–617.
22. Thoden JB, Firestine SM, Benkovic SJ, Holden HM. (2002) PurT-encoded glycinamide ribonucleotide transformylase. Accommodation of adenosine nucleotide analogs within the active site. *J. Biol. Chem.* **277**, 23898–23908.
23. Li S, Hill CP, Sundquist WI, Finch JT. (2000) Image reconstructions of helical assemblies of the HIV-1 CA protein. *Nature* **407**, 409–413.
24. Monaco-Malbet S, Berthet-Colominas C, Novelli A, et al. (2000). Mutual conformational adaptations in antigen and antibody upon complex formation between an Fab and HIV-1 capsid protein p24. *Structure* **8**, 1069–1077.

9

Computational Methods in Nanostructure Design

Replica Exchange Simulations of Self-Assembling Peptides

Giovanni Bellesia, Sotiria Lampoudi, and Joan-Emma Shea

Summary

Self-assembling peptides can serve as building blocks for novel biomaterials. Replica exchange molecular dynamics simulations are a powerful means to probe the conformational space of these peptides. We discuss the theoretical foundations of this enhanced sampling method and its use in biomolecular simulations. We then apply this method to determine the monomeric conformations of the Alzheimer amyloid-β(12–28) peptide that can serve as initiation sites for aggregation.

Key Words: Alzheimer amyloid-β peptide; biomaterials; conformational space sampling; molecular dynamics simulations; replica exchange algorithm.

1. Introduction

The self-assembly of normally soluble proteins or peptides into aggregate nanostructures is a process that is typically triggered in the cell under pathological conditions (heat shock, excessively high concentrations, etc.) *(1)*. The end product of aggregation is often a highly ordered fibrillar structure, enriched in β-sheet content *(2,3)*. The presence of these fibrils, known as *amyloid fibrils*, is a hallmark of a number of neurodegenerative diseases, including Alzheimer's disease and Parkinson's disease *(4)*. Interestingly, recent experiments suggested that even non-disease-related proteins can aggregate and form fibrils with a striking similarity to those produced by pathogenic proteins *(5)*. Even protein fragments and small peptides can form fibrils *(6)*. The implication of this finding is that aggregation is an inherent property of polypeptide chains.

From: *Methods in Molecular Biology, vol. 474: Nanostructure Design: Methods and Protocols*
Edited by: E. Gazit and R. Nussinov © Humana Press, Totowa, NJ

An intriguing application of peptide aggregation lies in the field of biomaterials. Indeed, peptides capable of aggregating into ordered fibrils could serve as building blocks for novel nanoscale biomaterials *(7)*. These nanostructures can be used in a number of applications, including drug delivery, biological surface engineering, as well as scaffolding for tissue repair, cell proliferation and differentiation, and templated assembly of metal nanowires ***(8–14)***.

To design biomaterials with specific properties, it is essential to understand the process of self-assembly. In particular, it is important to characterize the nature of the monomeric, unassembled states of these peptides and to determine from which conformations aggregation is initiated. In general, aggregating peptides do not possess a unique folded state but rather appear to be mostly unstructured, only populating a small fraction of ordered states. These different monomeric species could serve as different initiation sites for aggregation and give rise to different aggregation intermediates and hence to different fibril morphologies. The monomeric states of aggregating peptides are notoriously difficult to study experimentally as these peptides tend to form clusters below the concentrations required for nuclear magnetic resonance (NMR) studies ***(15)***. Moreover, ensemble-averaging experimental techniques (such as NMR) may not be appropriate for detecting structured conformational ensembles with low population.

Computer simulations offer a means of probing the conformation space of peptides at a level of atomic resolution often exceeding experimental capabilities ***(16–24)***. They are invaluable complements to experiment, allowing direct access to the entire conformational space sampled under a given set of conditions. In this chapter, we describe the use of molecular dynamics (MD) simulations to study the monomeric state of aggregating peptides. We focus on the use of an enhanced sampling technique known as replica exchange (REx) molecular dynamics. Because peptides tend to have energy landscapes dominated by high barriers and deep minima, the use of efficient sampling techniques (as opposed to a conventional constant temperature MD simulation) is essential to obtain a correct statistical description of conformational space.

We begin in **Subheading 2**. with a presentation of the theoretical foundations of MD and REx MD simulations. We include a pseudocode of an REx algorithm so that the reader can implement REx into an existing MD program. In **Subheading 3**., we describe the steps needed to carry out an REx simulation for a peptide in explicit solvent. An example is given for the simulation of the 12–28 fragment of the Alzheimer amyloid-β peptide ***(23)***. In **Subheading 4**., we elaborate on specific points and discuss alternate simulation approaches.

2. Materials

2.1. Theoretical Background

2.1.1. Molecular Dynamics Simulations

Molecular dynamics (MD) is a simulation method that can be used to determine the equilibrium and transport properties of classical many-body systems ***(25–31)***. For a system of interacting atoms or molecules, MD simulations follow and specify the time evolution of the system by numerically solving Newton's equations of motion under a given set of boundary conditions. For a classical system of N particles with positions and velocities denoted by the $3N$-dimensional vectors, $(\boldsymbol{r}_1(t), \ldots, \boldsymbol{r}_N(t))$ and $(\boldsymbol{\upsilon}_1(t), \ldots, \boldsymbol{\upsilon}_N(t))$, the corresponding Newton equations of motion are

$$m_i\ddot{r}_i = \boldsymbol{F}_i, \forall i \in [1, N] \tag{1}$$

where m_i and $\boldsymbol{F}_i$ are the mass of particle i and the force on particle i, respectively.

For *conservative* forces,

$$\boldsymbol{F}_i = -\frac{\partial V}{\partial \boldsymbol{r}_i} \tag{2}$$

and

$$E = K + V \equiv \frac{1}{2}\sum_{1=1}^{N} m_i \boldsymbol{\upsilon}_1^2 + V(\boldsymbol{r}_1,\ldots,\boldsymbol{r}_N) \tag{3}$$

where K is the kinetic energy, $V(\boldsymbol{r}_1, \ldots, \boldsymbol{r}_N)$ is the potential energy, and E is the total energy.

The latter is a *conserved quantity*:

$$\frac{\partial E}{\partial t} = 0 \tag{4}$$

The interatomic potential energy $V(\boldsymbol{r}_1, \ldots, \boldsymbol{r}_N)$ is given by an *empirical potential*. This is an analytical expression, typically a sum of pairwise terms in which the pair energy is a function of the interatomic distance (*see* **Note 1**).

A typical example of an empirical potential used in MD simulations has the following form ***(32)*** (*see* **Note 2**):

$$\begin{aligned} &\sum_{bonds} K_b(b-b_0)^2 + \sum_{angles} K_\theta(\theta_0-\theta_0)^2 + \\ &+ \sum_{dihedrals} K\chi(1+\cos(n\chi+\delta)) + \sum_{impropers} K_\phi(\phi-\phi_0)^2 + \\ &+ \sum_{non-bonded\ pairs} 4\in\left[\left(\frac{\sigma}{r}\right)^{12} - \left(\frac{\sigma}{r}\right)^{6}\right] + \frac{q_1 q_2}{\varepsilon_l r} \end{aligned} \tag{5}$$

where K_b, K_θ, K_χ, and K_ϕ are the bond, angle, dihedral angle, and improper dihedral angle energy constants, respectively; b, θ, χ, and ϕ are the bond length, bond angle, dihedral angle, and improper torsion angle, respectively, with the subscript zero representing the equilibrium values for the individual terms. Coulomb and Lennard-Jones 6–12 terms contribute to the nonbonded interactions; $\in$ is the Lennard-Jones well depth, and σ is related to the Lennard-Jones minimum distance by the equality $r_{min} = 2^{\frac{1}{6}}\sigma$. The various terms in the empirical potential (bond constants, etc.) are parameterized from *first principles* calculations or fit to experimental data.

The physical state of the system of interest at any time t is fully specified by solving **Eqs. 1** and **2** with $V(\boldsymbol{r}_1, \ldots, \boldsymbol{r}_N)$ defined in **Eq. 5**.

Newton's equations of motion (**Eq. 1**) are second-order ordinary different equations (ODE) in the variable t. They are solved numerically using integrators or *maps*. These involve *time-stepping* algorithms *(33)* in which solutions to Newton's equations of motion are calculated at subsequent times $t_{i+1} = t_i + \Delta t$ where Δt is the *time step* (*see* **Note 3**).

2.1.2. Statistical Mechanics

Statistical mechanics provides the link between the microscopic information obtained from MD simulations and the desired macroscopic thermodynamic properties of the system of interest *(34,35)*.

For a system of N particles, a microstate is defined as a point in the 6N-dimensional phase space:

$$\boldsymbol{x} = (\boldsymbol{p}_1, \ldots, \boldsymbol{p}_N, \boldsymbol{r}_1, \ldots, \boldsymbol{r}_N) \tag{6}$$

where $\boldsymbol{p}_i = m_i \dot{\boldsymbol{r}}_i$ is the momentum of particle i (*see* **Note 4**).

At thermodynamic equilibrium *(34–36)*, the macroscopic state of this same system is characterized by a fixed number of thermodynamics parameters, such as the number of particles N, the total energy E, the volume V, the pressure P, or the chemical potential μ. An extremely large number of microscopic states (6N–dimensional vectors, i.e., points $\boldsymbol{x}$ in phase space) are consistent with these fixed thermodynamics parameters. The collection of such microscopic states describes a particular type of statistical *ensemble*. Ensembles used in MD simulations include the *microcanonical ensemble* in which the total number of particles N, the volume V, and the energy E are held constant (*NVE* ensemble) (*see* **Note 5**), the *canonical ensemble* (*NVT* fixed), and the *NPT–ensemble*, in which N is fixed and the system is kept at constant temperature and pressure by contact with a thermostat and an external barostat, respectively (*see* **Notes 6** and **7**).

Given an ensemble of points (microscopic states) in the phase space, we define the *probability density* (or probability distribution) in the phase space

$\rho(x, t)$ such that, at any time t, $\rho(x, t)dx$ is the probability of observing a system in the phase space volume element dx. At thermodynamic equilibrium, the probability density is stationary, that is, $\delta\rho(x, t)\delta t = 0$ and $\rho(x, t) = \rho(x)$.

The ensemble average for a generic thermodynamic observable (physical quantity) $A(x)$ is given by:

$$\langle A \rangle = \int dx A(x)\rho(x) \quad (7)$$

The same thermodynamic observable can be calculated as a time average (over a time t_f) along the trajectory of the system in phase space:

$$\bar{A} = \frac{1}{t_f}\int dt A(p(t),\ r(t)) \quad (8)$$

The *ergodic hypothesis* states that a system, after a sufficiently long time, will visit all possible phase space points consistent with the fixed thermodynamics parameters. In addition, the time that the system will spend in a region of the phase space will be proportional to the probability density in that region. The ergodic hypothesis establishes the equivalence between ensemble average and time average as follows:

$$\lim_{t_f \to \infty} \bar{A}_{t_f} = \langle A \rangle \quad (9)$$

In the most basic formulation of MD, Newton's equations are solved numerically, and a succession of dynamical states is generated in accordance with the probability density of the microcanonical ensemble. The ensemble average in **Eq. 7** can then be replaced by the trajectory average in **Eq. 8** by invoking the ergodic hypothesis. As a result, several thermodynamic observables can be calculated as time averages over the simulation time length.

2.2. The Replica Exchange Algorithm

Molecular dynamics simulations of complex biomolecular systems at low temperatures tend to get trapped in multiple metastable states associated with local free-energy minima. Several enhanced sampling methods have been developed in recent years for improving the phase space sampling at low temperatures (*see* **Note 8**).

We focus here on the REx method, which has been successfully applied to simulations of biomolecular systems with rough energy landscapes ***(37–39)***.

The REx method relies on the construction of a generalized ensemble composed of M noninteracting replicas of the system of interest. The mth replica is in contact with its own heat reservoir and has an inverse temperature of $\beta_m = \frac{1}{k_B T_m}$ and a *canonical* probability distribution in configuration space:

$$\rho(\boldsymbol{c}_m,\beta_m) \propto e^{-\beta_m V(\boldsymbol{c}_m)} \tag{10}$$

where $\boldsymbol{c}_m$ describes the system's configuration, and $V(\boldsymbol{c}_m)$ is the potential energy.

A state (configuration) of the generalized ensemble is then specified by the set $\boldsymbol{C}=\{\boldsymbol{c}_1, \boldsymbol{c}_2, \ldots, \boldsymbol{c}_M\}$ and the probability distribution

$$\rho(\boldsymbol{C},\beta) \propto \prod_{m=1}^{M} \rho(\boldsymbol{c}_m,\beta_m) \tag{11}$$

where $\beta=\{\beta 1, \beta_2, \ldots, \beta_M\}$ spans over a range including both high and low temperatures.

The REx method involves running equilibrium MD simulations in the generalized ensemble as follows: (1) Each replica is simulated simultaneously and independently under canonical conditions for a given number of MD time steps. (2) The configurations of pairs of replicas at neighboring inverse temperatures are exchanged with a given probability. It can be easily shown that, to ensure that the equilibrium probability distribution $\rho(\boldsymbol{C}, \beta)$ will be approached, a correct choice for the exchange probability is ***(39)***

$$W(\boldsymbol{c}_m,\beta_m \mid \boldsymbol{c}_n,\beta_n) = \begin{cases} 1 & \mathit{for}\ \Delta < 0 \\ e^{-\Delta} & \mathit{for}\ \Delta > 0 \end{cases} \tag{12}$$

where $\Delta=(\beta_n - \beta_m)(V(\boldsymbol{c}_n) - V(\boldsymbol{c}_m))$.

The iteration of the entire exchange process (1 and 2) enables the various replicas to traverse the entire range of temperatures, so that replicas at low temperatures can easily escape from local minima. More generally, the exchange process increases the thermalization of every single canonical simulation. As a result, each replica will approach its canonical probability distribution.

The MD trajectories generated at different temperatures via the REx method may be combined using reweighting techniques such as the weighted histogram analysis method (WHAM) ***(40)***. Average values and fluctuations of different thermodynamic observables can be calculated over a given range of temperatures. Moreover, application of the WHAM equations on MD-REx data allows the accurate calculation of the potential of the mean force (PMF; *see* **Note 9**) along one or two reaction coordinates ***(41)*** (PMF profiles and PMF contour plots, respectively).

We present here a pseudocode for the REx algorithm. In this protocol, each node must maintain the following variables:

`tempByRank[]`: an array of temperatures, indexed by the rank of the node operating at that temperature, that is, `tempByRank[i] == T`, means that node i is operating at temperature T in the current iteration `tempByOrder[]`: an array of temperatures indexed by their ascending order, that is, `tempByOder[j] == T`, means that the `j`th temperature, in ascending order, is `T`

`myTempOrder`: a scalar holding the location of the temperature at which the node is operating in the `tempByOrder` array, that is, node `n` has `myTempOrder == k` means that node `n` is operating at temperature `tempByOrder[k]` in the current iteration

The invariant condition, which must be satisfied for all n at each iteration of the algorithm, is `tempByOrder[myTempOrder] == tempByRank[n]`.

Lines beginning with // are comments. The pseudocode, which is identical for each node in a parallel program using message passing interface (MPI) communication ***(42)***, is as follows:

```
// initialize my rank
myRank=MPIGetRank();
// initialize physics from program input
// initialize tempByRank and tempByOrder from program input
tempByRank[]=INPUT
tempByOrder[]=INPUT
// set temperature and myTempOrder
myTemp=tempByRank[myRank]
myTempOrder=myRank;
// main loop
for(iteration=0; iteration < max; iteration++){
// do physics
// do replica exchange
neighborRank=FindNeighbor(myTempOrder, iteration%2);
neighborTemp=tempByRank[neighborRank];
if( neighborRank != False ){
// communicate energies
// even rank neighbor sends first, odd receives and then
they swap
if( (myTempOrder%2)==0){
MPISend(myPEnergy, neighborRank);
neighborPEnergy=MPIReceive(neighborRank);
} else {
MPISend(myPEnergy, neighborRank);
neighborPEnergy=MPIReceive(neighborRank);}
// compute swap decision, symmetrically
if( SwapDecision(myPEnergy, neighborPEnergy) ){
// find location of my new temp in tempByOrder
for(i=0; i < numprocs; i++){
if(tempByOrder[i] == neighborTemp){
```

```
myTempOrder=i;}}
// update temperature
myTemp=neighborTemp;
tempByRank[myRank]=neighborTemp;}}
// now share new tempByRank among everyone
for( i=0; i < numProcs; i++){
MPIBroadcast( tempByRank[i], i);}
}
function FindNeighbor(myTempOrder, cycle){
if( cycle == 0){
// on even cycles do 0-1, 2-3, …
if( (myTempOrder%2) == 0 ){
neighborOrder=myTempOrder+1;
} else {
neighborOrder=myTempOrder-1;}
} else {
// on odd cycles do 1-2, 3-4, …
if( (myTempOrder%2) == 0 ){
neighborOrder=myTempOrder-1;
} else {
neighborOrder=myTempOrder+1;}}
if( (neighborOrder < 0) || (neighborOrder > = numProcs)){
// if neighborOrder is out of bounds, return False
return False;}
neighborTemp=tempByOrder[neighborOrder];
// search through tempByRank for neighborTemp
for(i=0; i < numProcs; i++){
if( tempByRank[i] == neighborTemp ){
return i;}}}
function SwapDecision(myPEnergy, neighborPEnergy){
// return True if the swap should be performed, False
   otherwise
}
```

See **Notes 10–14**.

2.3. Principal Components Analysis

The three-dimensional conformation of a protein arises from the mutual arrangements of quasi-rigid *secondary structures* and domains connected by flexible loops ***(43)***. The relative coordinates of these quasi-rigid elements define a *subspace* containing a reduced number of so-called *collective* coordinates. These collective coordinates can describe the large-scale collective motion (*see* **Note 15**) and conformational changes of a protein ***(44)***.

Principal components analysis (PCA) is one of the best-available methods for extracting collective coordinates from MD simulations ***(44–46)*** (*see* **Note 16**).

PCA is performed by diagonalizing the variance–covariance matrix $\boldsymbol{B}$ with elements that are defined as follows:

$$B_{ij} \equiv \left\langle \left(x_i - \langle x_i \rangle\right)\left(x_j - \langle x_j \rangle\right)\right\rangle \tag{13}$$

where the averages are over the instantaneous structures sampled during the simulation, and x_i is a mass-weighted atomic coordinate. A set of eigenvalues and eigenvectors is obtained by solving the standard eigenvalue problem:

$$\boldsymbol{B}W = W\Xi \tag{14}$$

where Ξ is the diagonal eigenvalue matrix, and W is the eigenvector matrix with an ith column that represents the axis of the ith collective coordinate in the conformational space. The eigenvalue Ξ_i represents the mean square fluctuation (MSF) along this axis.

Functionally relevant motions in proteins are related to a few collective coordinates, namely, the ones with the highest values of the eigenvalues Ξ_i (highest MSF). Therefore, the projection of the PMF (*see* **Note 9**) on these relevant collective coordinates is useful to relate free energy with a protein's conformational changes.

3. Methods

We describe an implementation of an REx MD simulation applied to the 12–28 fragment of the Alzheimer amyloid-β peptide *(23)*.

1. A force field, solvent model, and simulation program must be chosen. In this example, we use the OPLS force field *(47)*, TIP3P Water model *(48)*, and the GROMACS software *(49–51)* (*see* **Notes 17–19**).
2. The initial coordinates of a peptide can be obtained from the Protein Data Bank (www.pdb.org) or an extended all-*trans* conformation can be built if no NMR or X-ray crystal structures are available. In this example, we construct an extended conformation based on the amino acid sequence VHHQKLVFFAEDVGSNK.
3. The peptide must be solvated in a box of water. Here, we solvate the peptide in a cubic box filled with 7059 TIP3P *(48)* water molecules. The length of the simulation box (60.4 Å) was determined in short constant pressure simulations at $T = 280\,\text{K}$, which were equilibrated at a physiological external pressure of 1 atm (*see* **Note 20**).
4. Protonation states of all titrable residues must be set appropriately. Here, we set them to fit neutral pH levels, giving a zero total charge for the system.
5. To use a longer simulation time step of $2\,fs$, covalent bonds between heavy atoms and hydrogen were held constant (in the GROMACS protocol using the SETTLE algorithm; *52*) (*see* **Note 21**).
6. Nonbonded Lennard-Jones interactions were tapered starting at 7 Å and extending to 8 Å cutoff. Neighbor lists for the nonbonded interactions were updated every 10 simulation steps.

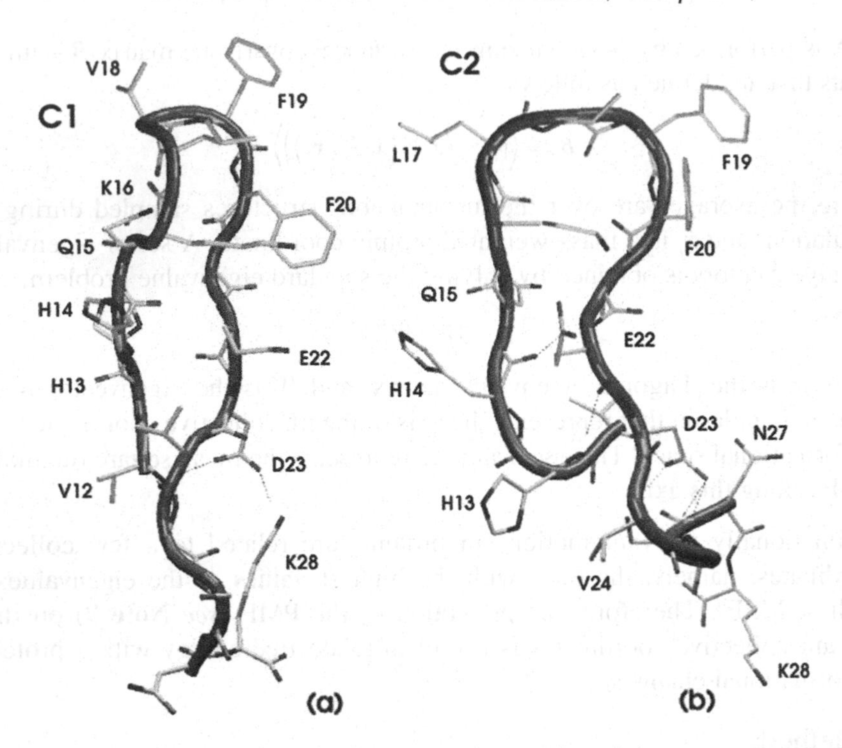

Fig. 1. Most representative conformations (centroids) of the two most populated clusters (**a**) C1 and (**b**) C2 obtained from the replica exchange molecular dynamics simulations of amyloid-β12–28 at T=280 K.

7. Electrostatic interactions were included using the reaction field *(53)* approach (*see* **Note 22**).
8. The temperature was controlled by the Nose-Hoover algorithm *(54)* with a 0.05 *ps* time constant.
9. For the REx simulations, the lowest temperature is chosen close to the temperature of interest. The highest temperature must be elevated enough so that conformational transitions are facilitated at these temperatures. An optimal temperature distribution scheme is an exponential one (*see* **Note 23**). Here, 60 replicas of the original system were considered at temperatures exponentially spaced between 280 and 600 K. Exchanges of replicas at adjacent temperatures were attempted every 500 simulation steps. The same time interval was used to periodically save atomic coordinates. The simulations were started from a random extended conformation, the same for all replicas, and equilibrated for about 4 *ns*. Subsequent equilibrium sampling runs were performed over 28 *ns*.
10. To classify all sampled conformations into groups of similar/dissimilar states, we performed a clustering analysis. The analysis was conducted according to the

GROMOS protocol **(55)**, using root mean square (RMS) deviations over C_α atoms of residues 2 to 16 as a measure of structural similarity. In our analysis, conformations with a mutual RMS < 2 Å were considered as belonging to the same cluster. The centroids (most representative structures) of the two most populated clusters are shown in **Fig. 1**.

11. Free-energy surfaces can be plotted as a function of a number of reaction coordinates. In **Fig. 2**, we plot the free-energy surface as a function of the principal components described in **Subheading 2.2.1**.
12. The monomeric structures can be used to infer possible assembled nanostructures. Two possible assemblies, based on the two centroids shown in **Fig. 1**, are given in **Fig. 3**.

4. Notes

1. Many-body terms have been introduced in recent years.
2. The form of the empirical potential shown here is for the CHARMM force field ***(32)***. Slight variations exist in other force fields ***(47,49,56)***.
3. There are size (number of atoms) and time (length of simulation) limitations for practical use of MD simulations. Currently, MD simulations can handle systems with 10^6 atoms for simulations times on the order of 10^{-6} s. The length of the simulations limits calculations to properties with relaxation times smaller than the simulation time. In **Subheading 2.2.**, we discuss the use of enhanced sampling techniques to overcome this limitation. To overcome the MD limitations related to system size, *periodic boundary conditions* can be used ***(33,57)***.
4. The conservation law is

$$H(\boldsymbol{x}) = \sum_{i=1}^{N} \frac{\boldsymbol{p}_i^2}{2m_i} + V(\boldsymbol{r}_1, \ldots, \boldsymbol{r}_N) = \text{Constant} = E \tag{15}$$

where $H(\boldsymbol{x})$ is the *Hamiltonian* of the system. A given microscopic state for the many-body system is therefore described as a point in a $6N$-dimensional space called *phase space*. Consequently, the time evolution of the many-body system can be pictured as a curve in the phase space. The flow in the phase space is determined by the time integration of the $6N$ Hamilton's equations:

$$\dot{\boldsymbol{r}}_i \frac{\partial H}{\partial \boldsymbol{p}_i} \qquad \dot{\boldsymbol{p}}_i = -\frac{\partial H}{\partial \boldsymbol{r}_i} \tag{16}$$

(or, equivalently, of the $3N$ Newton's equations), with the phase space point at $t = 0$ providing the initial conditions.

5. For a thermodynamic system with fixed number of particles N and volume V and following Hamilton's equations, the Hamiltonian is a constant of the motion (see **Eq. 15**) and equals the total energy of the system. This type of thermodynamic system is represented in statistical mechanics by the *microcanonical ensemble*

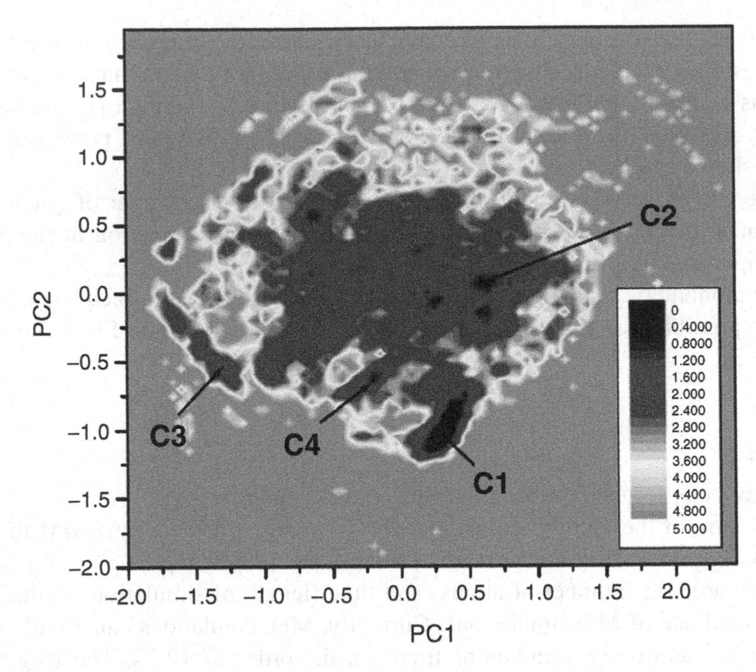

Fig. 2. Free-energy map (in units of k_BT) of the amyloid-β 12–28 peptide at $T = 280\,K$ projected onto the first two principal components PC1 and PC2. The centroid of cluster C1 was used as the reference state to compute the principal components.

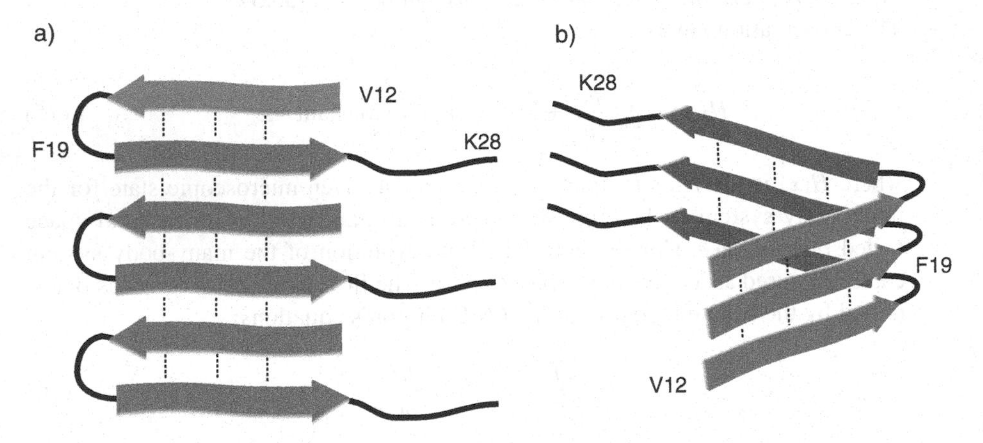

Fig. 3. Two possible models of amyloid-β 12–28 protofilaments compatible with the replica exchange molecular dynamics simulations. (**a**) In this model, monomeric β-hairpins (from cluster C1) align into larger antiparallel β-sheets through side-to-side self-assembly. Several β-sheets make up a protofilament. In model (**b**), loop-like amyloid β 12–28 monomers (from cluster C2) add up on top of each other to form two parallel β-sheets. Several of these β-sheet complexes self-associate, possibly through the exposed hydrophobic residues, to form a protofilament. In both models, the interstrand hydrogen bonds run parallel to the fiber axis.

(or *NVE* ensemble). The probability density for the microcanonical ensemble is defined as

$$\rho_{\mathrm{NVE}}(\boldsymbol{x}) \propto \delta(H(\boldsymbol{p}, \boldsymbol{r}) - E) \tag{17}$$

6. Laboratory experiments are frequently performed under conditions of constant temperature or constant pressure. A common ensemble is the *canonical ensemble* (*NVT*), an assembly of all microstates with fixed *N*, *V*, and *T*. The temperature of a physical system (unlike the total energy *E*) is directly observable and can be controlled by keeping the system in contact with a proper heat bath (thermostat). The probability density for the canonical ensemble is defined as

$$\rho_{\mathrm{NVT}}(\boldsymbol{x}) \propto e^{-\frac{E}{k_B T}} \tag{18}$$

where E is the total energy of the system, and k_B is the Boltzmann constant. The temperature T can be calculated from the equipartition theorem, which states that

$$\left\langle \frac{1}{2} \sum_{1=1}^{N} m_i \dot{\boldsymbol{r}}_i^{\,2} \right\rangle = \frac{3k_B T}{2} \tag{19}$$

where the left-hand side is the ensemble average of the total kinetic energy. Similarly, we can define the *NPT ensemble*, where *N* is fixed and the system is kept at constant temperature and pressure by contact with a thermostat and an external barostat, respectively.

7. For the calculation of the thermodynamic properties in the *NVT* and *NPT* ensembles, it is necessary to extend MD and modify the equations of motion ***(58)***. The modifications of the equations of motion may involve either deterministic or stochastic methods. We limit our analysis here to the extension of MD to include an external thermostat since MD simulations of biomolecules are typically carried out under canonical conditions (constant *NVT*). The two most sophisticated MD thermostats used in simulations of biological systems are the Nose-Hoover thermostat ***(54)*** and the Langevin thermostat ***(33)***. The former is entirely deterministic and based on the coupling of the coordinates and the velocities of all the particles to an additional variable (an additional degree of freedom representing a thermostat). This new *extended MD* is derived by modifying the Newton's equations of motion by addition of one extra degree of freedom and its conjugate momentum and has its theoretical foundations in *non-Hamiltonian* mechanics ***(58)***. The Langevin thermostat relies on the Langevin equation of motion:

$$m_i \ddot{\boldsymbol{r}}_i = \boldsymbol{F}_i - \xi \dot{\boldsymbol{r}}_i + \boldsymbol{F}' \tag{20}$$

where $\xi > 0$ is the damping factor, $\boldsymbol{F}'$ is a Gaussian random force, and the remaining symbols are equivalent to the ones defined in **Eq. 1**. The Langevin equation extends

Newton's equations of motion by introducing two nonconservative forces (a dissipative force and a random force) related such that the "fluctuation–dissipation" theorem is obeyed and the *NVT* sampling is guaranteed.

8. Most of the enhanced sampling methods used in biomolecular simulations are based on the concept of *generalized ensemble.* Generalized-ensemble methods rely on the generation of a random walk in energy space to overcome the multiminima problem and get a correct sampling of phase space. Two examples of such methods are the *multicanonical algorithm* ***(59)*** and *simulated tempering* ***(60)***.
9. The PMF is calculated from the reversed work theorem, which relates the radial distribution function with the change in Helmholtz free energy ***(34)***. This theorem has been extended to the calculation of the PMF along one or more generic reaction coordinates and has become of central importance in computational studies of biomolecular systems.
10. **Reference *61*** presents a UNIX socket code for REx implemented as a server (which makes exchange decisions) and clients (which perform MD). In contrast, our REx pseudocode relies on calls to the MPI ***(42)*** to accomplish communication between the nodes, all of which have identical tasks: They perform MD the majority of the time and make exchange decisions at predetermined intervals. The advantages of our approach are twofold. First, it relies on a well-established and well-maintained communication protocol (MPI), versions of which are deployed on most modern-day machines, to hide the details of synchronization between the nodes from the programmer. Second, it eliminates the need to maintain code for two kinds of behavior, server and client, by making all the nodes behave identically.
11. The efficiency of this implementation of REx relies on the fact that the temperatures are swapped among the nodes rather than the system coordinates. Communication is by far the most time-consuming activity in REx simulations. Therefore, making the size of the information communicated constant yields a tremendous improvement over an approach that communicates a variable amount of information that increases with the size of the system (i.e., the system coordinates).
12. We use pseudocode for three MPI calls, the prototypes for which are

```
MPISend(what, destination)
what=MPIReceive(source)
MPIBroadcast(what, source)
```

13. In this implementation of REx, each node alternately considers its left and right neighbors (in temperature order) as potential swap partners. That is, on even iterations swaps between (first and second), (third and fourth),... temperatures are considered, while on odd iterations swaps between (second and third), (fourth and fifth) ... temperatures are considered.

14. The function `SwapDecision(e1,e2)` must be symmetric (commutative) in its arguments, so that both nodes reach the same decision, to swap or not to swap, even though one is executing `SwapDecision(e1,e2)` and the other `SwapDecision(e2,e1)`.
15. This is any motion that involves a number of atoms moving in a concerted fashion.
16. Although the finite simulation times may limit the correct description of the slowest collective motions (i.e., the collective motions with timescale is of the same order of the simulation time length) *(62)*.
17. Other possible force fields include Amber *(56)*, GROMOS *(49)*, and CHARMM *(32)*.
18. Other possible water molecules include SPC *(63–65)*, TIP4P *(48)*, or polarizable water models *(66–68)*.
19. Other possible software include Amber *(69)*, CHARMM *(70)*, TINKER *(71)*, and NAMD (Nanoscale Molecular Dynamics) *(72)*.
20. An optimal simulation box is the truncated octahedron.
21. In CHARMM, the SHAKE *(73)* algorithm is used. Other variants include the RATTLE *(74)* algorithm.
22. A more accurate, albeit more costly, means to handle the electrostatics is to use particle mesh Ewald sums *(75)*.
23. In the event that the swap ratio is poor (below 10–15%) in a particular region (typically around a transition temperature), additional replicas can be manually added to the original exponential distribution. Alternately, a more optimal distribution and range of temperatures can be sought *(76)*. The number of temperatures needed for the simulation may limit the practical use of REx. Since the number of replicas scales with the square root of the number of particles in the system *(61)*, the "processor" cost for REx may become prohibitive for large systems.

References

1. Dobson CM. (2003) Protein folding and misfolding. *Nature* **426**, 884–890.
2. Serpell LC. (2000) Alzheimer's amyloid fibrils: structure and assembly. *Biochim. Biophys. Acta* **1502**, 16–30.
3. Tycko R. (2004) Progress towards a molecular-level structural understanding of amyloid fibrils. *Curr. Opin. Struct. Biol.* **14**, 96–103.
4. Selkoe DJ. (2003) Folding proteins in fatal ways. *Nature* **426**, 900–904.
5. Fandrich M, Fletcher MA, Dobson CM. (2001) Amyloid fibrils from muscle myoglobin. *Nature* **410**, 165–166.
6. López de la Paz M, Goldie K, Zurdo J, et al. (2002) *De novo* designed peptide-based amyloid fibrils. *Proc. Natl. Acad. Sci. U. S. A.* **99**, 16052–16057.
7. Tsai C, Zheng J, Aleman C, Nussinov R. (2006) Structure by design: from single proteins and their building blocks to nanostructures. *Trends Biotechnol.* **24**, 449–454.
8. Zhang S. (2002) Emerging biological materials through molecular self-assembly. *Biotechnol. Adv.* **20**, 321–329.
9. Zhang S. (2003) Fabrication of novel biomaterials through molecular self-assembly. *Nat. Biotechnol.* **21**, 1171–1178.

10. Rajagopal K, Schneider J. (2004) Self-assembling peptides and proteins for nanotechnological applications. *Curr. Opin. Struc. Biol.* **14**, 480–486.
11. Reches M, Gazit E. (2003) Casting metal nanowires within discrete self-assembled peptide nanotubes. *Science* **300**, 625–627.
12. Kol N, Adler-Abramovich L, Barlam D, Shneck R, Gazit E, Rousso I. (2005) Self-assembled nanotubes are uniquely rigid bioinspired supramolecular structures. *Nano Lett.* **5**, 1343–1346.
13. Reiches M, Gazit E. (2006) Designed aromatic homo-dipeptides: formation of ordered nanostructures and potential nanotechnological applications. *Phys. Biol.* **3**, S10–S19.
14. Scheibel T, Parthasarathy R, Sawicki G, Lin X-M, Jaeger H, Lindquist S. (2003) Conducting nanowires built by controlled self-assembly of amyloid fibers and selective metal deposition. *Proc. Natl. Acad. Sci. U. S. A.* **100**, 4527–4532.
15. Temussi PA, Masino L, Pastore A. (2003) From Alzheimer to Huntington: why is a structural understanding so difficult? *EMBO J.* **22**(3), 355–361.
16. Ma B, Nussinov R. (2002) Stabilities and conformations of Alzheimer's β-amyloid peptide oligomers ($a\beta_{16-22}$,$a\beta_{10-35}$): sequence effects. *Proc. Natl. Acad. Sci. U. S. A.* **99**(22), 14126–14131.
17. Ma B, Nussinov R. (2006) The stability of monomeric intermediates controls amyloid formation: Abeta25–35 and its N27Q Mutant. *Biophys. J.* **90**(10), 3365–3374.
18. Buchete N, Tycko R, Hummer G. (2005) Molecular dynamics simulations of Alzheimer's β-amyloid protofilament. *J. Mol. Biol.* **353**, 804–821.
19. Tsai H-H, Reches M, Gunasekaran K, Gazit E, Nussinov R. (2005) Energy landscape of amyloidogenic peptide oligomerization by parallel-tempering molecular dynamics simulation: significant role of asn ladder. *Proc. Natl. Acad. Sci. U. S. A.* **102**, 8174–8179.
20. Klimov DK, Thirumalai D. (2003) Dissecting the assembly of $a\beta_{16-22}$ amyloid peptides into antiparallel β sheets. *Structure* **11**, 295–307.
21. Gnanakaran S, Nussinov R, Garcia AE. (2006) Atomic-level description of amyloid β dimer formation. *J. Am. Chem. Soc.* **128**, 2158–2159.
22. Wei G, Shea J-E. (2006) Effects of solvent on the structure of the Alzheimer amyloid-β(25–35) peptide. *Biophys. J.* **91**, 1638–1648.
23. Baumketner A, Shea J-E. (2006) Folding landscapes of the Alzheimer amyloid-β(12–28) peptide. *J. Mol. Biol.* **363**, 945–957.
24. Baumketner A, Shea J-E. The structure of the Alzheimer amyloid 10–35 peptide probed through replica-exchange molecular dynamics simulations in explicit solvent. *J. Mol. Biol.* **366**.
25. Alder BJ, Wainwright TE. (1957) Phase transition for a hard sphere system. *J. Chem. Phys.* **27**, 1208–1290.
26. Rahman A. (1964) Correlations in the motion of atoms in liquid argon. *Phys. Rev.* **27**, A405–A411.
27. Verlet L. (1967) Computer "experiments" on classical fluids. I. thermodynamical properties of Lennard–Jones molecules. *Phys. Rev.* **159**, 98–103.

28. Barojas J, Levesque D, Quentrec B. (1973) Simulations of diatomic homonuclear liquids. *Phys. Rev. A* **7**, 1092–1105.
29. Stillinger FS. (1975) Theory and molecular models for water. *Adv. Chem. Phys.* **31**, 1–101.
30. Ryckaert JP, Bellemans A. (1975) Molecular dynamics of liquid normal butane near its boiling point. *Chem. Phys. Lett.* **30**, 123–125.
31. McCammon JA, Gelin BR, Karplus M. (1977) Dynamics of folded proteins. *Nature* **267**, 585–590.
32. MacKerell AD Jr, Bashford D, Bellott M, et al. (1998) All-atom empirical potential for molecular modeling and dynamics studies of proteins. *J. Phys. Chem. B* **102**, 3586–3616.
33. Allen M, Tildesley DJ. (1987) *Computer Simulation of Liquids*. Clarendon Press, Oxford, UK.
34. Chandler D. (1987) *Introduction to Modern Statistical Mechanics*. Oxford University Press, Oxford, UK.
35. Pathria RK. (1996) *Statistical Mechanics*. Butterworth-Heinemann, Oxford, UK.
36. Fermi E. (1956) *Thermodynamics*. Dover, New York.
37. Hukushima K, Nemoto K. (1996) Exchange Monte Carlo method and application to spin glass simulations. *J. Phys. Soc. Jpn.* **65**(6), 1604–1608.
38. Hansmann UHE. (1997) Parallel tempering algorithm for conformational studies of biological molecules. *Chem. Phys. Let.* **281**, 140–150.
39. Sugita Y, Okamoto Y. (1999) Replica-exchange molecular dynamics method for protein folding. *Chem. Phys. Lett.* **314**, 141–151.
40. Kumar S, Bouzida D, Swendsen RH, Kollman PA, Rosenberg JH. (1992) The weighted histogram analysis method for free energy calculation on biomolecules. I. the method. *J. Comp. Chem.* **13**, 1011–1021.
41. Roux B. (1995) The calculation of the potential of the mean force using computer simulations. *Comp. Phys. Comm.* **91**, 275–282.
42. Snir M, Otto S, Huss-Lederman S, Walker D, Dongarra J. (1996) *MPI: The Complete Reference*. MIT Press, Cambridge, MA.
43. Branden C, Tooze J. (1999) *Introduction to Protein Structure*, 2nd ed. Garland, New York.
44. Hayward S, Go N. (1995) Collective variable description of native protein dynamics. *Annu. Rev. Phys. Chem.* **46**, 223–250.
45. Hayward S, Kitao A, Go N. (1994) Harmonic and anharmonic aspects in the dynamics of bpti: a normal mode analysis and principal component analysis. *Protein Sci.* **3**, 936–943.
46. Kitao A, Go N. (1999) Investigating protein dynamics in collective coordinate space. *Curr. Opin. Struct. Biol.* **9**, 164–169.
47. Kaminski GA, Friesner RA, Tirado-Rives J, Jorgensen WL. (2001) Evaluation and reparametrization of the OPLS-AA force field for proteins via comparison with accurate quantum chemical calculations on peptides. *J. Phys. Chem. B* **105**, 6474–6487.

48. Jorgensen WL, Chandrasekhar J, Madura JD, Impey RW, Klein ML. (1983) Comparison of simple potential functions for simulating liquid water. *J. Chem. Phys.* **79**, 926–935.
49. Lindahl E, Hess B, van der Spoel D. (2001) GROMACS 3.0: a package for molecular simulation and trajectory analysis. *J. Mol. Mod.* **7**, 306–317.
50. Berendsen HJC, van der Spoel D, van Drunen R. (1995) GROMACS: a message-passing parallel molecular dynamics implementation. *Comput. Phys. Commun.* **91**, 43–56.
51. Monticelli L, Ash W. Replica exchange facility for GROMACS. University of Calgary, Calgary, Alberta, Canada.
52. Miyamoto S, Kollman PA. (1992) Settle: an analytical version of the shake and rattle algorithms for rigid water models. *J. Comp. Chem.* **13**, 952–962.
53. Tironi IG, Sperb R, Smith PE, van Gunsteren WF. (1995) Generalized reaction field method for molecular dynamics simulations. *J. Chem. Phys.* **102**, 5451–5459.
54. Nosé S. (1991) Constant temperature molecular dynamics methods. *Prog. Theor. Phys.* **103**, 1–46.
55. Daura X, Gademann K, Jaun B, Seebach D, van Gunsteren WF, Mark AE. (1999) Peptide folding: when simulation meets experiment. *Angew. Chem. Int. Ed.* **38**, 236–240.
56. Ponder J, Case D. (2003) Force fields for protein simulations. The Amber biomolecular simulation programs. *Adv. Protein Chem.* **66**, 27–85.
57. Frenkel D, Smit B. (1996) *Understanding Molecular Simulations*. Academic Press, San Diego, CA.
58. Martyna GJ, Tuckerman ME. (2000) Understanding modern molecular dynamics: techniques and applications. *J. Phys. Chem. B* **104**, 159–178.
59. Berg BA, Neuhaus T. (1992) Multicanonical ensemble: a new approach to simulate first-order phase transition. *Phys. Rev. Lett.* **68**(1), 9–12.
60. Marinari E, Parisi G. (1992) Simulated tempering–a new Monte Carlo scheme. *Europhys. Lett.* **19**, 451.
61. Nymeyer H, Gnanakaran S, Garcia AE. (2004) Atomic simulations of protein folding, using the replica exchange algorithm. *Methods Enzymol.* **383**, 119.
62. Balsera MA,Wriggers W, Oono Y, Schulten K. (1996) Principal component analysis and long time protein dynamics. *J. Phys. Chem.* **100**, 2567–2572.
63. Berendsen HJC, Postma JPM, van Gunsteren WF, Hermans J. (1981) Interaction models for water in relation to protein hydration. In: Pullman B, ed. *Intermolecular Forces*. Dordrecht, the Netherlands: Reidel, pp. 331–342.
64. Grigera JR, Berendsen HJC, Straatsma TP. (1987) The missing term in effective pair potentials. *J. Phys. Chem.* **91**, 6269–6271.
65. Glaettli A, Daura X, van Gunsteren WF. (2003) A novel approach for designing simple point charge models for liquid water with three interaction sites. *J. Comput. Chem.* **24**, 1087–1096.
66. Chen B, Xing J, Siepmann JI. (2000) Development of polarizable water force fields for phase equilibrium calculations. *J. Phys. Chem. B* **104**, 2391–2401.

67. Ren P, Ponder JW. (2003) Polarizable atomic multipole water model for molecular mechanics simulation. *J. Phys. Chem. B* **107**, 5933–5947.
68. Rick SW, Stuart SJ, Berne BJ. (1994) Dynamical fluctuating charge force fields: application to liquid water. *J. Chem. Phys.* **101**, 6141–6156.
69. Case D, Cheatham T III, Darden T, et al. (2005) The Amber biomolecular simulation programs. *J. Comp. Chem.* **26**, 1668–1688.
70. Brooks BR, Bruccoleri R, Olafson B, States D, Swaninathan S, Karplus M. (1983) CHARMM: a program for macromolecular energy minimization and dynamics calculations. *J. Comp. Chem.* **4**, 187–200.
71. Ponder J. (2003) *TINKER: Software Tools for Molecular Design*, 4.1 ed. Washington University School of Medicine, St. Louis, MO.
72. Phillips J, Braun R, Wang W, et al. (2005) Scalable molecular dynamics with NAMD. *J. Comp. Chem.* **26**, 1781–1802.
73. Ryckaert JP, Ciccotti G, Berendsen HJC. (1977) Numerical integration of the Cartesian equations of motion of a system with constraints: molecular dynamics of n-alkanes. *J. Comp. Phys.* **23**, 327–341.
74. Anderson H. (1983) Rattle: a "velocity" version of the shake algorithm for molecular dynamics calculations. *J. Comp. Phys.* **52**, 24–34.
75. Essmann U, Perera L, Berkowitz ML, Darden T, Lee H, Pedersen LG. (1995) A smooth particle mesh Ewald method. *J. Chem. Phys.* **103**, 8577–8593.
76. Trebst S, Troyer M, Hansmann UHE. (2006) Optimized parallel tempering simulations of proteins. *J. Chem. Phys.* **124**, 174903.

10

Modeling Amyloid Fibril Formation

A Free-Energy Approach

Maarten G. Wolf, Jeroen van Gestel, and Simon W. de Leeuw

Summary

Amyloid fibrils are structures consisting of many proteins with a well-defined conformation. The formation of these fibrils has been the subject of intense research, largely due to their connection to several diseases. We focus here on the computational studies and discuss these from a free-energy point of view. The fibrillogenic properties of many proteins can be predicted and understood by taking the relevant free energies into account in an appropriate way. This is because both the equilibrium and the kinetic properties of the protein system depend on its free-energy landscape. Advanced simulation techniques can be used to understand the relationship between the free-energy landscape of a protein and its three-dimensional structure and propensity to form amyloid fibrils. We give an overview of existing simulation techniques that operate at a molecular level of detail and that are capable of generating relevant free-energy values. The free energies obtained with these methods can be inserted into a statistical-mechanical or kinetic framework to predict mean fibril properties on length scales and time scales that are inaccessible by molecular-scale simulation methods.

Key Words: Amyloid fibrils; free energy; modeling; protein aggregation; proteins; simulation techniques; statistical mechanics.

1. Introduction

Protein aggregation plays a pivotal role in many naturally occurring processes. The cytoskeleton, for instance, contains fibrillar aggregates of tubulin and actin *(1–3)*, while capsid proteins combine to form the protective shell of rod-like viruses, such as tobacco mosaic virus *(4,5)*. Aggregates of protein molecules can vary strongly in size, shape, and function, and their shape and function are often closely related. However, an important class of fibrillar protein aggregates exists that does not perform a function but rather appears to disrupt natural

From: *Methods in Molecular Biology, vol. 474: Nanostructure Design: Methods and Protocols*
Edited by: E. Gazit and R. Nussinov © Humana Press, Totowa, NJ

processes. These amyloid fibrils have been implied in a number of diseases, many of which are life threatening. These include Alzheimer's disease (connected to amyloid fibril formation of the amyloid-β protein or Aβ), Parkinson's disease (linked to the aggregation of α-synuclein), and Huntington's disease *(6–9)*.

Under certain conditions, such as an appropriately chosen temperature or pH, many proteins form amyloid fibrils *(7,9–11)*. A schematic overview of the most important processes during amyloid fibril formation is given in **Fig. 1** for the Alzheimer's-related Aβ protein depicted in **Fig. 2**. These conditions may vary from protein to protein, as may some details of the fibril structure *(10,12)*. However, the overall structure and behavior of fibrils that originate from different proteins are often remarkably similar *(7,9,10,13,14)*. For instance, it has been observed that although the length of amyloid fibrils within a sample may vary considerably, their thickness is more or less constant *(7,9,10,14)*. Furthermore, they usually consist of several intertwined protofilaments

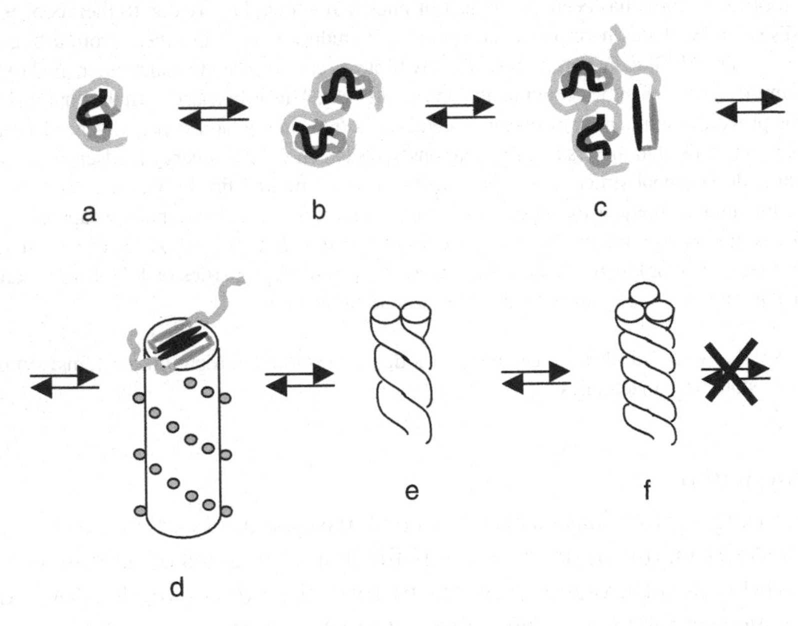

Fig. 1. Schematic view of amyloid Aβ protein fibril formation. (**a**) The monomer; (**b**) the dimer. Here, the darkness of the chain is a measure of the hydrophobicity of the amino acids. (**c**) A possible nucleus; (**d**) a (proto)fibril that consists of two ordered protofilaments. In the cross section, it is shown that the two protofilaments interact through their most hydrophobic residues. The gray circles indicate the locations of the hydrophilic residues that protrude from the body of the protofilament. (**e**) and (**f**) Amyloid fibrils that consist of two or three intertwining protofibrils. Here, the hydrophilic chains are omitted for clarity.

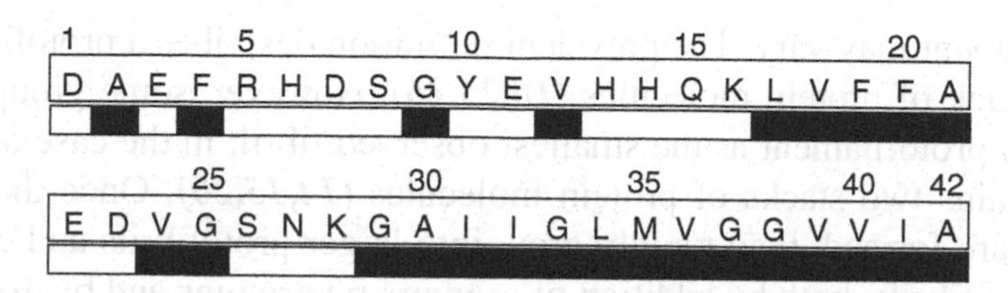

Fig. 2. The composition of $A\beta_{1-42}$ protein. Each amino acid is indicated by its one-letter abbreviation, and the numbers correspond to the position in the chain. Below the letters is a color code. Black indicates a nonpolar side chain (hydrophobic residue); white indicates a polar side chain. The distribution of residues with polar and apolar side chains along the molecule gives an area of the protein that is strongly hydrophobic near the C-terminus, a much more hydrophilic part of the molecule near the N-terminus, and a mixed area in between. This corresponds to the darkness and lightness used in Fig. 1.

connected by lateral interactions. Inside these protofilaments, the individual protein molecules generally exist in a well-defined conformational state that allows stabilization by intermolecular β-sheet-type bonds (so-called cross-β-sheets) in the axial direction ***(7,10,15,16)***. In addition to these common structural elements, many proteins have the mechanism by which they form amyloid fibrils in common: Amyloid fibril formation can be described in general terms as a process of nucleation and growth ***(7,9,17,18)***. As such, one would expect that a general theoretical treatment can shed light on the fibrillization of different protein molecules. While it is certainly possible to predict important general properties with this type of modeling ***(19)***, it turns out that subtle differences in the fibril preparation can have a major effect on the structure of the mature fibril ***(12,20)***. Hence, a more detailed understanding of the protein–protein interaction is required to describe these differences.

To obtain a detailed picture of amyloid fibril formation, this process has to be studied at several length scales. The protein monomer is the building block of amyloid structures, and its properties are therefore very important. In its monomeric form, a protein tends (on average) to adopt the conformation with the lowest (Gibbs) free energy, the native state. However, it also occurs that a protein molecule partially unfolds (in the literature, this is often referred to as *misfolding*) and becomes attached to another monomer. Subsequently, the dimer may grow into a nucleus. The identity of the nucleus for fibrillization is still a hotly debated topic ***(21–23)***. In fact, for some proteins, the possibility of a nucleus that consists of a single protein molecule has been discussed ***(24)***.

After nucleation, small elongated aggregates are formed that are stabilized by the cross-β-sheet interaction characteristic of the amyloid-type structure. These structures are called *protofilaments*. Note that in the literature, the definition of a protofilament may vary. The prevalent definition describes a protofilament as a single stack of protein molecules ***(10,25,26)***. However, some groups choose to define a protofilament as the smallest observed fibril; in the case of Aβ, this fibril contains two stacks of protein molecules ***(14,15,20)***. Once these protofilaments are formed, they rapidly grow into larger protofibrils and eventually into mature fibrils, both by addition of monomeric proteins and by assembly of protofilaments *(7)*. Apparently, a number of processes and (intermediate) structures are important in the formation of amyloid structures. Which processes are likely to occur and which conformations tend to be stable is described by the free-energy landscapes of the protein aggregates. In a first step to fully understanding amyloid fibril formation, it is therefore crucial that one can determine the free energies of the species that play a role in this process.

We review how free energies relevant to amyloid fibril formation can be obtained from molecular-scale modeling and subsequently applied to predict overall fibril properties. **Subheading 2**. contains a description of the concept of free energy in the context of protein aggregation. In **Subheading 3**., we discuss the computational techniques that can be applied to obtain the relevant free energies, with a focus on molecular-scale modeling. This type of modeling can provide insight into the detailed structural features of the fibrils. In **Subheading 4**., we review the results that have been reported so far. Particularly, we discuss the energetics of each step in the process of amyloid fibril formation. We start by considering the free-energy landscape of a single protein molecule. Subsequently, we review nucleation and the formation of protofilaments, and finally we briefly discuss fibril elongation. **Subheading 5**. provides a discussion of general, coarse-grained models that can predict trends and describe the common properties of many different amyloid fibrils. We discuss the approximations and input parameters that are required by this type of theory. As we show, the free energies obtained from molecular-scale modeling or experiment are essential to make useful predictions on the mesoscale with coarse-grained models.

2. Free Energy in Protein Aggregation

Transitions in biomolecular systems are conveniently discussed in terms of the free-energy landscape, which depicts changes in the free energy as transitions take place. In other words, the free energy is considered as a function of the relevant "reaction" coordinates describing the transition. The complexity of biomolecular systems is reflected in the free-energy landscape: It often takes the form of a rugged landscape with many mountain passes and valleys. Stable and metastable states of a biomolecular system can be identified as valleys in this landscape. The equilibrium population of these valleys is obtained directly from their relative depths. Since the full landscape contains the barriers between different states, it provides information on the rate of transition between these states as well. In this chapter, we mainly focus on (meta)stable states, that is, the valleys in the free-energy landscape.

The free-energy landscape differs from the potential energy landscape, which is investigated more readily in simulation studies. This landscape displays the potential energy of a biomolecular system as a function of the molecular degrees of freedom, that is, the particle positions. Although this information is generally very useful, a complete picture of transitions between different states of the biomolecular system can be obtained only by considering the free-energy landscape.

The free-energy landscape of a protein in solution can be readily manipulated by changing certain parameters of the system. Here, one can think of changing the properties of the proteins themselves, such as the amino acid sequence (a mutation), or of a change in the surroundings (e.g., in solvent composition or temperature) *(27)*. The dependence of the behavior of the proteins on such manipulations is due to the nature of the interaction of the proteins with themselves and each other: The different (parts of the) proteins are held together by weak physical interactions, such as hydrogen bonding, ionic interactions, and hydrophobic effects. In addition to the parameters mentioned, there are numerous other ways to change the free-energy landscape of a system that can result in a shift of the thermodynamic equilibrium and the kinetic properties *(27)*.

Unless detailed structural information is available, it is often difficult to predict *a priori* what effect a specific change can have on the properties of a protein system. It is possible, however, to discern if a particular parameter plays a role by studying whether the free-energy difference between two states ΔG changes as a result of a shift in its value *(24)*. If this is the case, and $\Delta\Delta G = \Delta G_{after} - \Delta G_{before} \neq 0$, then we may conclude that the parameter we changed is one that is relevant for the transition. For instance, if we introduce a single mutation in the protein and subsequently study the aggregation behavior, $\Delta\Delta G$ gives information about the importance of the amino acid that was replaced in the aggregation process. Because the change of the system may shift the free energy of either of the two states or of both, the interpretation of experimental $\Delta\Delta G$ data can be ambiguous. In computer simulations, visualization is straightforward, and it is sometimes possible to decide which of the two states the mutation affects.

Alternatively, but equivalently, we can describe the overall free energy of an amyloidogenic system by taking into account the free energy of each protein aggregate that is present. This is a function not only of the number and type of interactions that exist between the protein molecules but also of the entropy associated with the conformation of the protein molecules. This contribution is usually given in terms of the partition function of an aggregate. In addition, we need to include a mixing entropy term that describes the distribution of the proteins in the system (including the distribution of the protein molecules over the formed aggregates) ***(28)***. This type of description represents a good starting point for a statistical-mechanical description of amyloid formation as discussed in **Subheading 5**.

Often, the two components of the free energy described, entropy and interaction energy, act in opposite directions. On the one hand, the system strives for a minimum of interaction energy, leading to the formation of large aggregates. On the other hand, entropy works against assembly and favors the formation of many small aggregates. In some cases, however, structure formation allows for a net entropy gain due to an increase in the degrees of freedom of the surroundings (e.g., the solvent) ***(29)***. That this may be the case for at least some amyloid-forming proteins follows from the (somewhat counterintuitive) observation that fibril formation can increase with increasing temperature ***(30)***.

3. Calculation of Free Energy From Simulation

From the free-energy landscape, the relationships between relevant conformational states can in theory be deduced and used to explain amyloid formation. Computationally, a free-energy landscape can be obtained from the distribution along the relevant coordinates linking states *A* and *B*. This is because the free-energy difference ΔG between two states *A* and *B* in this landscape (in terms of the thermal energy $k_B T$, with T the absolute temperature and k_B Boltzmann's constant) determines the population of these two states N_A and N_B in thermodynamic equilibrium:

$$\beta \Delta G_{(A->B)} = \ln\left(\frac{N_A}{N_B}\right) \qquad (1)$$

Here, $\beta = 1/(k_B T)$ is the inverse of the thermal energy. There are two basic computational techniques that can in principle provide average distributions. In Monte Carlo (MC) simulations, the phase space of a system is generally sampled by an importance-sampling random walk. This involves repeatedly taking a step in a random direction in phase space from the current conformation to a new one and evaluating this step by a Metropolis criterion. In the Metropolis criterion, the step is accepted if $e^{-\beta \cdot \Delta E}$ is larger than a random number between 0 and 1, with ΔE the potential energy difference between the current and the new state. A step resulting in a negative ΔE (i.e., the energy of the new state is smaller than the energy of the old state) is therefore always accepted, while accepting a step accompanied by a positive ΔE depends on the random number. The acceptance probability P_{acc} in a Metropolis criterion is summarized as $P_{acc} = \min(1, e^{-\beta \cdot \Delta E})$. In this way, a representative ensemble of the system is obtained from which the populations of states *A* and *B* are derived and consequently the free-energy difference is determined.

In molecular dynamics (MD) simulations, the properties of interest are measured during a certain time interval, typically on the order of a hundred nanoseconds. The evolution of the system in time is determined by solving suitable equations of motion. To obtain the free-energy difference of **Eq. 1**, the populations of states *A* and *B* are measured. The average measured population can then be assumed equal to the equilibrium population of the system if the simulation time is sufficiently long (ergodic theorem). The details of the MC and MD techniques are beyond the scope of this chapter; we refer those interested to some seminal works on this subject ***(31,32)***.

The determination of an accurate free-energy landscape requires that the distribution be sampled adequately (i.e., the statistical error in the distribution must be small) along the relevant coordinates. Because the size and the complexity of a biomolecular system result in a very large and complicated phase space, adequate sampling requires a huge data set, which is not feasible with detailed standard MC or MD simulations. Although regular MD simulations cannot be used to fully explore the free-energy landscape of complex systems, their ability to describe the time dependence of the properties of the system makes them very popular as a way to test the stability of proposed fibril structures ***(16,33–35)***.

Many approaches have been proposed to solve the sampling problem, and some of these have been applied to study amyloid fibril formation. These approaches can be roughly divided into two types. The first entails a decrease of the complexity of the model system, sacrificing detail to enhance computability. This has led to the development of lattice models ***(36–38)*** and off-lattice coarse-grained models ***(39–41)***. An advantage of these models is that they may provide a general description of protein fibrillization, valid for more than one protein. On the other hand, due to their inherent approximations, they likely cannot capture the behavior of a specific protein in full detail. The second type of approach is to use more advanced simulation methods; these methods are discussed next and summarized in **Table 1**.

3.1. Replica Exchange Molecular Dynamics

In MD, the transition from one valley in the free-energy landscape to another is the result of thermal motion. However, since the barriers separating these valleys are generally higher than the thermal energy k_BT, transitions are a rare event, and only one free-energy minimum is sampled. Replica exchange molecular dynamics (REMD) or parallel tempering is a method devised to increase the number of transitions between the valleys in the free-energy landscape during MD simulations ***(31,42)*** while maintaining the elaborate sampling of the free-energy minima characteristic of MD.

In REMD, a number of identical systems (replicas) are simulated simultaneously but at different temperatures. At set time intervals, there are attempts to

Table 1
Summary of Advanced Simulation Techniques Used to Gain an Energetic Detailed Understanding of Fibril Formation

	Characteristics	Application Range[a]
Replica exchange molecular dynamics (REMD) *(42)*	A number of MD simulations of the same system are performed simultaneously but at different temperatures. Every *n* time steps, attempts are made to swap copies at different temperatures, which are accepted or rejected by a Metropolis criterion. Simulations at low temperature will explore the free-energy minima efficiently. The thermal motion in simulations of higher temperature facilitates barrier crossing. The phase space is sampled more extensively, and free-energy data can be extracted.	The conformational space of a polypeptide (25 residues maximum) in explicit solvent can be sampled exhaustively.
Umbrella sampling *(45)*	A path is selected between two states. A number of MD or MC simulations are performed, each with a different biasing potential to sample a specific region of the path exhaustively. The simulations can be unbiased and combined to obtain the distribution along the sampled path, and a free-energy profile associated with this path can be extracted.	The range depends on a number of factors: the size of the system; the number of individual simulations required; the sampling time required.
State free energy *(47,48)*	The absolute free energy of a specific state is calculated as a sum of three contributions: (1) the protein intramolecular interaction energy, (2) the effect of dissolving a protein, (3) the entropy.	Each state free energy can be calculated from a few picosecond MD. A free-energy landscape requires this for each possible conformation.

(continued)

Table 1 *(continued)*

	Characteristics	Application Range[a]
	Free-energy differences can be calculated easily, but they have a very large statistical error.	
	Free-energy landscapes can be devised but require calculating the absolute free energy of every possible conformation.	
Activation relaxation technique (ART) ***(51,53)***	Samples the minima in a system by transitions through saddle points on the energy landscape. The transitions are accepted or rejected by a Metropolis criterion.	In combination with the OPEP force field simulation, the conformational space of eight tetrapeptides can be exhaustively sampled, but four decamers are out of range.
	The entropy of the thermal vibrations of a minimum is neglected.	
	Efficiently samples the energy landscape of the system, showing low-energy conformations.	

MC, Monte Carlo; MD, molecular dynamics; OPEP, optimized potential for efficient peptide structure prediction in solution.

[a]Required simulation time generally increases rapidly with system size because properties take a longer time span to reach equilibrium, and simulations of large systems are slower than those of small systems.

swap the temperatures of different replicas, which are accepted or rejected by the following Metropolis criterion: $P_{acc} = \min\{1, e^{-\Delta\beta\cdot\Delta E}\}$, which also depends on the temperature difference. The replicas therefore perform a random walk through the temperature range used for the REMD. Replicas at elevated temperatures can cross energy barriers with relative ease, whereas low-temperature replicas explore the valleys in the free-energy landscape. The replicas in high-temperature simulations thus complement the low-temperature replicas, allowing for a more thorough sampling of configurational space ***(43)***. Using this method, the equilibrium distribution over the different minima is obtained, from which relevant free-energy differences can be calculated. More generally, from these equilibrium distributions any thermodynamic quantity can be calculated for any temperature using multiple-histogram reweighting techniques ***(42)***.

Although REMD can provide a detailed understanding of fibril formation, it is computationally very demanding, requiring weeks of computing on a parallel cluster to sample the conformational space of a small peptide (30 residues) in water adequately. Similarly, dimerization studies are limited to small (7-residue) peptides that form fibrils. For larger systems, it is impossible to sample the whole conformational space within a reasonable amount of time. However, this method can still be applied to scan for possible structures on the pathway to fibril formation. In this way, various structures that may play a role at the beginning of fibril formation, including interlocking β-sheet structures, have been identified for the fragment $A\beta_{10-35}$ (for $A\beta_{10-35}$, *see* **Fig. 2**) ***(44)***.

3.2. Umbrella Sampling

In many cases, only the free-energy difference between two states is of interest, and it suffices to sample only the relevant states adequately without sampling the rest of phase space. The free-energy difference can be calculated from the ratio of the populations of the two states (**Eq. 1**). However, due to barriers in the free-energy landscape, state *B* is generally not well sampled by the probability distribution around state *A* (see **Fig. 3**) and vice versa. Hence, the ratio N_A/N_B is either intractable ($N_B = 0$) or subject to a large statistical error ($N_B << N_A$). Equal sampling of states *A* and *B* in one simulation can be attained by introducing an extra potential (umbrella potential) to the system. The free-energy difference can then be calculated with sufficient accuracy ***(45)***.

A single simulation in which both state *A* and *B* are sampled is not very efficient. Therefore, a general umbrella sampling involves a path that connects two states, which is explored exhaustively. Since the free-energy difference

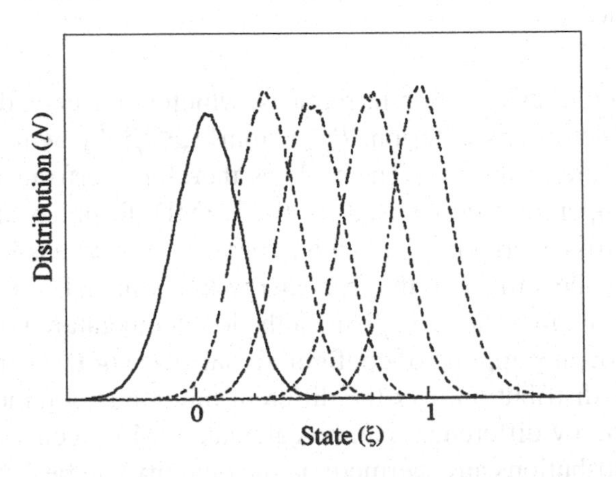

Fig. 3. Various windows resulting from an umbrella sampling. Probability distribution around state *A* (ξ = 1, solid line) does not sample state *B* (ξ = 0). Four biased simulations (dashed line) ensure accurate sampling of the path connecting states *A* and *B*.

is a state function, the path can be chosen at random, but a wise choice can shorten the calculations. To ensure efficient sampling of the path, a number of copies of the system are simulated. Each copy only differs by a biasing potential (umbrella potential) that enhances sampling around a specific point along the path (*see* **Fig. 3**). A given biased simulation results in a small sampling window that is sufficiently accurate to calculate the free-energy landscape* along that part of the path. To generate a free-energy landscape of the entire path, the various windows are combined by use of the weighted histogram analysis method (WHAM) ***(46)***.

Currently, an umbrella sampling method is under development in our group to calculate the free energy of filament elongation of the peptide KFFE. In this case, a convenient path is defined by the center-of-mass (COM) distance between the two last peptides in a filament parallel to the fibril axis. We introduce a harmonic potential as the biasing potential to sample a specific region of the path. This results in a free-energy landscape as a function of the COM distance, from which the free-energy difference between a protein attached to a filament and a free monomer can be obtained. This method can be applied to small peptides (up to 10 residues) to calculate the free-energy gain by the cross-β-sheet interactions.

3.3. State Free Energy

There is not always a simple way to connect two states of a system in a simulation to calculate the free-energy difference. When finding a good path is difficult, the free-energy difference between two relevant states can be obtained by considering the absolute free energy corresponding to the appropriate state. This state free energy can be approximated by considering the free energy as a combination of separate energy terms ***(47,48)***. For a protein in conformation *A*, this means

$$G_A = E_{protein} + E_{solvation} - TS_{protein} \tag{2}$$

where G_A is the calculated free energy of conformation *A*, $E_{protein}$ is the energy of the intramolecular interactions of the protein in conformation *A*, $E_{solvation}$ is the energy associated with protein–solvent interaction incorporating both potential energy and entropic contributions, T is the temperature, and $S_{protein}$ is the entropy of conformation *A*. A typical state free-energy calculation starts with a very short (around 100 ps to prohibit structural changes) all-atom MD simulation

* The free-energy landscape between two elements along a distance coordinate is often referred to as the effective pair potential or the potential of mean force (PMF).

of the relevant conformation to account for the fluctuations in the conformation due to thermal motion. A representative ensemble of structures from this MD simulation can then be used to calculate the average energy contributions to the free energy. The energy of the intramolecular interactions $E_{protein}$ is the molecular mechanical energy, consisting of bond, angle, dihedral, van der Waals, and electrostatic interactions, over this representative ensemble. The energy of solvation $E_{solvation}$ is usually calculated by considering the structures obtained from the MD simulations but taking the solvent into account implicitly. Implicit solvent models generally consist of three terms ***(48,49)***. Two arise from nonpolar contributions: the energy to create a cavity the size of the conformation in the solvent and the energy from van der Waals interactions between the solvent and the protein. The third term is a contribution from the solvent polarization by the protein. Finally, the entropy of the conformation $S_{protein}$ can be estimated from a quasi-harmonic or a normal-mode analysis of the MD simulation ***(47,48)***. The accuracy of this method is generally low due to the numerous assumptions and approximations made, but the computation to calculate the state free energy of one conformation is very fast, and results can be obtained within hours. The absolute free-energy values are then used to calculate the free-energy difference between relevant states. From these free-energy differences, the equilibrium populations can be calculated with, for example, a thermodynamic model ***(50)***, which is discussed in more detail in **Subheading 5**.

Although state free-energy calculations can give a relationship between two well-defined states, it does not provide any information about the importance of these conformations in the whole ensemble. The method can also be used to devise a free-energy landscape by performing an exhaustive search over all accessible conformations. This is only possible for small protein molecules (approximately 10 residues or less). Protein aggregates generally have too many degrees of freedom to be accurately described by this technique.

3.4. Activation Relaxation Technique

The activation relaxation technique (ART) ***(51)*** is an advanced MC technique that samples the minima in an energy landscape. Where standard MC usually takes a step of predescribed length in a random direction, ART crosses a saddle point in a random direction to find the next minimum in the energy landscape. A basic ART step consists of four stages: First, the energy landscape surrounding a minimum is sampled in a random fashion until a negative eigenvalue of the Hessian matrix is found. Such an eigenvalue indicates a saddle point in the direction of the eigenvector associated with it. Second, the system is directed toward this saddle point by a force aimed along this eigenvector, while the energy is minimized until all forces on the atoms vanish, indicating the saddle point has been reached. Third, the system is pushed over the saddle point and allowed to relax into the next minimum using standard minimization techniques. Finally, the new structure is accepted or rejected by a Metropolis criterion.

Applying the ART method in peptide simulations is often done in combination with the OPEP force field. The optimized potential for efficient peptide structure prediction in solution (OPEP) *(52)* is an approximate energy function using a hybrid level of description insofar as the main chain of each amino acid is taken into account explicitly, whereas each side chain is represented by a single bead with an appropriate van der Waals radius. In this potential, terms are incorporated that maintain stereochemistry, avoid steric overlap, allow for backbone hydrogen bonding, and treat the side-chain interactions based on their hydrophobicity or (partial) charge. Solvent effects are approximated by a continuous model (implicitly).

For a full description of the free energy, it is important to consider the entropy contribution. Since the OPEP energy function treats solvent implicitly and the ART method does not include thermal motion, it is impossible to account for the entropic contributions, and the sampled energy landscape is therefore a mixture of energy and free energy *(53,54)*. This means that the structure with the lowest energy does not necessarily represent a free-energy minimum. Nevertheless, the ART-OPEP technique generates folding mechanisms of helix and hairpin formation as well as native structures that match those found using more detailed models *(53)*, and less-structured peptides are found to match experimental constraints *(55)*.

Because ART-OPEP omits a number of contributions to the entropy, it can be used on slightly larger systems than, for instance, REMD. Besides investigating monomer phase space, the aggregation of two to eight small peptides (maximum seven residues) has been studied. However, exploring the whole phase space is not feasible when simulating more complex systems (e.g., assembly of four 15-residue peptides).

4. Understanding Fibril Formation in Terms of Free Energy

4.1. Fibrillogenic Character of Peptide Monomers

Application of the free-energy calculations discussed in Chapter 3 shows that any attempt to answer the question whether a peptide is fibrillogenic begins with a study of the monomer conformational space. For the majority of biologically active peptides at physiological conditions, the native state has a free energy significantly lower than any other conformation, mainly due to strong intramolecular interactions resulting in a well-defined, stable monomer structure *(56)*. The lack of these interactions as well as stress in the peptide chain due to intramolecular interactions can eliminate a clear minimum.

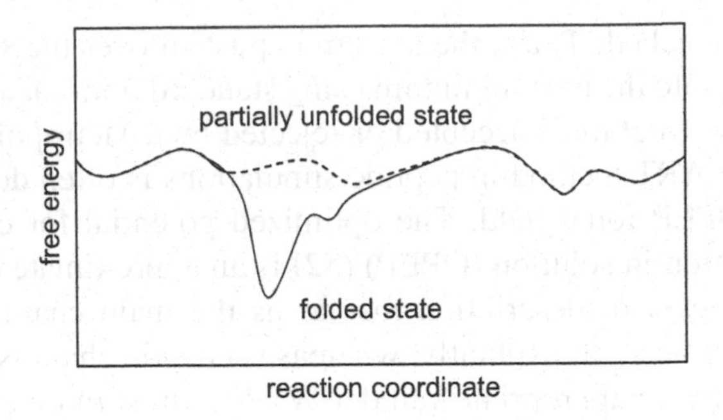

Fig. 4. Schematic view of a monomer free-energy landscape with a stable native state (solid line) and without a clear minimum (dashed line).

The absence of such a minimum results in many equivalent free-energy minima (*see* **Fig. 4**), some of which may have a low barrier to fibril formation. This is illustrated by results of REMD and ART-OPEP simulations of various fragments of Alzheimer's Aβ protein differing in sequence length. The $A\beta_{21-30}$ peptide does not form fibrils when dissolved (although it can exist in a stable fibril structure) ***(55)***. The structures corresponding to the free-energy minima found with REMD and ART-OPEP are in good agreement with the constraints deduced from nuclear magnetic resonance (NMR) spectroscopy. All minima are characterized by a large propensity to form a loop that is stabilized by a strong interaction between residues Val24 and Lys28 ***(54,55,57)***. The resulting deep free-energy minima explain the resistance of $A\beta_{21-30}$ solutions to form fibrils. In contrast, REMD studies of peptides that are prone to form fibril structures suggest that these peptides do not fold to a unique native state ***(58–60)***. The minima in the free-energy landscape are dominated by random-coil structures and an occasional secondary structure element. The great diversity in conformations these peptides adopt proves the absence of a global minimum and reflects their fibrillogenic properties.

Even if the free-energy landscape of a monomeric peptide possesses a deep minimum, a significant population of a partially unfolded conformation can also arise due to a local minimum in the free-energy landscape of a monomeric peptide. If this local minimum has intermediate stability (i.e., if the free energy lies between the top of the barrier and the global minimum), this can promote fibril formation ***(61)***. In this case, the formation of fibrils is accelerated because the partially unfolded structure has a lower threshold toward amyloid formation than the native state, while the barrier between the partially unfolded state and the native state ensures that peptides do not immediately revert back to the native state (*see* **Fig. 5**). The importance of this effect is clearly seen when

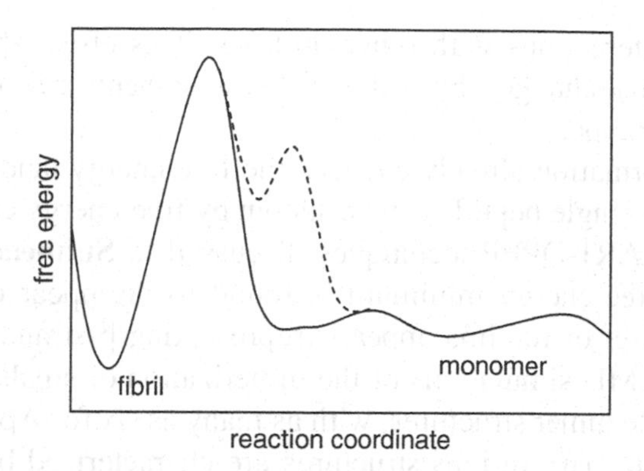

Fig. 5. Schematic view of a free-energy landscape with a barrier separating a monomer and a fibril state. An intermediate state between the monomer and the fibril (dashed line) facilitates fibril formation.

comparing Alzheimer's $A\beta_{25-35}$ and its Asn27–Gln† mutant, where the former forms fibril structures, while the latter does not. The complete conformational space has been systematically explored for both peptides by changing the ϕ and ψ, angles and for each conformation the state free energy was calculated *(61)*. The wild-type peptide shows many extended conformations of intermediate free energy, while the mutated peptide shows only structures with low and high free energies. A partially unfolded state with intermediate stability that tends to facilitate aggregate formation is thus effectively removed from the monomer structure pool by the mutation.

4.2. Nucleation, Protofibril Formation, and Fibril Elongation

Amyloid formation is well described by a nucleation-growth mechanism, and the formation of a nucleus is the first barrier on the path to fibril formation *(7,62)*. We define a nucleus as the starting structure from which a fibril can grow; however, at present the specific nature of the nucleus is still subject to debate *(21–24)*.

A structural transformation of the protein is required for it to go from a free monomer to a building block of a fibril. While a single molecule is unlikely to acquire the β-strand conformation characteristic of an amyloid fibril, this structural rearrangement is possible after aggregation because the free-energy landscape of the protein changes, showing other stable secondary structure

† Protein residue number 27 is mutated from asparagine to glutamine.

motifs due to interactions with other proteins. This effect shows an analogy with environmental changes, by which different solvents affect the shape of the free-energy landscape.

That dimer formation already changes the free-energy landscape describing the structure of a single peptide can be shown by free-energy calculations using the REMD and ART-OPEP techniques discussed in **Subheading 3**. ***(63–65)***. The monomer free-energy minimum is found to disappear on dimer formation, and a new set of minima appears, representing β-strands, among others. For instance, REMD simulations of the dimerization of small peptides show a wealth of possible dimer structures, with as many as six for $A\beta_{16-22}$ and four for the KFFE peptide. The various structures are characterized by different interaction types, such as backbone hydrogen bonding, hydrophobic clustering, and even water-mediated coulombic interactions between charges ***(64)***, as well as by different atoms involved in these interactions. An ART-OPEP study of $A\beta_{16-22}$ dimer and trimer formation showed that these structures assemble to antiparallel cross-β-sheets ***(65)***; however, the conformational space holds a number of structures with only slightly higher energies. These structures are characterized by out-of-register antiparallel cross-β-sheets or a parallel cross-β-sheet.

The change in the free-energy landscape due to interactions with different peptides is strongly dependent on the amino acid sequence of the peptide ***(63)***. Since every peptide is capable of forming cross-β-sheet hydrogen bonds, this sequence dependence can be explained by the contributions of the side-chain interactions. Here, the contributions to the free energy, the interaction energy and the entropy, act as opposing forces. On dimerization, very bulky hydrophobic side chains experience a larger intermolecular energy than small side chains, while their intramolecular energy is unchanged. In addition, peptides containing side chains with a lot of conformational freedom experience a reduction of this freedom on dimerization coupled to a sizable loss of entropy. This was clearly observed in experiments in which the peptides KFFE and KVVE were found to form fibrils, while the peptides KAAE and KLLE did not ***(66)***. REMD simulations showed that the intramolecular interaction energy does not change for any of these peptides during dimerization, but the strength of the intermolecular interaction showed the tendency KFFE > KVVE ~ KLLE > KAAE ***(63)***. The entropic contribution determines the difference in fibrillogenic nature of KVVE and KLLE. This is because leucine side chains experience larger freedom in solution than valines, resulting in a larger entropic reduction on dimerization. This results in the following tendency to form dimers: KFFE > KVVE > KLLE > KAAE. The considerations discussed here are based on a dimerization simulation, but similar effects are expected for fibril formation because similar interactions are involved in the fibrillization. REMD dimerization considerations and experimental observations of fibril formation show good agreement regarding the relative tendency to form dimers.

Although dimers are the first step in fibril formation, they do not necessarily represent a stable nucleus. The energy landscape of KFFE hexamers obtained with ART-OPEP shows two families of conformations, with one dominated by a β-barrel-like structure, and the other family showing double-layer cross-β-sheet structures. The formation of these two structural families is indifferent to the size of the assembly as simulations of seven and eight chains also show them *(67)*. However, besides these families the octamer also shows structures in which the cross-β-sheet interaction stretches over at most two or three peptides, and the heptamer shows amorphous aggregates, both with equivalent energies to the aforementioned structural families.

In contrast to small peptides, which form cross-β-sheet structures rapidly although not exclusively, larger peptides take more time to adopt these structures *(68)*. The vast conformational space available to random-coil oligomers prohibits a rapid cross-β-sheet formation. It also implies a very large entropic contribution to the free energy of this state, while the strong interactions present in cross-β-sheet structures indicate a large potential energy contribution. The exact free-energy difference between these two states is still unclear.

After the nucleus phase, growth into mature fibrils occurs. It is reasonable to assume that this happens mainly by addition of monomeric peptide units. Due to the large size of these systems, simulations on fibril elongation are challenging. The free-energy difference associated with fibril elongation of small peptides can be calculated through an umbrella sampling technique (Wolf et al., work in progress). In the case of reversible fibril elongation, these free-energy differences can also be obtained with experimental methods *(24)*. The average population of relevant states is dictated by the free-energy differences between these states (**Eq. 1**). In experiments, the average population of a specific state can be measured as the concentration of that state. In case of fibril elongation, the two relevant states are a monomer in solution or in a fibril. Because it is very difficult to measure the concentration of fibrils or elongation sites, it is often assumed that this concentration does not change, that is, that no new fibrils are formed. In this case, the monomer concentration is equivalent to the dissociation equilibrium constant and thus related to the free-energy difference *(24)*. Both the computational and the experimental techniques provide a way to calculate $\Delta\Delta G$ values involving fibril elongation of mutated peptides to distinguish the important interactions in the fibril.

5. Thermodynamic, Statistical-Mechanic, and Kinetic Models

In addition to the molecular simulation techniques discussed in some detail, a more coarse-grained description of amyloid fibril formation can be applied. The term *coarse grained* refers to a description in which molecular details are largely neglected, and a simplified model is instead outlined to predict the behavior of a system. Applying this type of theory can provide a general understanding of amyloid fibril formation that goes beyond any one system (e.g., a description that describes the behavior not only of Aβ protein but also of a whole family of fibrillogenic proteins). In addition, these techniques may be applied to systems (e.g., aggregates) that are too large to model on a molecular scale. Coarse-grained models can be statistical-mechanical (or thermodynamical) in nature and as such describe only the equilibrium properties of a system. Alternatively, they can strive to take the time dependence of any reaction explicitly into account (dynamic or kinetic modeling). Applying these techniques, we can predict the average dimensions of all species, as well as the fraction of protein molecules in each of the aggregated states, as a function of several external parameters, such as the overall protein concentration, the temperature, and (for kinetic or dynamic models) as a function of time. There is a marked difference between these models and those described in detail in **Subheading 3**. While the latter models focus on a detailed description of a particular system and can as such give a good estimate of relevant (microscopic) free energies, the models discussed in this section can take these, or experimentally determined, free-energy values and use them to predict the behavior of such a system on a mesoscopic or macroscopic length scale ***(19,50)***.

Many experimental studies of amyloid fibril formation focus on the time dependence of fibril formation ***(12,17,69)***. It is therefore not surprising that a host of kinetic models, coupled to reaction schemes, has been introduced to better understand the fibrillization process ***(12,17,22–24,69,70)***. These models describe fibrillization as a series of reaction steps, each with one or two corresponding reaction constants. They are coarse grained because they do not take molecular detail into account but rather focus on stoichiometry and (ir)reversibility of the reaction steps. However, because they are often designed to describe a very specific protein system, the models presented in the literature tend to be quite detailed and as such are likely applicable only for the protein and conditions for which they were originally set up. While the nucleation-and-growth mechanism is a common ingredient in many of these models, the details of the reaction path often differ from model to model. Important points of difference between the models are the roles of protofilaments and protofibrils, the presence of micellar intermediates and paranuclei, the identification of oligomeric species as on-pathway or off-pathway structures, and the size of the nucleus ***(12,17,69–71)***. To use a kinetic model to predict the properties of a particular amyloidogenic system, the reaction constants of each step between monomer and fibril must be known. In case of a reversible step, the equilibrium constant may be readily calculated from the free-energy difference between the reactants and the products. More generally, reaction constants depend on the height of the free-energy barrier between the products and reactants.

In some cases (e.g., for the Alzheimer's-related Aβ protein), the formation of amyloid fibrils has been found to be a reversible process ***(24)***. If this is the case, the system is under thermodynamic rather than kinetic control. This means that a thermodynamic or statistical-mechanical model may be expected to provide a good description of the amyloid fibril formation ***(19,50)***. In contrast, when considering irreversible assembly, the dynamics of the assembly need to be taken into account explicitly because the experimentally observed fibrillar state may not correspond to an equilibrium state. Statistical-mechanical techniques have long been applied to protein systems, and protein aggregation has been extensively studied theoretically by Oosawa and Asakura ***(1)***. However, these techniques have only recently been applied to amyloid-like systems ***(19,24,50,72,73)***. To describe amyloid fibril formation, a minimum of three energetic contributions must be considered: the change in free energy that results from an axial bond (β-sheet-type interaction), that from a lateral bond between protofilaments, and that from a conformational transition. In reality, more energetic parameters may play a role since more than two conformations may play a role during amyloid formation, and the formation of a fibril from protofilaments may involve more than one type of lateral bond ***(10,15)***. In the next paragraph, we discuss a statistical-mechanical model ***(19)*** to illustrate the application of these techniques.

In our statistical-mechanical description of amyloid fibril formation, we consider three species: monomeric proteins, which are in a non-β-strand conformation; protofilaments, which are linear aggregates (in other words, a single stack) of protein molecules; and fibrils, which consist of several laterally associated protofilaments. This is shown schematically in **Fig. 6**, where a "blob" indicates a protein molecule in a non-β-conformation, and a disk indicates one in a β-conformation. As shown in the figure, we do not restrict the conformation of the molecules in the protofilament state; that is, protein molecules in a protofilament can be in the (ordered) β-strand conformation characteristic of the mature fibril, or they can be in a disordered conformation. An exception to this rule is made for the first and last monomers of the protofilament, which we define as disordered. In our definition of a fibril, on the other hand, we distinguish two regions: the body of the fibril, which contains only proteins in the β-strand conformation, all of which are laterally associated to one protein in every other neighboring protofilament, and the fibril ends, which do not laterally associate and are disordered. Note that the lengths of these fibril ends may equal zero, leading to an entirely ordered fibril.

To describe the protofilaments theoretically, we use a two-state model based on the one-dimensional Ising model ***(74)***. In our model, every intermolecular

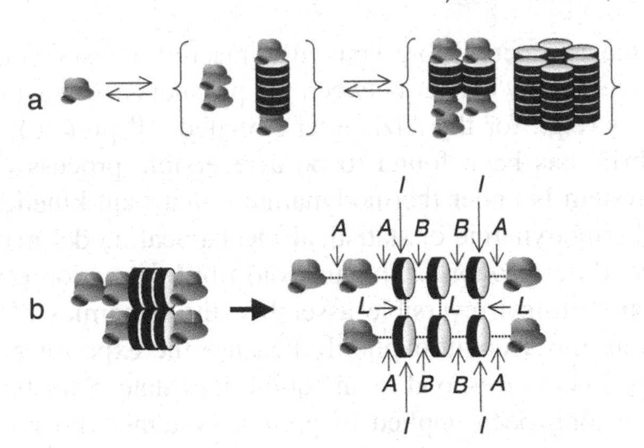

Fig. 6. (**a**) A general schematic picture of the aggregated states accounted for in our statistical-mechanical model. "Blobs" indicate disordered proteins, disks are proteins with a β-strand conformation. (**b**) Overview of the free-energy parameters associated with the different types of interactions inside an amyloid fibril.

interaction is characterized by one of two discrete binding free energies: the "β binding free energy" *B* if both molecules are in the β-strand conformation, and the "non-β binding free energy" *A* if this is not the case (*see* **Fig. 6b**). For the fibrils, this description is extended with a lateral-association free-energy parameter, designated *L*, in a similar way as was done in **ref.** ***72***. Finally, a free-energy parameter *I* was introduced that favors the formation of homogeneous fibrils by penalizing the presence of an interface between an ordered and a disordered region along the protofilament or fibril axis. With these free-energy parameters, we can calculate the partition functions of all species exactly (within the confines of the model). We can then determine the overall free-energy density (the Helmholtz free energy per unit volume) of the protein solution *ØF* using standard self-assembly theory ***(28,75,76)***. This free-energy density is given in units of thermal energy and as such is dimensionless. For solutions that are dilute enough that interaggregate interactions may reasonably be ignored, it is described by

$$\Delta F = \sum_{X} \rho(X)\left[\ln \rho(X) - 1 - \ln Q(X)\right], \tag{3}$$

where the summation is over all different aggregate types. Indicated by *X* in the equation, the aggregate type is defined by the number of protein molecules and the number of lateral interactions contained within the aggregate. ρ(*X*) is the dimensionless number density of the aggregates of type *X*. This corresponds to the number of aggregates of type *X* present per unit volume of protein solution. *Q*(*X*), finally, is the partition function of an aggregate of

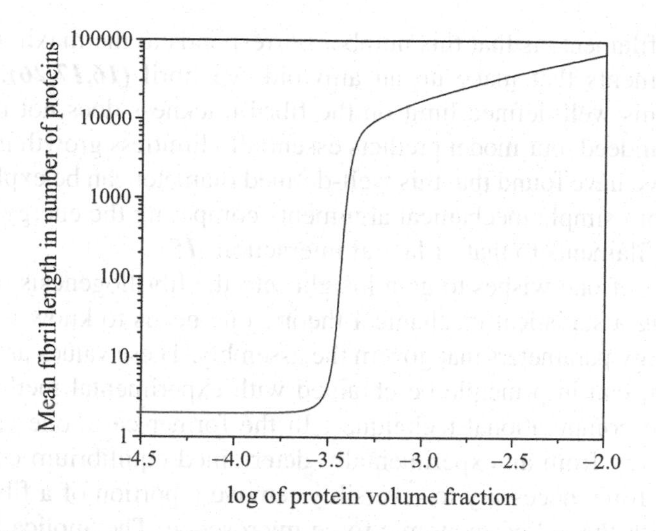

Fig. 7. Example of a prediction from the statistical-mechanical model ***(19)***. Depicted is the mean length of a fibril that consists of six protofilaments, given in terms of the number of protein molecules that are associated in the direction of the fibril axis. The mean length is plotted against the overall volume fraction of the protein molecules in solution. This curve is found for values of the free-energy parameters of $A = -2.1$, $B = -4.8$, $I = 2.3$, and $L = -3.8$ times the thermal energy. Similar curves are found for different values.

type X and contains the information on the conformational state of all protein molecules that make up the aggregate. Setting the functional derivative of the free-energy density with regard to the number density equal to zero, while imposing conservation of mass, allows us to determine the equilibrium properties of the protein solution.

Using the statistical-mechanical framework outlined, we can predict several properties of the amyloid fibrils ***(19)***. These include the mean fibril length and the fractions of protein molecules in any aggregated state as a function of the overall protein concentration. In **Fig. 7**, we give an example of the results of our model. Here, we show the mean length (measured in terms of the number of protein molecules in the direction of the fibril axis) of fibrils that consist of six protofilaments as a function of the total volume fraction of protein molecules. The figure shows that there is a strong dependence of the mean fibril length on the overall protein concentration. At low concentrations, the fibrils do not grow, and most of the material is present as monomers. At a critical concentration, there is a fairly sudden increase in the mean fibril length, followed by a slow rise of the mean fibril length. The reason we chose to look at fibrils consisting of six protofilaments is that this number corresponds to the maximum number of protofilaments that make up an amyloid Aβ fibril ***(16,17,26)***. While the reason for this well-defined limit on the fibril thickness does not follow from our model (indeed, our model predicts essentially limitless growth in the lateral direction), we have found that this well-defined diameter can be explained quite naturally from simple mechanical arguments comparing the energy of bending of the protofilaments to that of lateral interaction ***(15)***.

Of course, if one wishes to gain insight into the fibrillogenesis of a specific protein using a statistical-mechanical theory, one needs to know the values of the free-energy parameters that govern the assembly. These values are not readily available but can in principle be obtained with experimental methods or with atomic-scale computational techniques. In the former case, one can compute the free energy from an experimentally determined equilibrium constant ***(73)*** or from the force necessary to reversibly remove a portion of a fibril ***(77)***, for instance, with the aid of an atomic force microscope. The application of computer simulation to determine free energies has been described in some detail here. For an excellent example of the interplay between thermodynamic and computational modeling, refer to **ref. *50***.

6. Concluding Remarks and Prospects

In this review, we highlight the role of free energy in amyloid fibril formation. We start with a focus on computational methods by which relevant free-energy differences are obtained and discuss their application in light of amyloid formation, and we close with a discussion of novel coarse-grained models that require these same energies as input parameters. Given appropriate values of the local free energies, a statistical-mechanical model can predict overall (measurable) mesoscopic properties of amyloid fibrils.

We expect that the use of computer simulations to understand complex systems will continue to increase significantly in the future because the capability of computers still grows every year, and the development of more advanced simulation techniques provides more efficient methods to harvest this power. The size of the systems studied with computational methods will thus increase, and we anticipate that in the next few years, this will lead to significant advances in understanding amyloid formation. The conformational space and the free-energy landscape of small fibrillogenic proteins will be understood in molecular detail. In addition, the early aggregation steps of more complex systems requiring significant conformational changes can be studied. Finally, the full nucleus formation of small fibrillogenic peptides will be investigated in molecular detail. The use of these simulations will also allow testing of new therapeutic agents and to elucidate their interaction with the fibrillogenic protein.

Acknowledgments

We thank the Netherlands Organization for Scientific Research NWO for funding (grant 635.100.012, program for computational life sciences). We are also grateful to Jaap Jongejan and Jon Laman for critically reading the manuscript and many useful discussions.

References

1. Oosawa F, Asakura S. (1975) *Thermodynamics of the Polymerization of Protein*. Academic Press, New York.
2. Korn ED. (1982) Actin polymerization and its regulation by proteins from non-muscle cells. *Physiol. Rev.* **62**, 672–737.
3. Downing KH, Nogales E. (1998) Tubulin and microtubule structure. *Curr. Opin. Cell Biol.* **10**, 16–22.
4. Kegel WK, van der Schoot P. (2006) Physical regulation of the self-assembly of tobacco mosaic virus coat protein. *Biophys. J.* **91**, 1501–1512.
5. Klug A. (1983) From macromolecules to biological assemblies (Nobel lecture). *Angew. Chem. Int. Ed. Engl.* **22**, 565–582.
6. Chiti F, Dobson CM. (2006) Protein misfolding, functional amyloid, and human disease. *Annu. Rev. Biochem.* **75**, 333–366.
7. Rochet J-C, Lansbury PT. (2000) Amyloid fibrillogenesis: themes and variations. *Curr. Opin. Struct. Biol.* **10**, 60–68.
8. Serpell LC. (2000) Alzheimer's amyloid fibrils: structure and assembly. *Biochim. Biophys. Acta* **1502**, 16–30.
9. Thirumalai D, Klimov DK, Dima RI. (2003) Emerging ideas on the molecular basis of protein and peptide aggregation. *Curr. Opin. Struct. Biol.* **13**, 146–159.
10. Khurana R, Ionescu-Zanetti C, Pope M, et al. (2003) A general model for amyloid fibril assembly based on morphological studies using atomic force microscopy. *Biophys. J.* **85**, 1135–1144.
11. Selkoe DJ. (2002) Deciphering the genesis and fate of amyloid β-protein yields novel therapies for Alzheimer disease. *J. Clin. Invest.* **110**, 1375–1381.
12. Goldsbury C, Frey P, Olivieri V, Aebi U, Müller SA. (2005) Multiple assembly pathways underlie amyloid-β fibril polymorphisms. *J. Mol. Biol.* **352**, 282–298.
13. Nelson R, Eisenberg D. (2006) Recent atomic models of amyloid fibril structure. *Curr. Opin. Struct. Biol.* **16**, 260–265.
14. Tycko R. (2004) Progress towards a molecular-level structural understanding of amyloid fibrils. *Curr. Opin. Struct. Biol.* **14**, 96–103.
15. van Gestel J, de Leeuw SW. (2007) The formation of fibrils by intertwining of filaments: model and application to amyloid Aβ protein. *Biophys. J.* **92**, 1157–1163.
16. Guo J-T, Wetzel R, Xu Y. (2004) Molecular modeling of the core of Aβ amyloid fibrils. *Proteins* **57**, 357–364.
17. Pallitto MM, Murphy RM. (2001) A mathematical model of the kinetics of β-amyloid fibril growth from the denatured state. *Biophys. J.* **81**, 1805–1822.

18. Walsh DM, Lomakin A, Benedek GB, Condron MM, Teplow DB. (1997) Amyloid β-protein fibrillogenesis—detection of a protofibrillar intermediate. *J. Biol. Chem.* **272**, 22364–22372.
19. van Gestel J, de Leeuw SW. (2006) A statistical-mechanical theory of fibril formation in dilute protein solutions. *Biophys. J.* **90**, 3134–3145.
20. Petkova AT, Leapman RD, Guo Z, Yau W-M, Mattson MP, Tycko R. (2005) Self-propagating, molecular-level polymorphism in Alzheimer's β-amyloid fibrils. *Science* **307**, 262–265.
21. Kisilevsky R. (2000) Review: amyloidogenesis-unquestioned answers and unanswered questions. *J. Struct. Biol.* **130**, 99–108.
22. Pellarin R, Caflisch A. (2006) Interpreting the aggregation kinetics of amyloid peptides. *J. Mol. Biol.* **360**, 882–892.
23. Powers ET, Powers DL. (2006) The kinetics of nucleated polymerizations at high concentrations: amyloid fibril formation near and above the "supercritical concentration." *Biophys. J.* **91**, 122–132.
24. Wetzel R. (2006) Kinetics and thermodynamics of amyloid fibril assembly. *Acc. Chem. Res.* **39**, 671–679.
25. Lührs T, Ritter C, Adrian M, et al. (2005) 3D structure of Alzheimer's amyloid-β(1–42) fibrils. *Proc. Natl. Acad. Sci. U. S. A.* **102**, 17342–17347.
26. Serpell LC, Blake CCF, Fraser PE. (2000) Molecular structure of a fibrillar Alzheimer's Aβ fragment. *Biochemistry* **39**, 13269–13275.
27. Chiti F, Stefani M, Taddei N, Ramponi G, Dobson CM. (2003) Rationalization of the effects of mutations on peptide and protein aggregation rates. *Nature* **424**, 805–808.
28. Cates ME, Candau SJ. (1990) Statics and dynamics of worm-like surfactant micelles. *J. Phys. Condens. Matt.* **2**, 6869–6892.
29. Makhatadze GI, Privalov PL. (1995) Energetics of protein structure. *Adv. Protein Chem.* **47**, 307–425.
30. Stine WB, Dahlgren KN, Krafft GA, LaDu MJ. (2003) In vitro characterization of conditions for amyloid-β peptide oligomerization and fibrillogenesis. *J. Biol. Chem.* **278**, 11612–11622.
31. Frenkel D, Smit B. (2002) *Understanding Molecular Simulation: From Algorithms to Applications*. Academic Press, San Diego, CA.
32. Allen MP, Tildesley DJ. (1987) *Computer Simulation of Liquids*. Clarendon Press, Oxford, UK.
33. Ma B, Nussinov R. (2002) Stabilities and conformations of Alzheimer's β-amyloid peptide oligomers ($A\beta_{16-22}$ $A\beta_{16-35}$ and $A\beta_{10-35}$): sequence effects. *Proc. Natl. Acad. Sci. U. S. A.* **99**, 14126–14131.
34. Röhrig UF, Laio A, Tantalo N, Parrinello M, Petronzio R. (2006) Stability and structure of oligomers of the Alzheimer peptide $A\beta_{16-22}$: from the dimer to the 32-mer. *Biophys. J.* **91**, 3217–3229.
35. Zanuy D, Ma B, Nussinov R. (2003) Short peptide amyloid organization: stabilities and conformations of the islet amyloid peptide NFGAIL. *Biophys. J.* **84**, 1884–1894.

36. Oakley MT, Garibaldi JM, Hirst JD. (2005) Lattice models of peptide aggregation: evaluation of conformational search algorithms. *J. Comp. Chem.* **26**, 1638–1646.
37. Dima RI, Thirumalai D. (2002) Exploring protein aggregation and self-propagation using lattice models: phase diagram and kinetics. *Protein Sci.* **11**, 1036–1049.
38. Harrison PM, Chan HS, Prusiner SB, Cohen FE. (2001) Conformational propagation with prion-like characteristics in a simple model of protein folding. *Protein Sci.* **10**, 819–835.
39. Favrin G, Irbäck A, Mohanty S. (2004) Oligomerization of amyloid Aβ16–22 peptides using hydrogen bonds and hydrophobicity forces. *Biophys. J.* **87**, 3657–3664.
40. Jang H, Hall CK, Zhou Y. (2004) Thermodynamics and stability of a β-sheet complex: molecular dynamics simulations on simplified off-lattice protein models. *Protein Sci.* **13**, 40–53.
41. Urbanc B, Cruz L, Yun S, et al. (2004) *In silico* study of amyloid β-protein folding and oligomerization. *Proc. Natl. Acad. Sci. U. S. A*, **101**, 17345–17350.
42. Sugita Y, Okamoto Y. (1999) Replica-exchange molecular dynamics method for protein folding. *Chem. Phys. Lett.* **314**, 141–151.
43. Cecchini M, Rao F, Seeber M, Caflisch A. (2004) Replica exchange molecular dynamics simulations of amyloid peptide aggregation. *J. Chem. Phys.* **121**, 10748–10756.
44. Jang S, Shin S. (2006) Amyloid β-peptide oligomerization *in silico*: dimer and trimer. *J. Phys. Chem. B* **110**, 1955–1958.
45. Torrie GM, Valleau JP. (1977) Non-physical sampling distributions in Monte-Carlo free-energy estimation—umbrella sampling. *J. Comp. Phys.* **23**, 187–199.
46. Roux B. (1995) The calculation of the potential of mean force using computer-simulations. *Comput. Phys. Commun.* **91**, 275–282.
47. Kollman PA, Massova I, Reyes C, et al. (2000) Calculating structures and free energies of complex molecules: combining molecular mechanics and continuum models. *Acc. Chem. Res.* **33**, 889–897.
48. Vorobjev YN, Hermans J. (1999) ES/IS: estimation of conformational free energy by combining dynamics simulations with explicit solvent with an implicit solvent continuum model. *Biophys. Chem.* **78**, 195–205.
49. Feig M, Brooks CL. (2004) Recent advances in the development and application of implicit solvent models in biomolecule simulations. *Curr. Opin. Struct. Biol.* **14**, 217–224.
50. Tiana G, Simona F, Broglia RA, Colombo G. (2004) Thermodynamics of β-amyloid fibril formation. *J. Chem. Phys.* **120**, 8307–8317.
51. Malek R, Mousseau N. (2000) Dynamics of Lennard-Jones clusters: a characterization of the activation-relaxation technique. *Phys. Rev. E* **62**, 7723–7728.
52. Derreumaux P. (2000) Generating ensemble averages for small proteins from extended conformations by Monte Carlo simulations. *Phys. Rev. Lett.* **85**, 206–209.

53. Mousseau N, Derreumaux P, Gilbert G. (2005) Navigation and analysis of the energy landscape of small proteins using the activation-relaxation technique. *Phys. Biol.* **2**, S101–S107.
54. Chen W, Mousseau N, Derreumaux P. (2006) The conformations of the amyloid-β (21–30) fragment can be described by three families in solution. *J. Chem. Phys.* **125**, 084911.
55. Lazo ND, Grant MA, Condron MC, Rigby AC, Teplow DB. (2005) On the nucleation of amyloid β-protein monomer folding. *Protein Sci.* **14**, 1581–1596.
56. Bryngelson JD, Onuchic JN, Socci ND, Wolynes PG. (1995) Funnels, pathways, and the energy landscape of protein-folding—a synthesis. *Proteins* **21**, 167–195.
57. Baumketner A, Bernstein SL, Wyttenbach T, et al. (2006) Structure of the 21–30 fragment of amyloid β-protein. *Protein Sci.* **15**, 1239–1247.
58. Baumketner A, Shea J-E. (2006) Folding landscapes of the Alzheimer amyloid-β(12–28) peptide. *J. Mol. Biol.* **362**, 567–579.
59. Baumketner A, Shea J-E. (2007) The structure of the Alzheimer amyloid β 10–35 peptide probed through replica-exchange molecular dynamics simulations in explicit solvent. *J. Mol. Biol.* **366**, 275–285.
60. Wei G, Shea J-E. (2006) Effects of solvent on the structure of the Alzheimer amyloid-β(25–35) peptide. *Biophys. J.* **91**, 1638–1647.
61. Ma B, Nussinov R. (2006) The stability of monomeric intermediates controls amyloid formation: Aβ25–35 and its N27Q mutant. *Biophys. J.* **90**, 3365–3374.
62. Kodali R, Wetzel R. (2007) Polymorphism in the intermediates and products of amyloid assembly. *Curr. Opin. Struct. Biol.* **17**, 48–57.
63. Baumketner A, Shea J-E. (2005) Free energy landscapes for amyloidogenic tetrapeptides dimerization. *Biophys. J.* **89**, 1493–1503.
64. Gnanakaran S, Nussinov R, Garcia AE. (2006) Atomic-level description of amyloid β-dimer formation. *J. Am. Chem. Soc.* **128**, 2158–2159.
65. Santini S, Mousseau N, Derreumaux P. (2004) *In silico* assembly of Alzheimer's $A\beta_{16-22}$ peptide into β-sheets. *J. Am. Chem. Soc.* **126**, 11509–11516.
66. Tjernberg L, Hosia W, Bark N, Thyberg J, Johansson J. (2002) Charge attraction and β propensity are necessary for amyloid fibril formation from tetrapeptides. *J. Biol. Chem.* **277**, 43243–43246.
67. Melquiond A, Mousseau N, Derreumaux P. (2006) Structures of soluble amyloid oligomers from computer simulations. *Proteins* **65**, 180–191.
68. Boucher G, Mousseau N, Derreumaux P. (2006) Aggregating the amyloid $A\beta_{11-25}$ peptide into a four-stranded β-sheet structure. *Proteins* **65**, 877–888.
69. Modler AJ, Gast K, Lutsch G, Damaschun G. (2003) Assembly of amyloid protofibrils via critical oligomers—a novel pathway of amyloid formation. *J. Mol. Biol.* **325**, 135–148.
70. Lomakin A, Teplow DB, Kirschner DA, Benedek GB. (1997) Kinetic theory of fibrillogenesis of amyloid β-protein. *Proc. Natl. Acad. Sci. U. S. A.* **94**, 7942–7947.
71. Bitan G, Kirkitadze MD, Lomakin A, Vollers SS, Benedek GB, Teplow DB. (2003) Amyloid β-protein (Aβ) assembly: Aβ 40 and Aβ 42 oligomerize through distinct pathways. *Proc. Natl. Acad. Sci. U. S. A.* **100**, 330–335.

72. Nyrkova IA, Semenov AN, Aggeli A, Bell M, Boden N, McLeish TCB. (2000) Self-assembly and structure transformations in living polymers forming fibrils. *Eur. Phys. J. B* **17**, 499–513.
73. O'Nuallain B, Shivaprasad S, Kheterpal I, Wetzel R. (2005) Thermodynamics of Aβ(1–40) amyloid fibril elongation. *Biochemistry* **44**, 12709–12718.
74. Ising E. (1925) Beitrag zur Theorie des Ferromagnetismus. *Z. Phys.* **31**, 253–258.
75. van der Schoot P, Michels MAJ, Brunsveld L, Sijbesma RP, Ramzi A. (2000) Helical transition and growth of supramolecular assemblies of chiral discotic molecules. *Langmuir* **16**, 10076–10083.
76. van Gestel J, van der Schoot P, Michels MAJ. (2005) Growth and chirality amplification in helical supramolecular polymers. In *Molecular Gels: Materials With Self-Assembled Fibrillar Networks*. Dordrecht, the Netherlands: Springer, pp. 79–94.
77. Kellermayer MSZ, Grama L, Karsai A, et al. (2005) Reversible mechanical unzipping of amyloid β–fibrils. *J. Biol. Chem.* **280**, 8464–8470.

11

Computer Modeling in Biotechnology

A Partner in Development

Aleksei Aksimentiev, Robert Brunner, Jordi Cohen, Jeffrey Comer, Eduardo Cruz-Chu, David Hardy, Aruna Rajan, Amy Shih, Grigori Sigalov, Ying Yin, and Klaus Schulten

Summary

Computational modeling can be a useful partner in biotechnology, in particular, in nanodevice engineering. Such modeling guides development through nanoscale views of biomolecules and devices not available through experimental imaging methods. We illustrate the role of computational modeling, mainly of molecular dynamics, through four case studies: development of silicon bionanodevices for single molecule electrical recording, development of carbon nanotube-biomolecular systems as in vivo sensors, development of lipoprotein nanodiscs for assays of single membrane proteins, and engineering of oxygen tolerance into the enzyme hydrogenase for photosynthetic hydrogen gas production. The four case studies show how molecular dynamics approaches were adapted to the specific technical uses through (i) multi-scale extensions, (ii) fast quantum chemical force field evaluation, (iii) coarse graining, and (iv) novel sampling methods. The adapted molecular dynamics simulations provided key information on device behavior and revealed development opportunities, arguing that the "computational microscope" is an indispensable nanoengineering tool.

Key Words: Biosensors; carbon nanotubes; coarse-grained modeling; DNA sequencing; empirical force field; high-density lipoprotein; high-throughput simulations; hydrogenase; molecular dynamics; multiscale modeling; nanodisc; nanopore; oxygen migration pathways; polarization; protein engineering; tight-binding method.

1. Introduction

Biotechnology exploits biological processes to create novel technological solutions, either through manufacturing synthetic devices that can directly interact with cellular machinery or by altering the design of biological

From: *Methods in Molecular Biology, vol. 474: Nanostructure Design: Methods and Protocols*
Edited by: E. Gazit and R. Nussinov © Humana Press, Totowa, NJ

molecules, such as proteins. Biotechnology has the potential to revolutionize medicine and offer novel technological solutions. Computer modeling can greatly accelerate the process of designing synthetic biodevices or engineering biological machines, but modeling tools and methods are significantly less developed than those available in the mainstream life sciences. This chapter provides an overview of the computational methods and tools we recently developed with the aim of dramatically improving the capabilities of computer modeling in biotechnology. The development efforts focus on the following areas: silicon bionanodevices, carbon nanotube-biomolecular systems, lipoprotein assemblies, and protein engineering for H_2 production.

1.1. Silicon Bionanodevices

Miniature devices in which single biomolecules directly interact with silicon electric circuits can already be manufactured and used for biomedical applications. Modeling such devices is challenging as no ready-to-use methods and software exist. We have integrated molecular dynamics (MD) simulations of biomolecules and continuum electrostatic models of silicon-based synthetic materials into a coherent computational method that allows researchers to relate changes in biomolecular structures to electronic processes in silicon semiconductor devices and vice versa.

1.2. Carbon Nanotube-Biomolecular Systems

Unique optical properties of carbon nanotubes (CNTs) allow changes in the conformation of adjacent biomolecules to be detected by spectroscopy. To aid the rational design of biosensors or drug development platforms operating using optical characteristics, the coupling between the biomolecular structure and the optical spectra of CNTs has to be established. We have developed a quantum mechanical/molecular mechanical (QM/MM) method for simulating CNT in complex with biomolecules and suggest a new semiempirical method for computing optical spectra of nanotube-biomolecular systems.

1.3. Lipoprotein Assemblies

Nanodiscs are nanometer-size protein–lipid particles being developed as platforms in which to study membrane proteins. Each nanodisc particle contains two scaffold proteins that encircle a small lipid bilayer, effectively producing a soluble platform in which membrane proteins can be embedded in a "native" environment. To optimize the self-assembly procedure, characterize the structure of assembled nanodiscs, and further improve on the design of the scaffold proteins, a variety of coarse-grained (CG) methods and tools have been developed by us that benefit the development of not only nanodiscs but also other bionanotechnology applications.

1.4. Protein Engineering for H_2 Production

Engineering the gas migration rates in gas-binding proteins such as hydrogenases requires the ability to precisely characterize the migration pathways taken by the gas molecules. Methodologies to comprehensively describe the location, energy profiles, and kinetic transport rates of gas migration pathways in proteins are being developed and integrated into the protein design process.

Most of the tools and methods described in this chapter are available in the form of plug-ins or modules to the publicly available software Visual Molecular Dynamics (VMD) *(1)* and Nanoscale Molecular Dynamics (NAMD) *(2)*. This review assumes that the reader has basic knowledge of biomolecular modeling and of MD methodology in particular. For more information on the last subjects we refer the reader to recent and classic reviews *(1–3)* and books *(4,5)*.

2. Silicon Nanopores for Sequencing DNA

Knowledge of an individual's genetic makeup can lead to the prediction and treatment of many diseases by means of personal genomic medicine and pharmacogenomics *(6)*. For the sequencing of a patient's genome to become a common medical routine, the development of an inexpensive, high-throughput genome-sequencing technique is necessary. An efficient sequencing technique is also required for fundamental medical research, in particular to identify genes critical in the development of human cancers *(7)*. The National Institutes of Health recently set an ambitious goal of reducing the cost of complete genome sequencing 10,000-fold to a mere $1000 within a 10-year period *(8,9)*. This goal constitutes a scientific and engineering challenge that can be met with the help of silicon nanotechnology. It is already possible to manufacture a nanochip with features comparable in size to a single nucleotide and that is sensitive enough to measure a single nucleotide's electric field *(10,11)*.

The general idea behind the DNA-sequencing device is to measure the highly localized electric field of a DNA molecule *(12,13)*. A metal-oxide semiconductor capacitor membrane with a nanometer-size pore confines the DNA, as shown in **Fig. 1**, slowing its movement and restricting its conformational fluctuations. At the same time, the conducting layers of this membrane serve as electrodes and register the electrostatic potential of the nucleotides as they pass through the nanopore. Given the difference in size and charge distribution of the DNA nucleotides adenine, cytosine, guanine, and thymine, the electrical signal recorded by the device during DNA translocation can, in principle, be used to identify the nucleotide sequence (*see* **Fig. 1**). Through integrated circuit techniques, the nanopore capacitor can be combined with an amplifier and other signal-processing circuitry, increasing the signal-to-noise ratio and temporal resolution of the device. Although researchers have succeeded in manufacturing operational nanopore capacitors and demonstrated the feasibility of recording the

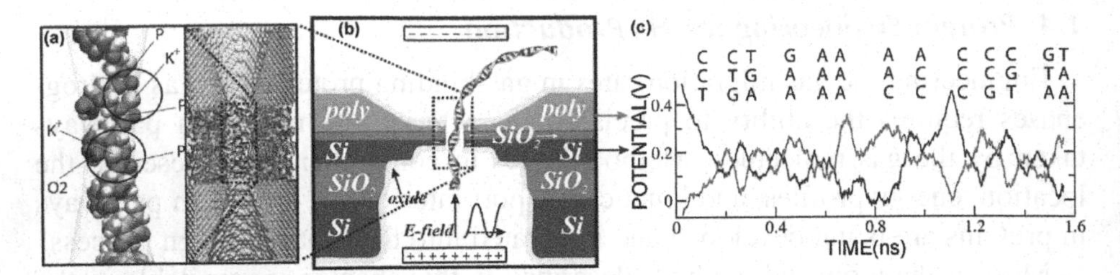

Fig. 1. Nanopore device for sequencing DNA. (**a**) Atomic-scale model of a DNA strand confined to a nanopore in a synthetic membrane. (**b**) A schematic view of the DNA sequencing device. (**c**) The correspondence between the nucleotide sequences and the electrostatic potentials recorded by the device as predicted by a molecular dynamics simulation.

electrical signatures of DNA molecules ***(12–19)***, relating the measured signal to a particular nucleotide sequence remains a major challenge.

Experiments allow one to measure DNA's electrostatic potential but not to visualize DNA's location and conformation in the pore at the time of the measurement. To decipher the electrostatic signals produced by DNA, its trajectory in the pore must be characterized in atomic detail. This cannot be achieved using current experimental techniques. Interpretation of the experimental data therefore requires molecular modeling, which can provide complete information about the DNA's position, conformation, and the electric field it generates. A simulation of the entire system, consisting of the DNA, the electrolytic solution, and the nanopore device, will allow elaboration of a protocol for deciphering DNA sequences based on measured electrical signals. MD simulations have already estimated the pore sizes that can provide optimal confinement for single-stranded and double-stranded DNA ***(16)***. The electric field signals, as well as the current blockades produced by the DNA translocation through nanopores, were characterized by all-atom MD ***(17)*** and multiscale methods ***(12,13)***. While these studies have already provided deep insight into the correlation between the DNA translocation and the resulting electrical signals, further development of the MD methodology is necessary to design a measuring protocol for identifying individual nucleotides.

2.1. Modeling Silicon-Biomolecular Systems

Historically, methodologies for MD simulation of biomolecular systems and inorganic materials have been developed independently. As a result, they use incompatible empirical force fields to describe interatomic interactions; thus, the interaction between DNA and a nanopore's synthetic membrane cannot be well described by any existing force field. To relate the sequence of DNA nucleotides passing through a nanopore in a capacitor membrane to the electric signals induced at the plates of the capacitor, a number of modeling tools and methods that enable microscopic simulations of nanodevices comprising both inorganic and biomolecular components need to be developed. MD simulations of such systems require a molecular force field, compatible with the Amber ***(20)*** and CHARMM ***(21–23)*** biomolecular force fields, that accurately reproduces the interaction between biomolecular and inorganic components. Other needed tools and methods include procedures for building and simulating crystalline and amorphous inorganic materials, atomless representation of synthetic surfaces, and tools for computing electrostatic potentials.

Another challenge that needs to be addressed is accounting for electronic processes occurring in synthetic nanodevices and for the mutual interactions between the nanodevice and the biomolecular system. A foundation of a multiscale methodology, which relies on a continuum description of the electronic processes inside the synthetic device while using an all-atom MD description for biomolecules, has been developed ***(13,19)***. Using this methodology, the electrostatic potential induced by the translocation of DNA through the nanopore device can be computed. Further development of the multiscale methodology is necessary to account for dynamic correlations between the continuum description of the electric field in a synthetic membrane and the all-atom MD description of DNA's conformations in the solution.

2.1.1. Building Atomic-Scale Models of Inorganic Nanodevices

To provide the inorganic models needed for all-atom simulations of nanobiodevices, we have implemented a method for building inorganic structures that can be used in conjunction with existing models of biomolecular structures. The nanostructures are constructed using a set of common operations that include replication of the inorganic unit cell, shaping the nanodevice, and specifying the connectivity of the inorganic atoms. This method was deployed to build nanopores in crystalline Si_3N_4 ***(15,17)*** and amorphous SiO_2 ***(24)*** membranes and to investigate electric field-driven permeation of water, ions, and DNA ***(12,15–18,24)***. To generate amorphous SiO_2 nanopores, MD of heating-annealing cycles were carried out starting from the crystalline SiO_2 structure ***(24)*** (*see* **Fig. 2b**).

The method for building nanodevice structures was implemented as a plug-in in VMD. This INORGANIC STRUCTURE BUILDER plug-in can generate molecular structures for crystalline lattices of biotechnologically relevant inorganic materials such as Si, SiO_2, Si_3N_4, Au, and graphite (*see* **Fig. 2a**). Researchers can select a material and its crystalline modification from a list of structures already in the database, specify the required size and shape of the system, and set desired boundary conditions. A structure will then be created for the system

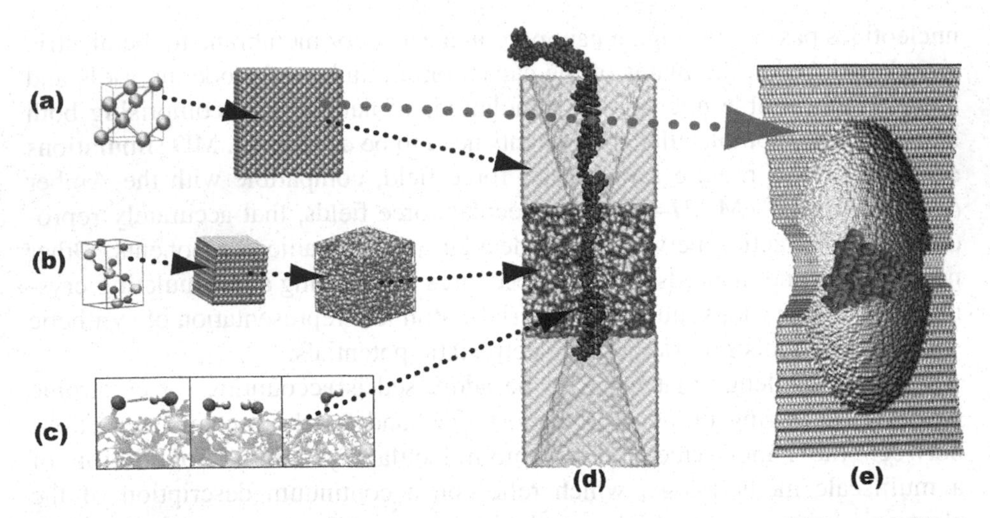

Fig. 2. Building hybrid biomolecular/silicon structures: (**a**) a crystalline polysilicon structure; (**b**) an amorphous silica structure; (**c**) the silica surface is refined and chemically modified; (**d**) a completed model of DNA in a silicon nanopore (water and ions not shown); (**e**) the same procedures can be used to build a silicon nanoreactor with a trapped biomolecule.

by replicating the crystal unit cell. To generate an amorphous SiO_2 device, a researcher can choose to perform a heating-annealing cycle or to build the structure using the blocks of amorphous SiO_2.

2.1.2. Modeling Amorphous SiO_2 Surfaces

Cutting a crystalline or amorphous structure to the desired shape often leaves unsaturated bonds at the surface. In aqueous media—or in wet air due to condensation—silanol (–SiOH) or siloxane (–SiO–) functional groups are formed at the surface of silica, with their protonation states depending on the pH of the environment. The same applies to silicon, which is covered by a SiO_2 layer when in contact with air or water. In silicon nitride (Si_3N_4), terminal primary and secondary amino groups (–NH_2 and –NH) are formed, as well as silanol groups, due to the partial oxidation of the surface. The surface protonation and the resulting net surface charge influence the adsorption of biomolecules and ions to the inorganic surface and hence are critical to modeling the nanodevice.

A ubiquitous feature of silicon-based materials is the presence of an amorphous SiO_2 layer at the surface. The SiO_2 layer coats the area of the silicon-based device that is most exposed to water and is responsible for the material's surface properties. We have implemented a method of modeling amorphous SiO_2 surfaces *(24)* by which a set of intermediate MD annealing steps is performed that mimics experimental annealing. As input, this method requires a raw nanodevice structure (constructed from an amorphous SiO_2 template) and the heating/quenching rates of the annealing cycle. To mimic the atomic arrangement of real SiO_2 surfaces, the nanodevice structure is annealed by employing the force field glassff 2.01 ***(25)***. This force field has been targeted for SiO_2 glasses and takes into account two- and three-body nonbonded interactions. The method yields surfaces that compare well to the surfaces produced experimentally. This method can be used to build silicon-based nanodevices such as nanopores ***(15)***, nanowires ***(26)***, and nanofluidics ***(27)*** systems. The SiO_2 glass force field was recently implemented in a single-processor version of NAMD.

The method described reproduces the structural features of amorphous SiO_2 surfaces with great accuracy. Furthermore, the resulting surfaces can be refined to produce other chemically modified SiO_2 forms common in experiments ***(28)***. We have developed another method to introduce hydroxyl groups at the exposed surface ***(24)***. Starting from a template structure, hydroxyl groups are generated by breaking siloxane bonds (–SiO–) at the surface. Each broken –SiO– bond produces dangling Si and O atoms, which are subsequently converted into a pair of hydroxyl groups ***(24)***, allowing the hydroxyl concentration to be precisely controlled. The method can be adapted to introduce small peptides or other organic molecules instead of the hydroxyl groups. Such surfaces grafted with organic molecules are common in biotechnology applications, such as DNA microarrays ***(29)*** and chromatographic resins ***(28)***.

2.1.3. Force-Field Development for Silicon-Based Materials

A necessary condition for MD simulations to be useful as a predictive tool in nanotechnology is the availability of robust force fields that can accurately describe interactions at the biomolecular/inorganic interface. Currently, several microscopic force fields exist for biomolecular systems and for inorganic systems separately, but not for composite biomolecular/inorganic systems. Toward the goal of developing a hybrid biomolecular/inorganic force field, we have developed two force fields parameterized to match key macroscopic properties of inorganic materials: their dielectric response in the bulk ***(15,17)*** and their hydrophobicity at the surface ***(24)***. Furthermore, the force fields developed are CHARMM compatible and can thus be deployed in conjunction with biomolecular models.

The dielectric properties of a nanodevice can influence not only the electrostatic force exerted on a nearby molecule but also the sensitivity of the device to the external electrostatic field. We have developed a straightforward method for matching the dielectric properties of inorganic materials by means of harmonic restraints applied to the atoms of the inorganic nanodevice. The magnitude of the harmonic restraints controls the displacement of the charged atoms and thereby the dielectric constant ***(15,17)***. For example, we have implemented MSXX, a

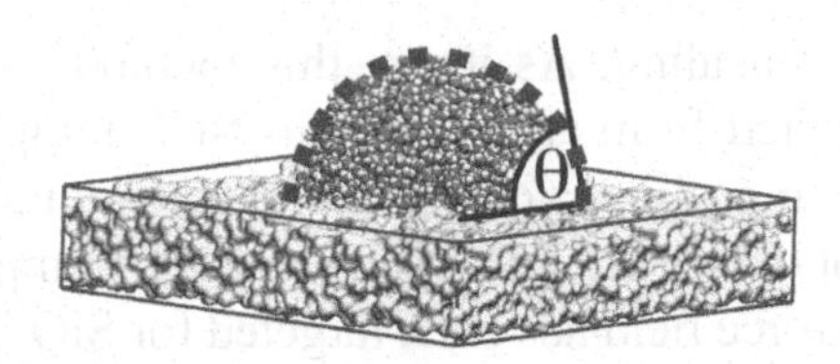

Fig. 3. Surface hydrophobicity reproduced using molecular dynamics (MD) simulations. The figure shows a water droplet resting on top of a rectangular slab of SiO_2. The intermolecular interactions of the SiO_2 slab have been tuned to reproduce the water contact angle θ observed in experiments.

force field for MD simulations of Si_3N_4 derived from ab initio calculations on small silicon nitride clusters ***(30)***. For this purpose we have brought MSXX into a CHARMM-compatible form and refined it to reproduce the experimental dielectric constant of Si_3N_4 (7.5 at 300 K). The resulting force field was used to study DNA translocation through Si_3N_4 nanopores ***(15–17,31)***.

The hydrophobicity of silicon-based surfaces is another macroscopic property critical to modeling of a nanodevice. We have implemented a method initially proposed by Werder et al. ***(32)*** to select force field parameters that reproduce surface hydrophobicity well. The hydrophobicity is quantified by measuring the water contact angle, i.e., the angle between the tangent of a water droplet and the solid surface (*see* **Fig. 3**). This method has already been employed to parameterize intermolecular interactions with amorphous SiO_2 surfaces ***(24)***.

The development of a CHARMM-compatible silicon/biomolecular force field will allow researchers to perform realistic quantitative simulations of interactions between biomolecules and silicon-based materials, including crystalline and amorphous SiO_2 and Si_3N_4. Preliminary work has already led to a force field that adequately describes silica–water interaction and correctly reproduces the experimental data on water/silica contact angles ***(24)***. Nevertheless, the force fields developed have not yet been calibrated to describe interactions with organic molecules. It is therefore necessary to extend the force fields for silicon-based compounds to accurately reproduce experimentally measured quantities characterizing the interactions between inorganic surfaces and biomolecules, such as the adhesion energy. Further development of the silicon/biomolecular force field will rely on experimental adsorption data for small organic molecules ***(33)*** and biomolecules such as DL-tryptophan ***(34)***, cholesterol ***(34)***, and polynucleotides ***(35)*** on silicon surfaces. These data will be used to calibrate the force field parameters, ensuring their accuracy for bionanotechnological applications.

2.1.4. Implicit Models of Synthetic Surfaces

Many properties of a nanodevice can be more easily represented by a mathematical function defined at the nanodevice surface than through an all-atom description. For example, if a nanodevice material is inert, a simple mathematical function can capture the steric restraints imposed by the shape of a nanodevice. Such atomless representation of inorganic objects dramatically reduces the degrees of freedom of the system, thereby increasing the speed of the simulation. To enable this kind of atomless representation to be deployed in MD simulations, we have implemented the TclBC Forces module in NAMD *(17,36)*.

Previously, arbitrary geometry-based constraints were applied through a scripting interface of NAMD (Tcl Forces) *(2)*. Such implementation required all atomic coordinates to be sent to a single processor, where the forces due to the user-defined geometry restraints would be computed, imposing a significant bottleneck on the speed of the parallel simulation. Since the force on each atom depends only on its position relative to the known boundary, NAMD does not need to exchange atom coordinates between processors, thus permitting the performance bottleneck to be removed by distributing the calculation script to all processors. One of the first applications of TclBC Forces investigated the voltage-driven translocation of DNA through an atomless model of a synthetic nanopore *(18)* (*see* **Fig. 4**) and of the α-hemolysin channel *(37)*.

2.1.5. Visualization of the Electrostatic Potential

Computational studies of nanodevices require not only the construction of atomic-scale models and force fields enabling microscopic simulations of the device, but also the implementation of new tools to analyze the data resulting from simulations. Since nonbonded interactions at the scale of several nanometers are dominated by electrostatic forces, developing a tool to analyze such forces is of paramount importance. To study the dynamics of ions and charged molecules in a synthetic nanopore *(16)*, we developed the PMEPOT plug-in for VMD, which allows one to compute electrostatic potential maps averaged over an MD trajectory.

The PMEPOT plug-in integrates the particle-mesh Ewald algorithm of NAMD with the graphical user interface of VMD, enabling the efficient calculation and a visual display of long-range electrostatics. By averaging the electrostatic potential over a 1- to 20-ns MD simulation, one can obtain an averaged electrostatic potential map of the system that can be displayed in VMD.

The PMEPOT plug-in was used to visualize the distribution of the electrostatic potential in Si_3N_4 nanopores (shown in **Fig. 5**) *(16,18)*, in the α-hemolysin channel *(37)*, in the mechanosensitive channel of small conductance (MscS) *(38)*, as well as in the potassium *(39)* and ceramide *(40)* channels.

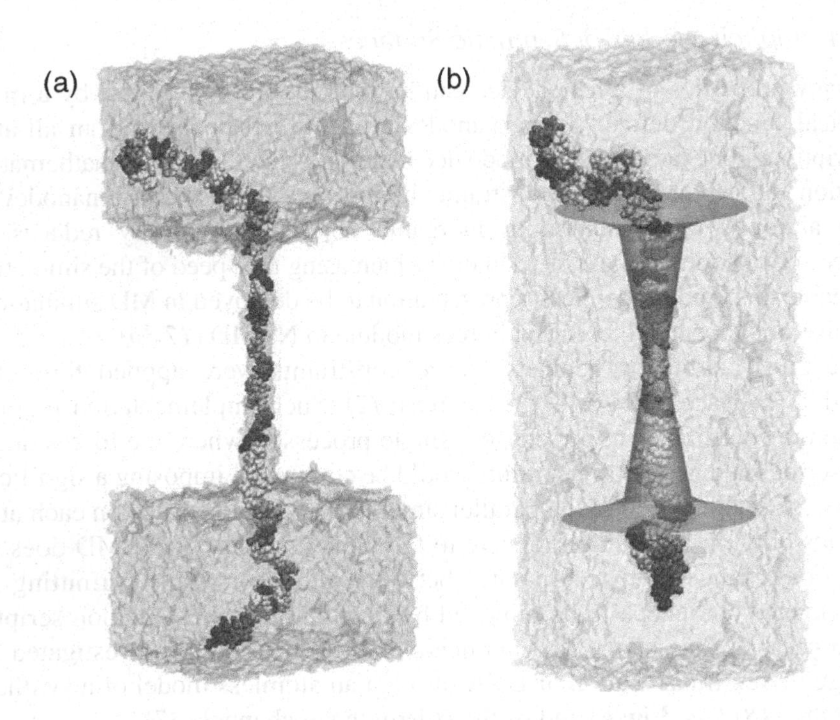

Fig. 4. Implicit model of a nanodevice. (**a**) TclBC Forces in NAMD were used to study the translocation of single-stranded DNA through a "phantom" pore *(18)*. (**b**) TclBC Forces define a nanopore boundary (conical surface) that restrains only the DNA strand, allowing water and ions to move freely *(18)*.

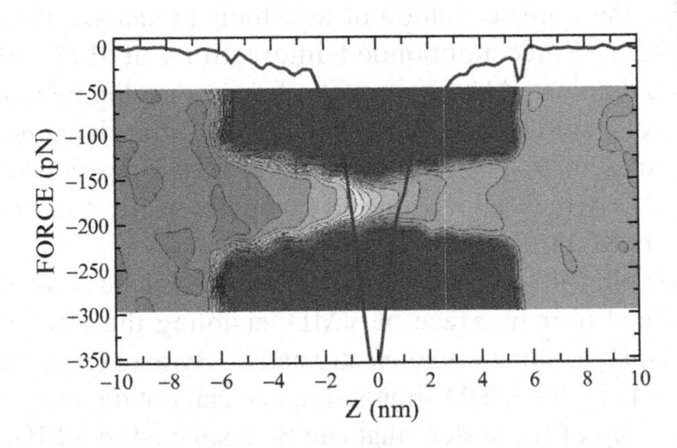

Fig. 5. Visualization of the electrostatic potential. The chart plots the electrostatic force acting on a probe charge in a nanopore (black line) calculated using the PMEPOT plug-in of VMD. The background image illustrates the distribution of the electrostatic potential inside the nanopore. The gradient of the electrostatic potential drives permeation of charged species, such as ions or DNA, through the nanopore.

2.1.6. Multiscale Modeling of DNA-Semiconductor Systems

To design a synthetic nanopore device for DNA sequencing, researchers have to establish a one-to-one correspondence between the atomic-scale details of DNA's translocation through the pore and the electrical signals measured by the device. The signal generation in the silicon domain, which consists of dielectric and semiconductor layers, cannot be described within the framework of an MD methodology. Semiconductor physics ***(41,42)*** must be used for modeling the silicon domain, while an all-atom MD methodology is appropriate for the biomolecular domain simulations. Fortunately, the characteristic times of electronic processes in the silicon domain are much shorter than the time needed to produce a noticeable change in the distribution of charges in the biomolecular domain. On the other hand, it is usually not necessary to calculate the distribution of the electric fields in the silicon domain with the same spatial resolution as in the biomolecular domain. The difference in temporal and spatial scales makes it possible to handle the two domains using different methods.

Modeling nanodevices built on semiconductors ***(13,19)*** exceeds the capabilities of MD methods and requires integration of computational electronics methods into MD. A multiscale method to describe a metal-oxide semiconductor (Si-SiO_2-Si) (**Fig. 6**) nanopore has been developed ***(13)***. In this method, the conformations of DNA are described using MD while the remaining parts of the system (i.e., the electrolytic solution and the nanopore) are described using a continuum Poisson electrostatic model ***(13,19)***. Snapshots extracted from MD simulations are input to a self-consistent Poisson solver. The solver relies on Boltzmann statistics to describe the distribution of ions in the electrolytic solution and on Fermi-Dirac statistics to describe the electrons and holes in the silicon materials ***(13)***.

The multiscale method was used to determine the feasibility of using a Si-SiO_2-Si nanopore to sequence DNA with single-base resolution. The first application of this method provided estimates for the magnitude of the electric signals produced by DNA translocation through a 1-nm diameter nanopore in a capacitor membrane ***(13,19)***. The maximum recorded change in the potential caused by the DNA translocation was estimated to be about 35 mV, the maximum voltage signal due to the DNA backbone alone to be about 30 mV, and that due to a DNA base alone to be about 8 mV. Subsequently, we demonstrated that the effect of a single base mutation on the voltage trace varied from 2 to 9 mV, which is experimentally detectable ***(19)***. We also demonstrated that the nanopores fabricated on metal-oxide semiconductor membranes can be used to accurately count the number of nucleotides in a DNA strand ***(19)***. Currently, MD simulations and Poisson electrostatics calculations are carried out sequentially using NAMD and a custom code. The simulation method does not take into account the electrostatic feedback force due to the charge redistribution in the synthetic component of the device.

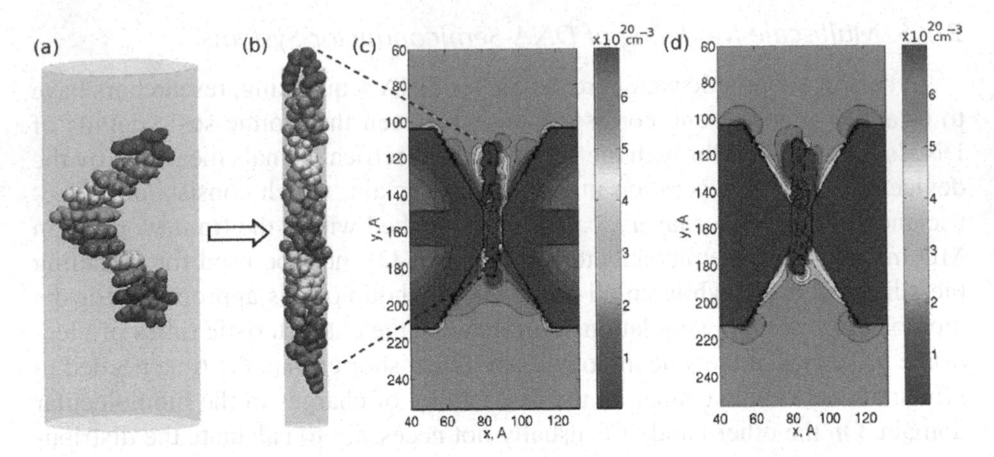

Fig. 6. Multiscale simulation of electric signal induced by DNA in a solid-state nanopore. Determined through molecular dynamics (MD), the conformation of DNA in a nanopore is related to the electrostatic potential in the semiconductor material surrounding the nanopore using a self-consistent Poisson solver *(13)*. (**a**) The conformation of a DNA strand in a phantom pore at the beginning of an MD simulation. Starting from this conformation, the phantom pore radius is gradually shrunk. (**b**) The conformation of the DNA strand in a 1-nm diameter pore. (**c**) Contour plot of negative charge concentration in the capacitor membrane and the electrolyte solution. (**d**) Same as in (**c**) for the positive charge concentration.

In the future, the feedback force will be included in the calculations. In addition to solid-state nanopores, the multiscale method can be applied to a variety of other systems, such as nanowires *(26)* and nanofluidics *(27)*.

2.2. Outlook

The rapid development of bionanotechnology has given rise to an abundance of novel devices in which biomolecules coexist and interact with silicon compounds or, more generally, inorganic nanostructures. In many devices, the electric field generated by nearby charged biomolecules is key to a device's functionality. For example, biofunctionalized thin-film silicon resistors *(43)* enable the direct detection of small peptides and proteins. Label-free electric detection of DNA hybridization is achieved by virtue of ion-sensitive field effect transistors *(44,45)*, enzyme field effect transistors *(46)*, and porous silicon sensors *(47,48)*. Amperometric biosensors use the glucose oxidase (GOx) enzyme immobilized on silicon nanowires that act as both substrate and electron-transfer mediator for high-sensitivity measurement of glucose concentrations *(49)*. Finally, a mechanical force can be directly applied to DNA and proteins using the tip of an atomic force microscope (AFM) *(50–52)*.

Further development of the computational methods and tools will enable design of new, more sensitive, and efficient bionanoelectronic devices arising from a better understanding of the correlations between the atomic-scale interactions at the device–biomolecular interface and the electronic processes in the device. Tools for atomic-scale modeling will benefit researchers seeking to optimize inorganic nanodevices used to adsorb, bind, sense, or transport biomolecules. These tools will allow one to obtain atomic-scale details of interactions between the biomolecules and inorganic surfaces in an MD simulation, thus providing insight into the processes involved. For example, a researcher will be able to simulate the adsorption of DNA probe oligonucleotides on silicon-based microarrays for DNA detection ***(53)***. Using a simulation of the interaction between an AFM tip and the surface of a virus capsid ***(50,54)*** or a liposome ***(55)***, a researcher will be able to increase the accuracy of the force measurements and AFM image resolution.

When the all-atom simulation tools are combined with the multiscale simulation tools, they can be used to study an even wider class of bionanotechnological systems. Thus, researchers can use the multiscale simulation tool to elucidate atomic-scale details of biomolecular systems based on electrical signals recorded by electronic bionanodevices and thereby optimize the nanodevices, improve their sensitivity, and extract more information from the recorded electrical signal. Further examples of the bionanodevices that can be simulated using the multiscale simulation tools, in conjunction with the all-atom simulation tools described, include biosensors based on various types of field effect transistors ***(44,45,56,57)***, amperometric glucose biosensors made of carbon nanofibers/silicon oil composites ***(58)*** or silicon nanowires ***(59)***, peptide and protein detectors using thin-film silicon resistors ***(43)***, and amperometric lactate biosensors based on CNTs immobilized on silicon/indium tin oxide substrate ***(60)***.

3. Carbon Nanotube Biosensors

Carbon nanotubes (CNTs) ***(61)*** are hexagonal lattices of carbon atoms rolled up into seamless cylinders. Due to their unique electronic, thermal, chemical, optical, and mechanical properties ***(62)***, they are of great interest for many nanotechnological applications. In particular, CNTs can be used as biosensors to detect various biomolecules ***(63)***. By decorating CNTs with biomolecules such as proteins ***(64)*** or DNA segments ***(65)***, they can be made selective to specific targets that, on binding to the decorated CNTs, alter the CNTs' optical spectra in experimentally detectable ways. Since CNTs fluoresce mainly in the near-infrared region, in which human tissues are transparent, and since decorated CNTs are easily assimilated into living cells, CNT biosensors are particularly well suited to be embedded in the human body.

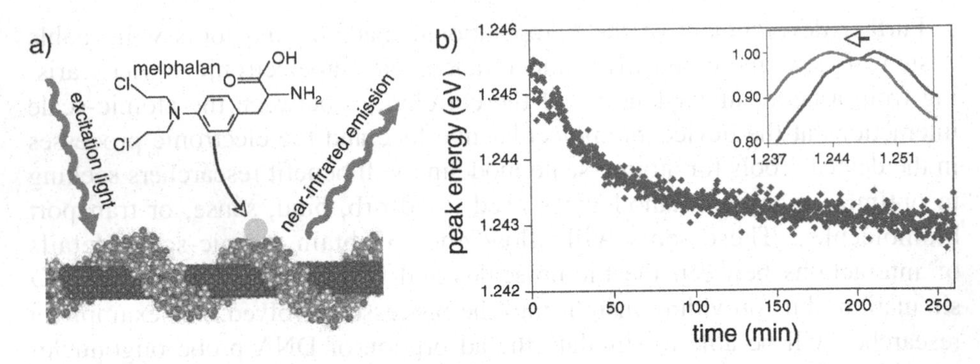

Fig. 7. Carbon nanotube (CNT)-based optical sensor. (**a**) Sensor for nucleic acid–drug interaction. A chemotherapeutic agent, melphalan (a drug used to treat ovarian and breast cancer), damages DNA wrapped around a carbon nanotube. (**b**) Energy peak of the carbon nanotube's near-infrared emission. The inset shows the emission spectra before (right curve) and after (left curve) DNA damage. DNA damage can be detected in real time. (Courtesy of the Strano group at the University of Illinois, Urbana-Champaign)

An example of ligand-specific biosensors is illustrated in **Fig. 7**. By monitoring the emission spectra of a DNA-nanotube complex, changes in the DNA's structure due to the action of a chemotherapeutic agent, melphalan, can be detected. A similar approach is effective in characterizing protein–DNA interactions, making CNT-DNA systems an attractive platform for the development of new drugs. Researchers have already demonstrated the feasibility of using nanotube–biomolecule interactions to sense various targets ***(64,65)***. However, the opportunities for rational design of nanotube biosensors are limited by insufficient understanding of nanotube–DNA–target interactions at the atomic level.

Although it has been experimentally demonstrated that CNTs can operate as optical biosensors ***(64,65)***, the coupling between changes in the biomolecular structure and the resulting shifts in the optical spectra remains poorly understood. First, the conformations of the DNA and proteins decorating the CNTs are not known. Second, the influence of the CNT chirality (the "twist" introduced in a CNT when it is rolled into a cylinder) and of the DNA sequence on the properties of the DNA-nanotube assemblies is not clear. Even less is known about the effect of the biological environment on the optical spectra of CNTs, making it difficult to interpret the spectral changes observed in experiments ***(64,65)*** in terms of the underlying interactions. Due to the difficulty of manipulating and tracking CNT sensors, the information provided by experiment is limited. Molecular modeling, however, can provide detailed information at the atomic level about the interactions between CNTs and biomolecules. It can serve as a tool for the rational design of such sensors, for example, guiding the engineering of DNA molecules to incorporate emission quenchers that can increase the signal-to-noise ratio of the sensor.

3.1. Modeling Carbon Nanotube-Biomolecular Assemblies

Many challenges currently facing research on CNT-based applications arise from a lack of adequate description of the interactions between nanotubes and their environment. Molecular modeling can provide detailed information about these interactions. For example, MD simulations have been applied to investigate the transport of water ***(66–69)***, ions ***(70)***, polymers ***(71)***, and nucleic acids ***(72,73)*** through CNTs as well as the insertion of a short single-walled CNT (SWNT) into a cell membrane ***(74,75)***. However, traditional MD studies use semiempirical force fields, which do not account for the high polarizability of the nanotube walls. *Ab initio* methods based on density functional theory, which take into account the polarizability of SWNTs, have been employed to study water and proton transport across SWNTs ***(76,77)***, but the computational cost of these methods is prohibitive for large systems such as SWNT-DNA complexes.

The simplest and most well-studied CNTs are SWNTs, which consist of one carbon cylinder. SWNTs have highly delocalized π-electrons that respond strongly to external fields ***(62)***, bringing about a polarizability that existing classical MD force fields treat inadequately. We have developed an empirical tight-binding QM model that accounts for the polarization effect of armchair SWNTs ***(78–82)***, the class of CNTs that has the highest symmetry. The tight-binding model reproduces results of more computationally expensive quantum chemistry calculations for both SWNT-water systems ***(79,80)*** and ion-SWNT systems ***(81)***.

To address the challenges in the nanotube-based applications, we have been developing a computational methodology for simulating SWNTs in complex with biomolecules. In the first series of studies, SWNTs were described as neutral cylinders with zero polarizability ***(69)***, while a classical MD force field was used to describe the interactions of the nanotubes with water. To determine the distribution of fixed partial atomic charges along a nanotube of a finite length, quantum chemistry calculations were carried out ***(80)***. Following that, a self-consistent tight-binding (SCTB) method was proposed that takes into account the polarization of an armchair nanotube in an external electric field ***(80)***. The efficacy of the polarizable model was investigated for two nanotube-water systems ***(79,80)*** and a potassium ion-nanotube complex ***(81)***.

3.1.1. Classical Model of Carbon Nanotubes

Classical MD, although not capable of accounting for the polarizability effects, is a popular method for investigating the interactions of biomolecules

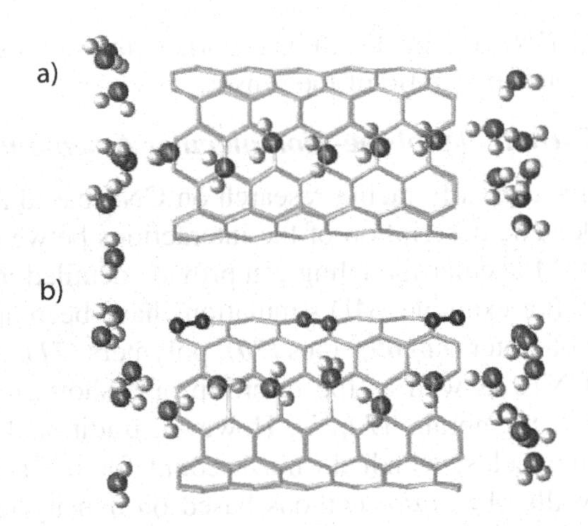

Fig. 8. Orientation of water molecules inside (**a**) a neutrally charged carbon nanotube and (**b**) a nanotube with charges at the ends. Uncharged carbon atoms are shown in licorice representation, while positively and negatively charged atoms are shown as white and black spheres, respectively.

with CNTs *(66–68,70–73,83)*. To evaluate the efficacy of this method, we have built an atomic model of an armchair SWNTs and conducted a series of classical MD simulations, investigating the transport of water and protons through the nanotube *(69)*. In these simulations, SWNTs were arranged into a hexagonally packed conformation, which prevents large movements of the SWNTs and creates a two-dimensional barrier for water and proton conduction. The simulations showed that water molecules inside a neutrally charged SWNT adopt a unipolar ordering along the SWNT axis as shown in **Fig. 8a**, and that modifications of the nanotube's atomic partial charges can induce a bipolar ordering of water molecules inside the nanotube (**Fig. 8b**). Well-organized arrangements of water molecules inside a nanotube were also found in other simulations *(66–68)* and later confirmed by neutron-scattering experiments *(84)*. Proton conduction through the SWNT was also investigated using the theory of network thermodynamics *(85–87)*, which assumes that the protons are transported through a water file connected via hydrogen bonds.

3.1.2. Electrostatics of Finite-Length Armchair Nanotubes

The atomic partial charges on nanotubes can significantly affect the interactions between nanotubes and biomolecules (especially charged molecules such as DNA). Various schemes have been adopted by the different force fields in use today to introduce such charges into calculations *(20,22,88)*. The restricted electrostatic potential (RESP) fitting scheme *(89,90)* has been used to determine the atomic partial charges on an armchair SWNT.

Table 1
RESP Charges for the 16-Å Single-Walled Carbon Nanotube (SWNT) Shown in Fig. 9

Method	H	C1	C2	C3	C4	C5	C6
RESP-α	0.1383	−0.1768	0.0327	0.0238	−0.0230	0.0065	−0.0016
RESP-β	0.1359	−0.1751	0.0441	−0.0012	−0.0012	−0.0012	−0.0012

RESP-α stands for standard restricted electrostatic potential (RESP) fitting, and RESP-β denotes a simplified RESP fitting scheme (see text for explanation).

The RESP fitting scheme employs a grid-fitting procedure that parameterizes the atomic partial charges to reproduce the electrostatic potential computed from density functional theory (RESP-α). The calculations have provided accurate values of the atomic partial charges for SWNTs shorter than 16 Å *(80)*. **Table 1** provides an example of fixed atomic partial charges for the 16-Å armchair SWNT shown in **Fig. 9**.

However, calculations based on density functional theory are too expensive for obtaining RESP charges for longer SWNTs. Since the results from RESP-α showed that atomic partial charges are nonzero mainly at the nanotube boundary (*80*,; **Table 1**), we developed a less computationally expensive RESP model (RESP-β), which constrains atomic partial charges of nonboundary carbon atoms to be identical. Despite the small number of parameters to be determined in RESP-β, the resulting atomic partial charges of SWNTs were found to be comparable with those obtained from RESP-α (*80*; **Table 1**). These charges serve as the basis for the development of a polarizable SWNT model.

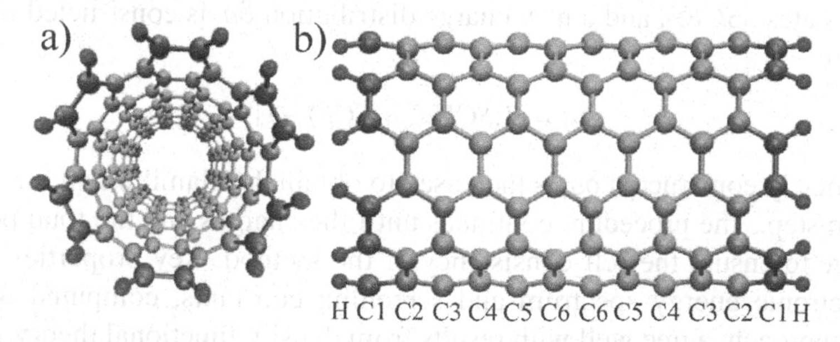

Fig. 9. A 16-Å armchair **single-walled carbon nanotube** (SWNT). (**a**) Top view and (**b**) side view. The SWNT contains 12 carbon sections (C1, C2, …, C2, C1). The SWNT ends are saturated with hydrogen atoms (H). The nanotube is colored according to the absolute values of the fixed atomic partial charges shown in **Table 1**. Dark is more charged, and faint is more neutral.

3.1.3. Polarizable Model of Armchair Nanotubes

The polarizability of SWNTs plays a critical role in SWNT–biomolecule interactions. Therefore, a polarizable SWNT model is needed to account accurately for the contribution of the SWNT electrons to intermolecular forces. Since in typical MD simulations the forces acting on simulated atoms need to be updated very frequently (e.g., at 1-fs time intervals), the model not only has to be accurate but also computationally efficient.

We have developed a semiempirical QM method, the self-consistent tight-binding (SCTB) method, to determine the induced partial charges arising from the polarizability of SWNTs at low computational cost *(78)*. In the SCTB method, the induced charge distribution is obtained by iteratively solving the single-electron tight-binding Hamiltonian,

$$\Sigma_{<i,j>}\gamma_{ij}\, a^{\dagger}_{i} a_{j} + \Sigma_{I}(U^{ext}_{i} + \Sigma_{j} U^{int}_{ij}\delta q_{j}) a^{\dagger}_{i} a_{i} \tag{1}$$

The first term on the right side of **Eq. 1** is the energy arising from the hopping of electrons between SWNT atoms. In this term, pairs $<i,j>$ include up to third-nearest neighbors; $a^{\dagger}_{i}$ and a_{i} are fermion (particle) creation and annihilation operators, respectively; and $\gamma_{i,j}$ is the probability for electrons to hop between the first-, second-, and third-nearest neighboring atoms, derived in **ref.** ***91***. The second term, U^{ext}, is the interaction energy between the atomic partial charges on the SWNT and the external field. The third term is the Coulomb interaction energy stemming from the induced charges on the SWNT atoms, with U^{int} the electron–electron Coulomb interaction potential and δq_{j} the induced charge at atom j. The induced partial charges δq_{j} are calculated through a charge refinement loop shown in **Fig. 10**. An initial charge distribution of SWNT atoms, δq_{j}, is obtained through RESP calculations; the Hamiltonian in **Eq. 1** is constructed based on δq_{j}; the Hamiltonian is diagonalized to obtain the self-consistent field ground states, $|SCF>$, and a new charge distribution δq_{j} is constructed from the equation

$$\delta q_{j} = e[\langle SCF|a^{\dagger}_{j} a_{j}|SCF\rangle - 1] \tag{2}$$

The newly constructed δq_{j} is then used to obtain the Hamiltonian for the next iteration step. The procedure continues until the charges and the total potential converge to ensure the self-consistency of the method. Key properties such as the electronic energy spectrum and screening constants, computed with the SCTB approach, agree well with results from density functional theory calculations despite the relative computational simplicity of the SCTB model *(79,80)*.

The polarizable SWNT model has been applied to three systems: an armchair SWNT with one water molecule placed on the SWNT axis *(80)*, an armchair SWNT with six water molecules inside the nanotube *(78,79)*,

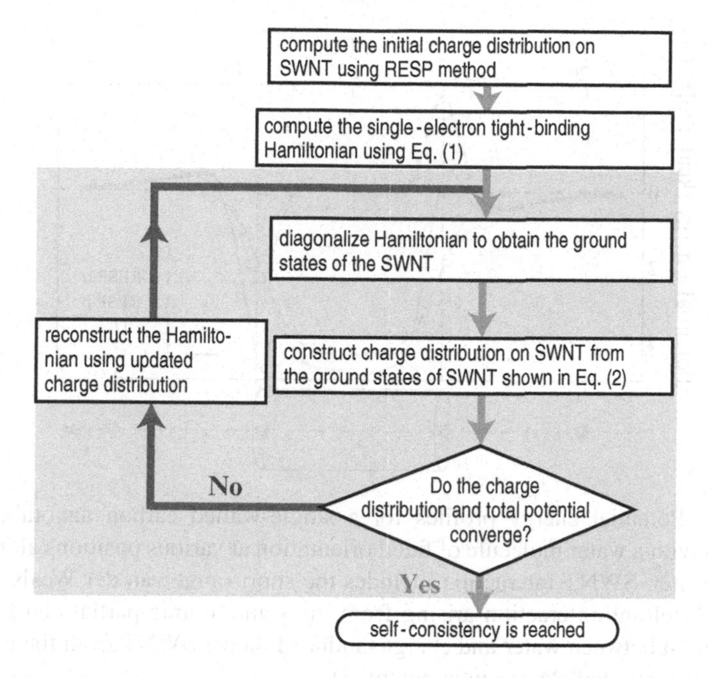

Fig. 10. Flowchart of the self-consistent calculation of induced charges and SCTB Hamiltonian *(80)*. The shaded area in gray is the atomic partial charges refinement loop. The initial charge distribution (δq_j in **Eq. 1**; see text) is taken from **restricted electrostatic potential** (RESP) calculations. The total potential matrix, the sum of the external potential U^{ext}, and the electron–electron Coulomb interaction from the non-uniform charge distribution U^{int}, is computed to construct the Hamiltonian in **Eq. 1**.

and an armchair SWNT with a potassium ion moving along the SWNT axis *(81)*. In the first study, the polarizable SWNT model was applied to compute the interaction energies between the SWNT and a water molecule at various positions along the nanotube axis *(80)*. These interaction energies, shown in **Fig. 11**, include the short-range van der Waals interaction U_{vdW}; the Coulomb interaction between the water molecule and the static atomic partial charges U_0, and the interaction between water and induced charges on the SWNT U_{ind}. One notices that U_0 is significant at the SWNT's entrance and exit, while the interaction between the water and the induced charges U_{ind} stabilizes the water molecule inside the SWNT.

In the second study, the polarizable SWNT model was applied to a system consisting of an armchair SWNT filled with six water molecules *(79)* (a conformation identical to that in **Fig. 8a**). Substantial screening of the water dipoles by the induced charges observed in this study agrees with density functional theory calculations *(76)*.

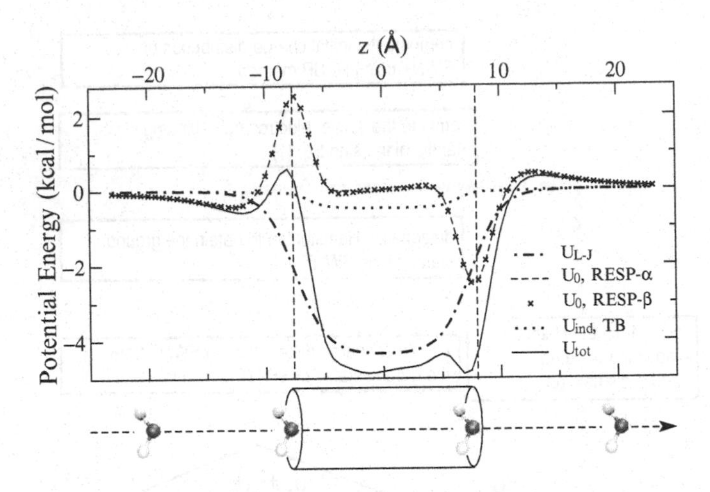

Fig. 11. Potential energy profiles for a single-walled carbon nanotube (SWNT) interacting with a water molecule of fixed orientation at various positions along the tube axis. The water–SWNT interaction includes the short-range van der Waals interaction U_{vdW}, the Coulomb interaction arising from the static atomic partial charges U_0, and the interaction between water and charges induced on the SWNT from the polarization U_{ind}. (RESP, restricted electrostatic potential).

In the third study, the polarizable SWNT model was applied to investigate the dynamics of a potassium ion inside the nanotube *(81)*. The SCTB calculations were carried out on the fly between integration steps of the MD routine. The main results obtained using the polarizable model, such as the ion's oscillation frequency, agree well with those calculated from Car-Parrinello MD *(82)*.

3.1.4. Implementation of the Polarizable Model in NAMD

To account for a nanotube's electronic degrees of freedom in MD simulations, work is in progress to integrate the polarizable SWNT model *(80,81)* into NAMD. The key aspect of such integration is to continuously update the atomic partial charges of the SWNT segment during an MD simulation, employing the method introduced in **ref.** ***81***. In the present simulation setup, the atoms of the SWNTs are held fixed, and the SWNT's π-electrons, which contribute the most to the SWNT polarizability, are treated by the QM model; the remainder of the system is described classically. In each MD integration step, the combined QM/MM module is called via a NAMD Tcl interface to diagonalize the QM Hamiltonian and thus obtain the SWNT induced charge δq_j for each atom j (see **Eq. 1**).

A problem arising from the QM treatment of SWNTs is that the QM formulation introduces an energy loss *(92)*. This energy loss arises from the (nonanalytical) dependence of the induced charge δq on the coordinates of the MM subsystem, which is not accounted for in a conventional MD algorithm. To conserve the total energy of the system, a force correction term can be introduced in the MM subsystem ***(92)***. In the simple case in which the classical subsystem consisted of a single potassium ion ***(81)***, the loss was avoided through inclusion of a term $\partial (\delta q_j)/\partial \mathbf{r}$ that accounted for the dependence of δq_j on the position $\mathbf{r}$ of the external charge (ion). With an analytical expression for $\partial (\delta q_j)/\partial \mathbf{r}$ not being available, the quantity was calculated by numerical differentiation. In the intended applications, there will be thousands of charges acting on the SWNT, making such a numerical approach unfeasible.

Since the correction force is small, the energy loss is currently compensated for by the Langevin thermostat feature of NAMD with an appropriate choice of the coupling constant. The results of such QM/MM calculations should be accurate enough to reveal, for example, the ordering of water around SWNT walls or the wrapping of DNA around nanotubes.

The studies of nanotube-water ***(79,80)*** and nanotube-ion ***(81)*** systems have validated the polarizable SWNT model ***(80,81)*** as a viable option for the accurate and efficient description of nanotubes in MD simulations. Integration of the polarizable SWNT model into NAMD yields a combined QM/MM description, which permit studies providing guidance for many practical problems related to nanotube-based technology.

3.2. Outlook

Applications of nanotube-biomolecular assemblies have been proposed in several areas of biotechnology, including biosensors, drug development platforms, and drug delivery systems. CNTs enclosed by biomolecules such as proteins ***(93–95)*** and DNA ***(96–98)*** have been shown to enter cells, suggesting applications as drug delivery devices. DNA interacts strongly with SWNTs to form a stable DNA-SWNT complex that effectively disperses and separates bundled SWNTs in an aqueous solution ***(99)***. CNTs are being used as gas sensors by measuring the change in the conductivity of CNTs when exposed to a small amount of certain gas molecules ***(63)***. Researchers are developing experimental techniques to adsorb and immobilize proteins and other biomolecules onto CNT surfaces to make chemical sensors such as immunosensors ***(100)***. A system containing a nanotube threaded through a nanopore has been devised for high-throughput DNA sequencing ***(101)***. CNT-based AFM probes have been designed for direct determination of haplotypes of DNA fragments ***(102)*** and for studying the structure of proteins ***(103)***. Researchers have been able to target and destroy cancer cells using bundles of CNTs, called *nanobombs*, by irradiating CNTs embedded inside cells with a laser beam. The nanobombs emit heat when irradiated, causing destruction of the cells ***(104,105)***. In many of the applications, computer modeling has the potential to greatly improve the rational design of CNT-biomolecular systems, image their internal organization, and predict their response to external factors.

Modeling SWNT–biomolecule interactions poses various challenges. First, electronic and optical properties of nanotubes are highly dependent on the nanotube chirality ***(62)***, thus requiring a flexible theoretical and computational approach that applies to all types of nanotubes. Second, the high polarizability of nanotubes needs to be modeled accurately and efficiently. Third, nanotube sensor applications require an efficient way to characterize the nanotube optical spectrum, taking its environment into account.

We have been focusing on one type of nanotube chirality, the armchair nanotube ***(69)***, and developed a polarizable model for it ***(78–80)***. Based on this work, we plan to integrate the polarizable SWNT model into NAMD ***(2)***; improve the efficiency of the present QM/MM polarization module, making it suitable for long simulations and large SWNT-biomolecule systems; and adapt the polarizable SWNT model to nanotubes of all chiralities. Furthermore, we seek to apply our semiempirical methods for calculating optical spectra of biological chromophores ***(106–109)*** to SWNT-biomolecule systems.

4. Nanodiscs for the Study of Membrane Proteins

Membrane proteins provide a way for cells and organelles to communicate and interact with the exterior environment. They control the passage of molecules and lipids across the cell membrane and are of great importance to human health, for example, as targets of most modern drugs ***(110–113)***. Unfortunately, membrane proteins can be difficult to study experimentally due to their tendency to aggregate in solution when removed from their native membranes. Detergent micelles and liposomes are conventionally used to study membrane proteins ***(114)***. However, the use of such systems often inactivates the proteins due to the nonnative environment.

Fortunately, nature has provided a system that can be engineered into a platform for studying membrane proteins, namely, discoidal high-density lipoproteins (HDLs) ***(115)***. Engineered forms of discoidal HDLs, called nanodiscs ***(116–119)***, are being developed as an alternative platform for embedding membrane proteins (**Fig. 12**) ***(120–134)***. Even though HDLs and, by extension, nanodiscs have been studied for decades and the implications of HDL for coronary heart disease are well documented ***(115)***, the structure of the primary protein component apolipoprotein A-I (apo A-I) and the transitions from lipid-free to nascent discoidal to mature spherical HDL particles are not well characterized. To date, only two high-resolution crystal structures ***(135,136)***, both in a lipid-free state, exist for apo A-I.

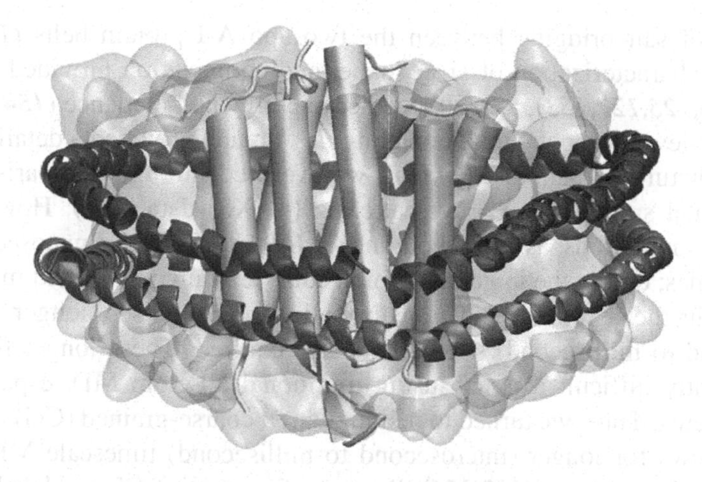

Fig. 12. Nanodisc with a bacteriorhodopsin monomer. Nanodiscs are discoidal protein-lipid systems used as platforms for embedding membrane proteins. The amphipathic membrane scaffold proteins (shown in black) wrap around a lipid bilayer (shown in transparent gray), effectively shielding the hydrophobic lipid tail groups from the aqueous environment. The bacteriorhodopsin (shown in gray) is embedded in the lipid bilayer.

Several models of the structure of discoidal HDL particles have been suggested, and although most experimental evidence points to a belt model ***(137–143)*** in which the amphipathic proteins wrap around a lipid bilayer in a belt-like manner, the issue is not yet settled. Even less is known about the formation of these discoidal protein-lipid particles, their transformation into spherical particles on the incorporation and esterification of cholesterol, or the incorporation of membrane proteins into nanodiscs. Computational modeling provides a means to design and test structural models of nanodiscs without need for costly and lengthy experiments. Such modeling should eventually reveal how nanodiscs assemble and how HDL particles form and eventually become spherical, providing dynamic information that is not generally available from experimental techniques such as nuclear magnetic resonance (NMR) and crystallography.

As a first step toward understanding the structure and assembly of nanodiscs as well as of HDL particles in general, all-atom simulations were first performed (in 1997) on the picket fence model of discoidal HDL, in which the helices of the protein are oriented parallel to the lipid acyl chains ***(144)***, using then-available experimental data ***(145–148)*** and structure prediction methods. All-atom simulations were later done (in 2005) on the double-belt model discoidal HDL, in which the protein helices are oriented perpendicular to the lipid acyl chains ***(149)***, relying on the Borhani crystal structure ***(135)***, evidence of salt bridging between the two apo A-I protein belts ***(150–152)***, extensive characterization of size, shape, and composition provided by nanodiscs ***(116,125,127,153)***, as well as other experimental evidence ***(154)***.

Atomic-level computational modeling has already provided a detailed image of the structure of nanodiscs, which was verified through comparisons with experimental small-angle X-ray scattering (SAXS) data ***(149)***. However, all-atom MD simulations, as carried out in **ref. *149***, are limited to nanosecond simulation times; due to the large macromolecular rearrangements and movements involved in the assembly and transformation of nanodiscs, longer timescale (microsend to millisecond) simulations are required. These longer timescales are currently difficult to reach using traditional all-atom MD, especially for large systems. Thus, we turned to residue-based coarse-grained (CG) modeling, which allows for longer (microsecond to millisecond) timescale MD simulations of such processes ***(155,156)***. Recent refinement of CG models has led to the development of CG lipid-water models that have been successfully used to investigate and assemble various lipid aggregates ***(157–162)***. The CG models proved to be useful in describing the assembly and dynamics of systems driven by mainly hydrophobic/hydrophilic interactions, including bilayers, micelles, and liposomes. Due to the amphipathic structure of membrane scaffold proteins and the fact that nanodisc assembly is driven by hydrophobic interactions, the behavior of lipoprotein systems will be primarily influenced by factors similar to those important in pure lipid assembly and, thus the systems are ideally suited for CG methods.

4.1. Computational Approach to Engineering Nanodiscs

To study the nanodisc system using CG methods, we needed to develop a number of methods and tools for constructing, parameterizing, simulating, and visualizing CG systems. Some of the tools were implemented in the visualization and simulation software VMD and NAMD, respectively. The CG residue-based model for nanodiscs required the implementation of the CG protein-lipid model into NAMD ***(155)***, the extraction of protein backbone parameters from all-atom simulations ***(156)***, and development of reverse CG methods to reconstruct an all-atom model from a CG description.

4.1.1. Imaging Nanodiscs Using All-Atom MD Simulations

All-atom MD simulations can provide the most detailed picture of nanodevices such as nanodiscs. We performed all-atom MD simulations on nanodisc particles in which the membrane scaffold proteins were wrapped around a nanometer-size lipid-bilayer in a belt-like manner, producing a detailed atomic image of a nanodisc ***(149)***. The aim was to test if the double-belt model nanodiscs were stable at least over the nanosecond simulation period, which indeed they were ***(149)***.

Comparisons with experimentally obtained small angle X-ray scattering (SAXS) patterns yielded further validation of the double-belt model ***(149)***.

Once a detailed atomic-level image of a nanodisc had been produced ***(149)***, we turned our attention toward understanding the process by which nanodiscs self-assemble. A detailed understanding of the nanodisc assembly process will permit the optimization of experimental nanodisc assembly procedures but could also lead to valuable information on nanodisc structure since the aggregation process self-selects a structure, eliminating bias. As stated, the nanodisc assembly process occurs on a much longer timescale than is attainable for all-atom MD, so we chose to use a CG modeling technique.

4.1.2. Coarse-Grained Protein-Lipid Model in NAMD

The CG models simplify the description of a system, reducing the overall system size by mapping clusters of atoms onto CG beads. Recent advances in CG modeling have resulted in CG lipid models that allow the study of lipid assembly on a micrometer length scale and on a millisecond timescale. In addition to the reduction in system size, CG models use a combination of short-range electrostatics and a minimal number of bead types to improve computational efficiency ***(157–159,162,163)***.

One of the first CG lipid models to successfully show assembly of lipids was developed by Klein and coworkers ***(158,159)***. The CG model was parameterized to mimic structural features resulting from all-atom simulations of DMPC (**dimyristoylphosphatidylcholine**) bilayers. The successful assembly of lamellar and inverted hexagonal phases of DMPC from an initially random distribution of lipids demonstrated the ability of the model to describe the dynamics of lipid assembly. Recent refinements and extensions to the CG model have led to the description of transmembrane-peptide-induced lipid sorting ***(164)***, lipid perturbation due to hydrophobic mismatch surrounding a nanotube ***(165)***, insertion of a model pore into a lipid bilayer ***(75)***, and the effects of anesthetics on model membranes ***(166)***. Other models that use a similar CG method have recently been developed by Marrink ***(157,167–171)***, Stevens ***(160,163)***, and Hilbers ***(162,172)*** and their coworkers.

The CG lipid models proved to be useful in describing the assembly and dynamics of systems driven mainly by hydrophobic/hydrophilic interactions, including bilayers, micelles, and liposomes. Due to the amphipathic structure of membrane scaffold proteins and the interaction with lipids in nanodiscs, the behavior of lipoprotein systems is primarily influenced by factors similar to those important in pure lipid assembly, making the systems ideally suited for CG methods. Therefore, we implemented in NAMD ***(2)*** the Marrink CG lipid model ***(157)*** developed originally for use with the MD program GROMACS ***(173)***.

The Marrink CG model was initially chosen due to its simple design, requiring only minimal CG bead types, a discrete set of nonbonded interactions, and use of potential energy functions usable in NAMD. Groups of atoms are represented by point-like beads, interacting through effective potentials. On average, the mapping is about 10 atoms per CG bead. Despite the similarities in potential energy functions, minor modifications to NAMD were required to allow for the use of the cosine-based angle potential energy term conventionally employed in GROMACS, as opposed to the harmonic angle term employed in NAMD. A series of lipid bilayer and micelle assembly simulations were done to verify that NAMD's implementation produced results comparable to Marrink's CG simulations on GROMACS ***(155,157)***. The implementation of Marrink's CG lipid model also required the conversion of GROMACS-style parameter and topology files to CHARMM-style files, which are accepted by NAMD. The CG MD simulations proved to be up to three orders of magnitude faster than equivalent all-atom simulations ***(155)***.

After successfully implementing Marrink's CG lipid model into NAMD, we extended it to also include proteins to be able to simulate nanodiscs ***(155,156)***. The residue-based CG model for proteins was created by assigning a set of two CG beads, a backbone bead and a side-chain bead, to each amino acid residue (except for glycine, which only has a backbone bead and no side-chain bead). All the CG backbone beads were of identical type (i.e., a neutral bead with hydrogen bond donor and acceptor properties), while the side chain beads were defined based on the general properties of the corresponding amino acid. Nonbonded interactions used the same finite set of interaction energies as in the original CG lipid model (i.e., attractive, semiattractive, intermediate, semi-repulsive, and repulsive interaction energies) ***(157)***. The equilibrium bond and angle values for the CG protein beads were extracted from all-atom simulations of preformed nanodiscs ***(149)***.

Using the newly developed CG model, we were able to simulate nanodisc stability and assembly. In simulations of preformed nanodiscs (in the double-belt model), overall structural features (known from experiments; ***116,127***) at various temperatures were reproduced and remained stable ***(155)***. Assembly simulations revealed the formation of discoidal particles, but the secondary structure of the scaffold protein was quite disordered, in disagreement with experimental evidence suggesting that the secondary structure of apo A-I is primarily helical ***(174,175)***. Therefore, to improve on the CG protein force field, an inverse Boltzmann technique was used to extract the parameters for the protein dihedrals from all-atom simulations to better represent the protein's secondary structure ***(156)***.

The membrane scaffold proteins in nanodiscs are known to be primarily helical ***(174,175)***, for which reason it was important that the residue-based CG protein model was able to reproduce an α-helical secondary structure well. Therefore, a dihedral force was applied to the protein backbone beads to maintain an improved helical secondary structure. We implemented a Boltzmann inversion procedure ***(176–180)***, allowing for the extraction of the force constant for the protein backbone dihedral from all-atom simulations ***(156)***. In the Boltzmann inversion procedure, one assumes that the dynamics of the system is governed by a potential $V(x)$, where x represents the degrees of freedom under consideration, $V(x)$ being chosen to reproduce a sampled equilibrium probability distribution $p(x)$ of configurations over x according to $p(x) = Z^{-1}e^{-V(x)/(k_BT)}$ (k is the Boltzmann constant, T is the temperature, and Z is the partition function). By running an all-atom simulation at constant temperature, one can determine the probability distribution of configurations over x. The inversion of this probability distribution according to $V(x) = -k_BT \ln[p(x)] + const$ provides the shape of the effective potential. The resulting dihedral potential lead them indeed to a well ordered double belt disk as demonstrated now.

4.1.3. Coarse-Graining Tools

To effectively use the developed CG protein-lipid model, a variety of scripts were written and released with VMD 1.8.5. The scripts collectively are called CGTools and allow for the setup of a CG system by converting an all-atom model to a CG model (**Fig. 13**). The conversion is done by placing a CG bead at the center of mass of the group of atoms it represents. CGTools also assigns the correct CG bead type depending on the amino acid or lipid group being

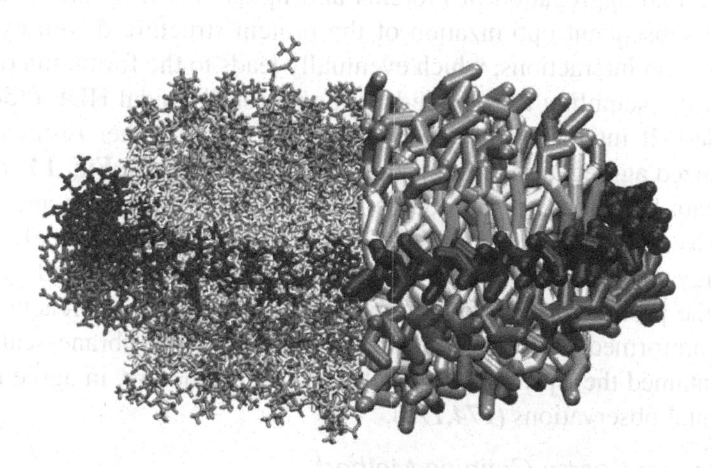

Fig. 13. Nanodisc: all-atom versus coarse-grained (CG) description. The CGTools released with VMD 1.8.5 allows for the conversion of an all-atom model to a CG model. An all-atom representation is shown on the left and the converted CG model on the right.

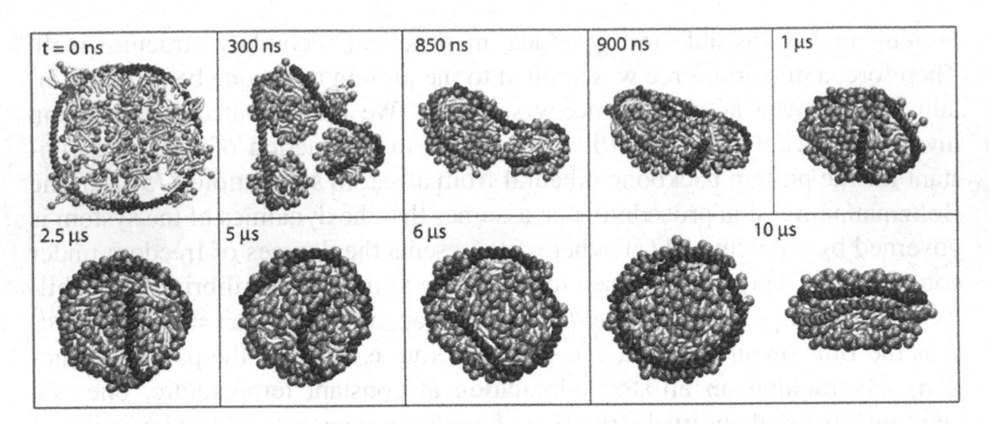

Fig. 14. Self-assembly of a nanodisc. The coarse-grained protein-lipid model developed and implemented in NAMD was used to assemble nanodiscs from an initially random distribution of lipids around two membrane scaffold proteins. Note the long simulation time of 10 μs.

replaced. In addition to successfully converting an all-atom structure to a CG structure, CGTools can also solvate and ionize the CG system.

Molecular dynamics simulations using the residue-based CG protein-lipid model ***(155,156)*** were performed to study the assembly of nanodiscs. A resulting 10-μs simulation (**Fig. 14**) ***(156)*** showed that assembly proceeds in two steps: an initial aggregation of proteins and lipids driven by the hydrophobic effect and subsequent optimization of the protein structure driven by specific protein–protein interactions, which eventually leads to the formation of a nanodisc particle resembling the double-belt model for discoidal HDL ***(156)***. Since the double-belt model was not assumed *a priori* but rather resulted from a self-organized aggregation process, the simulation shown in **Fig. 14** constitutes an important validation of the double-belt model for nanodiscs and discoidal high-density lipoproteins. Analysis of the assembly process revealed that the total energy and solvent-accessible surface area both converged after 10 μs and that the protein–protein interaction energies were slowly reaching values found in preformed nanodisc simulations ***(156)***. The membrane scaffold proteins maintained their primarily helical secondary structure, in agreement with experimental observations ***(174,175)***.

4.1.4. Reverse Coarse-Graining Method

It is desirable to reconstruct an all-atom model from a CG model to allow for atomic-level details to be captured. Such a reverse CG method, in which an all-atom structure is obtained, is often also needed to interpret experimental

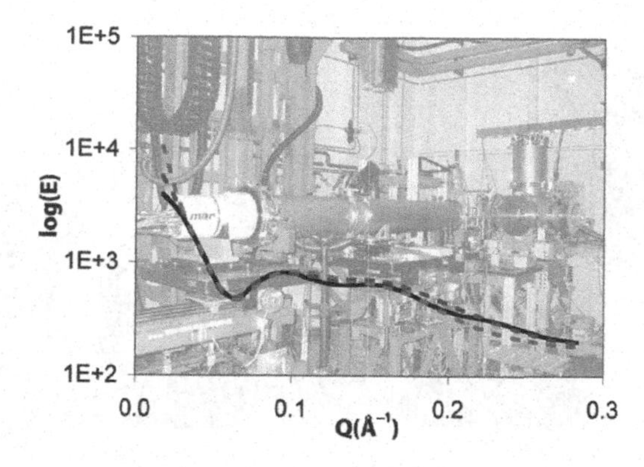

Fig. 15. Comparison of experimental and theoretical **small-angle X-ray scattering** (SAXS) curves for Nanodiscs. Experimentally measured curves of **dipalmitoylphosphatidylcholine** (DPPC) and **dimyristoylphosphatidylcholine** (DMPC) nanodiscs are shown as dashed lines. The black line represents the calculated SAXS curve obtained from coarse-grained simulations of a nanodisc.

data. We developed a procedure to systematically recast a CG system into an all-atom model *(181)*. The recasting is done by mapping the center of mass of the group of atoms represented by a CG bead to the bead's location. The resulting all-atom system is then energy minimized and equilibrated with the center of mass of the atom groups restrained to generate a realistic all-atom system. The reverse CG methodology was initially used to calculate theoretical SAXS curves (**Fig. 15**) for comparison with those measured experimentally for investigating the disassembly of nanodiscs with cholate *(181)*.

4.2. Outlook

The methods described here should permit the study of the discoidal-to-spherical transition of HDL, provide guidance in redesigning membrane scaffold proteins, and make possible the optimization of nanodisc assembly. Moreover, the development of CG models of synthetic materials together with the residue-based CG protein-lipid model *(155,156)* can be employed to investigate a variety of biotechnology applications and devices, such as the use of micelles and liposomes to solubilize membrane proteins *(114)* and to deliver drugs and genetic materials *(182–187)*. As delivery vehicles, micelles and liposomes need to fuse with biological membranes to perform their desired function. The fusion process is driven by the same hydrophobic/hydrophilic interactions operative in nanodisc assembly.

Supported bilayers, also used to study membrane proteins *(188)*, likewise could be coarse-grained with models developed for nanodiscs. In addition, two

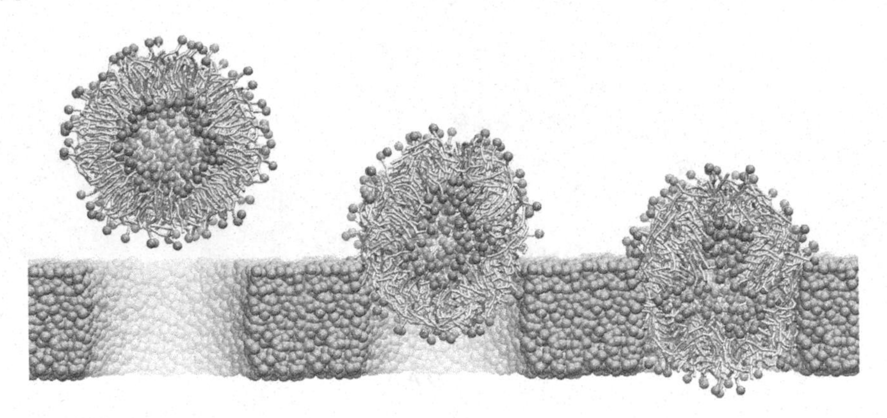

Fig. 16. Coarse-grained simulation of the assembly of a liposome with a silicon nanopore.

of the other systems described in this work, nanotubes and silica nanopores, could benefit from CG simulations. Nanotubes are being developed for use as optical sensors and targeted chemotherapeutic drug delivery vehicles. Both of these target biotechnology applications require that the nanotube be inserted into a membrane, a process ideally suited to CG modeling. A CG silica model, to be developed in order to describe the insertion of a liposome containing a membrane protein into a silica nanopore (**Fig. 16**), can be used in conjunction with CG models for proteins, lipids, and detergents to describe a variety of biotechnology applications, such as nanofluidics ***(189)***, nanoelectronics ***(190)***, nanostructures ***(191)***, and DNA microarray technologies ***(29)***.

Beyond biotechnology applications, the CG models developed here can also be used for applications in which long timescales and large systems need to be simulated. For example, our residue-based CG model is currently in use to investigate the formation of tubular membranes from flat membranes by means of N-BAR domains ***(192)***, proteins that reshape cellular membrane.

5. Design of an O_2-Tolerant Hydrogenase

With the world's oil reserves dwindling, the development of a viable new alternative energy fuel has become an urgent priority. Hydrogen gas (H_2), a renewable resource that boasts zero pollution, is a promising alternative to gasoline. One method of producing H_2 under development is by means of the unicellular green algae *Chlamydomonas reinhardtii*. Because *Chlamydomonas* has the natural ability to couple photosynthetic water oxidation to the generation of H_2 through the hydrogenase enzyme ***(193,194)*** (*see* **Fig. 17**), it could potentially be used to produce H_2 commercially ***(195–197)***. Such a means of H_2

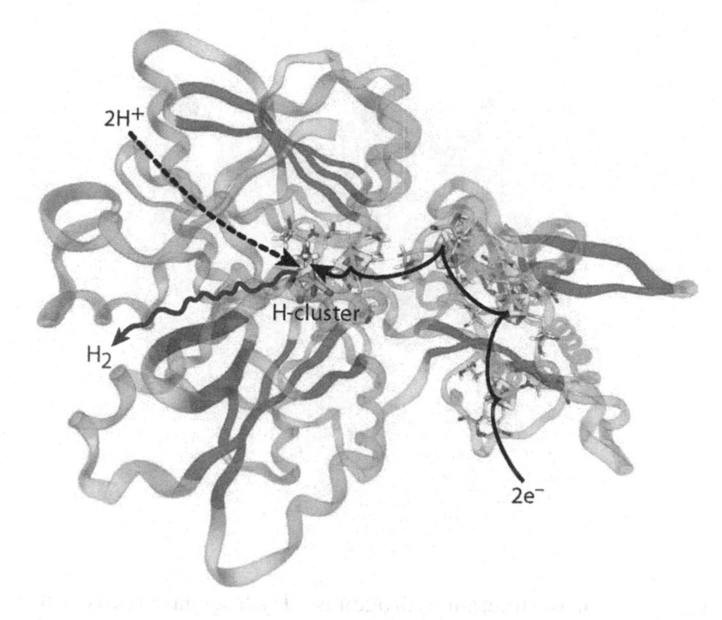

Fig. 17. The H_2 production reaction. The hydrogenase enzyme (shown in ribbons) generates H_2 from hydrogen ions (H^+) and electrons (e^-) coming from photosynthetic water oxidation. O_2 inactivates this reaction by binding to the H-cluster's active site irreversibly.

production would be affordable and efficient, requiring only water and sunlight, with up to 10% of incident sunlight energy converted.

While *Chlamydomonas* microalgae hold promise as a source of H_2, there remains one major hurdle preventing their widespread use. A single O_2 molecule can irreversibly bind to the hydrogenase's buried active site, inhibiting its activity and leading to enzyme degradation. For this reason, current H_2 production methods using microalgae must operate under anaerobic conditions, an expensive and impractical proposition. To make microalgae hydrogen production practical, work is being conducted that aims to engineer hydrogenase to prevent O_2 from making its way to the H_2 production site of the enzyme (as illustrated in **Fig. 18**), thus improving its resistance to O_2. Specifically, hydrogenases from the *Chlamydomonas* green algae, as well as an already more O_2-tolerant hydrogenase of known atomic structure from *Clostridium pasteurianum* ***(198)***, are being mutated and tested for their O_2 sensitivity. Incorporating the engineered hydrogenase back into *Chlamydomonas* would create a new strain of microalgae that could efficiently produce H_2 in open air, drastically reducing the cost of H_2 production.

Several studies suggested that the protein matrix plays an important role in O_2 accessibility inside proteins ***(199,200)***. To find the mutations that confer O_2

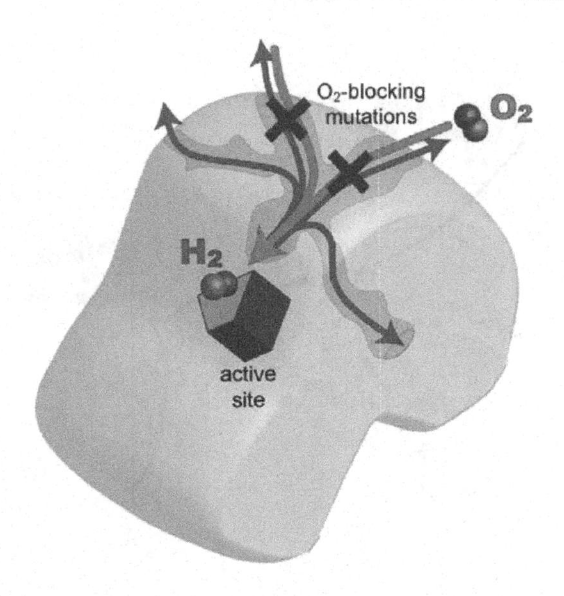

Fig. 18. Designing an O_2-tolerant hydrogenase. Hydrogenase is easily deactivated by O_2 molecules that reach its active site by permeating across the protein matrix. Since it has been found that H_2, due to its smaller size, can travel through more pathways than O_2, it should be possible to block O_2 by the selective mutation of O_2-pathway-lining residues while still allowing the H_2 product to exit the protein.

tolerance to hydrogenase, the entry and transport mechanisms of O_2 inside the protein matrix must be well understood. Furthermore, if the mutations are to block existing O_2 pathways, then the locations of these pathways, the energy barriers encountered by O_2, and the kinetic transport rate must also be reliably measured. Once the effect of a given set of mutations and, most importantly, the reasons for its success or failure are well understood, an optimized mutation strategy can be developed and followed. Specific information about the role of mutations, beyond their overall effect on macroscopic rates, is not accessible experimentally, and simulation is needed to understand the atomic-level mechanism that ultimately makes the mutations successful.

5.1. Methods for Probing Protein Gas Migration

Computer-assisted protein engineering is an important and challenging branch of biotechnology for which there are no ready-made solutions. Designing a hydrogenase that is O_2-tolerant requires, at the very least, the ability to accurately characterize O_2 pathways inside proteins. However, despite extensive research on specific aspects of gas migration in proteins, such as the myoglobin geminate recombination process ***(201–205)*** and numerous MD simulations targeting gas migration in various proteins ***(206–210)***, the information collected by the scientific community on gas conduction properties has mostly been of a narrow scope and is too limited for finding O_2 pathways in hydrogenase. Preliminary work has led to substantial advances in mapping gas pathways by means of the temperature-controlled locally enhanced sampling (LES) method ***(211,212)***, maximum volumetric solvent-accessible maps ***(212)***, and implicit ligand sampling ***(213)***.

5.1.1. Temperature-Controlled Locally Enhanced Sampling

The first step in improving hydrogenase's O_2 tolerance is to find the location of its O_2 and H_2 migration pathways. Because gas diffusion inside proteins is a random process involving the exploration of a constantly fluctuating protein matrix by small ligands, simulating gas permeation events one gas molecule at a time is very inefficient and provides poor statistics. A traditional method for finding gas migration pathways is to significantly increase the sampling of pathways using the LES method. With LES, additional copies of the gas ligand are created that are invisible to each other, yet every copy interacts simultaneously with a unique copy of the protein ***(209)***. This allows researchers to obtain statistics in a single simulation run that would otherwise require dozens to thousands of separate runs. While searching for gas migration pathways inside hydrogenase, it was found that the traditional implementation of LES produced results that were inconsistent with those from non-LES simulations. Notably, in LES simulations, the replicated particles experience artificially low energy barriers ***(214)*** and, consequently, follow unphysical or unrealistic pathways inside the protein.

To correct for this inadequacy, a modification of the algorithm, called temperature-controlled LES (TC-LES) ***(212)***, was developed. TC-LES scales the mass and Langevin temperature of the replicated particles in such a way that it offsets the artifacts introduced by the standard LES algorithm. Because TC-LES adjusts for errors introduced by having large numbers of ligand copies, TC-LES simulations can use over a thousand LES replicas instead of the previous limit of about a dozen. When applied to hydrogenase, TC-LES led to the discovery of previously unknown O_2 pathways inside hydrogenase ***(211,212,215)*** that could not be detected from solely looking at the protein's static structure. Especially intriguing was the discovery of two completely distinct modes of transport for both O_2 and H_2 inside hydrogenase ***(212)***. While H_2 placed inside the protein diffused radially outward as it would in a roughly uniform medium, every replicated LES O_2 molecule moved in unison, and in spurts, along well-defined and repeatable pathways. The motion of the O_2 replicas suggests that O_2 migrates in hydrogenase by following pathways consisting of series of transiently forming cavities in certain well-defined areas of the protein. The discovery of the O_2 permeation mechanism in hydrogenase was important in that it led to the development of more advanced gas pathway mapping techniques, such as maximum volumetric solvent-accessible maps and the implicit ligand sampling method described in **Subheading 5.1.3**.

5.1.2. Volumetric Solvent-Accessible Maps

The TC-LES method described only finds a subset of all the pathways, determined by both the initial gas positions and by the specific random events occurring in the simulation. This situation is not acceptable for protein design, for which one requires an easily automated procedure that will reliably identify *all* the pathways in a given protein or mutant. To address the challenge of locating all pathways, the maximum volumetric solvent-accessible map method was developed (*see* **Fig. 19**). Since gas molecules migrate inside proteins by taking advantage of transiently forming cavities, gas pathways can be mapped, in principle, by looking for such cavities as they occur when there is no gas in the protein. This is precisely what is accomplished by the volumetric solvent-accessible maps.

In the method, a short (1- to 5-ns) simulation of the studied protein is performed in the absence of any gas ligand. Then, for each step of the simulation trajectory, every internal cavity inside the protein matrix is found based on a triangulation of the protein's instantaneous coordinates. Finally, for every trajectory step, a three-dimensional (3D) volumetric map is created that contains, at each grid point, the radius of the largest sphere that contains the point and does not intersect any of the protein's atoms ***(212)***. By using a volumetric approach, it becomes possible to combine the maps from different steps of a

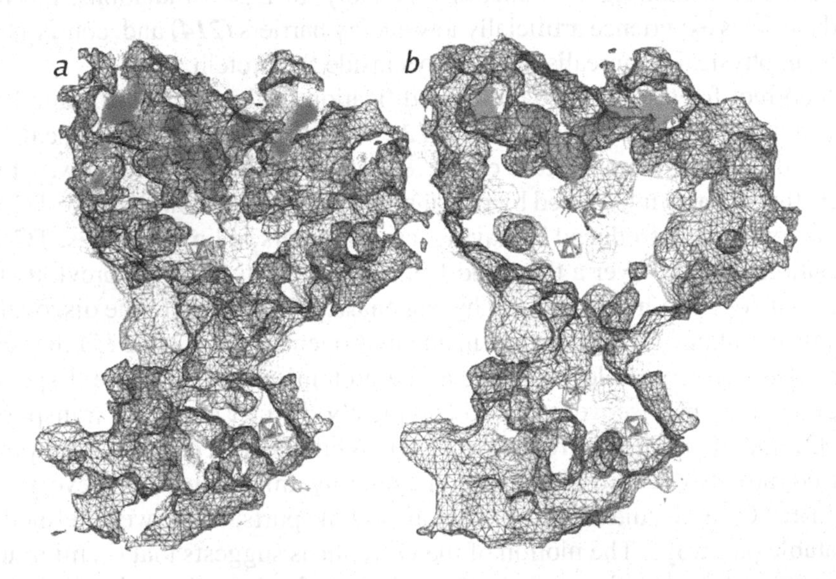

Fig. 19. Maximum volumetric solvent-accessible maps of hydrogenase. Maximum volumetric solvent-accessible maps are displayed as black wire frame for (**a**) H_2 and (**b**) O_2 in hydrogenase. The positions of the H_2 and O_2 ligands from the **temperature-controlled locally enhanced sampling** (TC-LES) simulations are displayed as a cloud of dots to highlight the excellent match between the two approaches.

simulation trajectory and create a map for an entire trajectory by averaging the maps or by keeping the minimum or maximum values at each grid point. While the maps generated from individual trajectory steps typically do not reveal the location of gas pathways any more than what is typically gleaned from a traditional static cavity search, by combining the maps of an entire protein trajectory and storing the maximum cavity size that occurs over time, the location of gas migration pathways can be found.

A detailed comparison of the gas migration pathways in hydrogenase using both TC-LES and the maximum volumetric solvent-accessible maps revealed an excellent match between the two methods ***(212)***. The volumetric approach, however, requires fewer computational resources and will find all the pathways in the protein as opposed to just a subset of the pathways.

5.1.3. Implicit Ligand Sampling

The maximum volumetric solvent-accessible map method does not provide all the data that is needed for a complete description of the gas migration pathways. Missing from this method is the ability to quantify the free-energy barriers experienced by migrating gas molecules. For this purpose, a new method, called implicit ligand sampling ***(213)***, was implemented. The implicit ligand sampling method relies on the fact that small gas ligands interact weakly with the host protein; therefore, their effect on the protein can be treated as a perturbation to the protein's equilibrium dynamics in the absence of the ligands. As a result, the implicit ligand-sampling approach only requires a 5- to 10-ns equilibrium simulation of the protein to infer all the pathways and energy barriers for any small gas molecule migrating inside it.

By using an energy perturbation scheme related to the commonly used "free energy perturbation" technique ***(216,217)***, "ghost" gas molecules, arranged in a regularly spaced grid, are placed inside the protein for each trajectory step of the equilibrium simulation. For each grid point and time step, the interaction energies between the inserted gas ligand and the protein coordinates are computed using many conformations of the ligand. From the distribution of measured energies, the potential of mean force (PMF) (i.e., the free energy of inserting a gas ligand at a specific point) can be computed at every point in space, creating a complete 3D PMF map. From this map, all the gas migration pathways inside the protein as well as the free-energy profile experienced by a gas molecule traveling along them can be recovered. The O_2 PMF map for sperm whale myoglobin, computed using implicit ligand sampling, is shown in **Fig. 20**. Since this method is applied to precomputed trajectories of the protein, individual PMF maps for different gas molecules such as O_2, carbon monoxide, nitric oxide, and xenon can be computed using few additional computational resources. Implicit ligand sampling has already been applied to reproduce

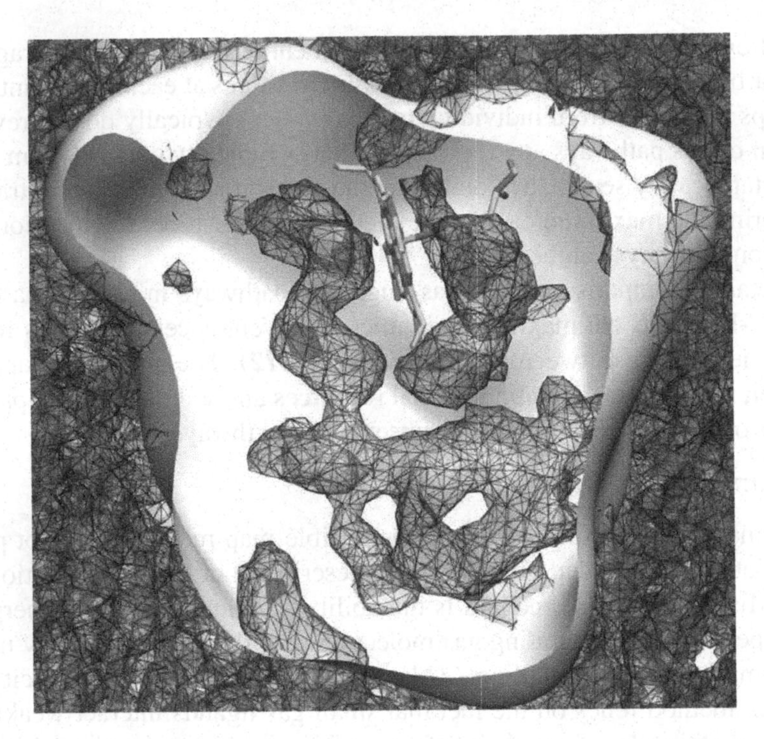

Fig. 20. O_2 **potential of mean force** (PMF) map of myoglobin. Two isosurfaces of the PMF for O_2 molecules, computed using implicit ligand sampling, are displayed inside sperm whale myoglobin. Areas that are relatively easily accessible to O_2 and serve as migration pathways are indicated by a wire-framemesh, while regions of the protein that are very favorable to O_2 and serve as holding pockets are shown in darker regions of the mesh.

experimental xenon-binding site locations ***(213,218)***, to map O_2 and carbon monoxide pathways inside sperm whale myoglobin ***(213)*** and other globins ***(219)***, to find O_2 permeation channels across AQP1 aquaporin ***(220)***, and to describe the O_2 pathways inside hydrogenase ***(221,222)***.

5.1.4. High-Throughput Simulation and Automation

Designing a protein to exhibit some desired property, or a nanodevice to work as a particular sensor, requires the testing of many variations to find the one that functions best. For example, the design of a hydrogenase mutant that is O_2 tolerant requires the screening and testing of a large number of mutant structures. Currently, a major obstacle to performing such screening is the relatively small number of simulations that can be easily set up, submitted, monitored, and analyzed concurrently without the aid of tools to automate the process. Each task is simple enough to be reliably performed by software, yet has a large enough number of discrete steps to be time consuming, distracting, and error prone for a human researcher. Since the tasks of creating, running, and analyzing results from a simulation require a significant time overhead, especially when performed on remote computer systems, the manual management required to maintain a large number of simultaneous simulations can quickly become prohibitive, resulting in unnecessarily long delays in obtaining simulation results. To enable studies requiring a large number of simulations in the future, a grid-based automation engine for NAMD simulations called NAMD-G *(223)* has been developed collaboratively with the National Center for Supercomputer Applications (NCSA).

Computer grids are collections of geographically separated computers linked by high-speed networks that can be treated in many respects as a single large computer. An internal version of NAMD-G already accomplishes several initial objectives regarding simulation automation *(223)*. Once invoked, NAMD-G submits a series of linked NAMD-G simulations to a grid-enabled remote supercomputer. NAMD-G then performs the steps needed to bring the simulation to completion, shown schematically in **Fig. 21**. NAMD-G handles all necessary authentication, so that the user does not need to keep track of log-ins and passwords for remote supercomputers and local clusters. Authentication is accomplished using the certificate-based Globus tool kit (http://www.globus.org), developed as part of the National Science Foundation (NSF) TeraGrid project, which provides a unified interface for high-performance computational resources. NAMD-G also performs all file transfer tasks automatically, including transferring the initial data to the remote computers, backing up the simulation's output to remote mass storage, and retrieving the output onto the user's local machine for subsequent analysis. Simulations almost always involve many steps, such as preequilibration, equilibration, and a production run, each of which may require the conditional submission of many remote jobs due to time limits imposed on typical supercomputer jobs. NAMD-G takes care of the dependencies between the various simulation steps, and on completing each job, it begins the next appropriate task in the simulation. Throughout the process, NAMD-G keeps the user informed regarding simulation progress and job failure through e-mail messages and status-monitoring commands. Since NAMD-G can monitor and fully automate a simulation from start to finish, it can be used as a foundation for future applications requiring arbitrarily large numbers of simulations (**Fig. 22**) *(223)*.

NAMD-G has already been employed in the study of the effect of mutations on hydrogenase's O_2 tolerance *(221,222)*. In addition, NAMD-G has also been used in the characterization of the gas migration pathways for a dozen monomeric globins, including various myoglobin *(213)*, invertebrate hemoglobins, and leghemoglobins *(219)*. For each globin, a 10-ns simulation was submitted using NAMD-G and then analyzed using the implicit ligand-sampling analysis described (shown in **Fig. 23**). Despite their very conserved secondary structures,

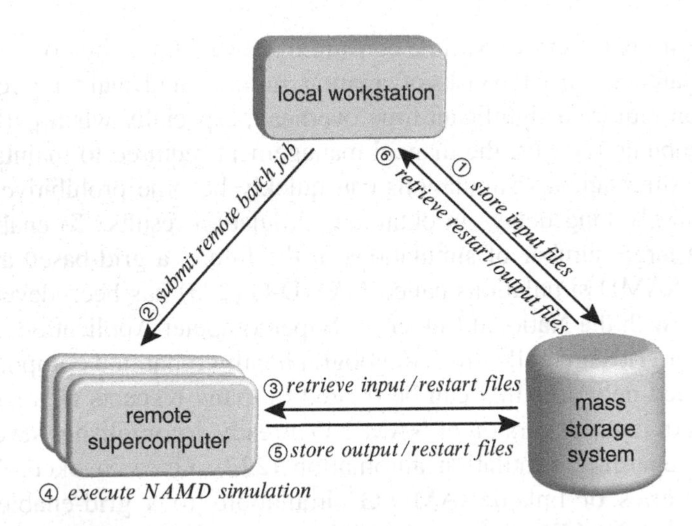

Fig. 21. NAMD-G overview. Many basic tasks normally done by the scientist are managed by NAMD-G. Tasks 2–6 are repeated until the simulation is complete.

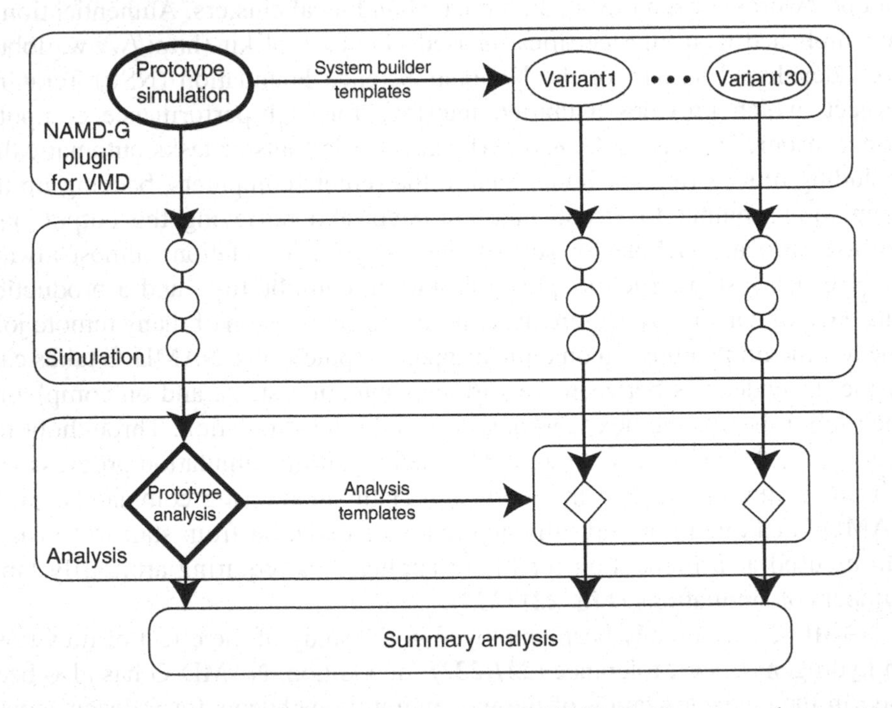

Fig. 22. A typical automated simulation using NAMD-G. The researcher creates a prototype system, which NAMD-G processes to obtain all the variant systems. NAMD-G executes all the simulations and finally initiates postprocessing analysis steps for each result.

the different types of monomeric globins exhibit a distinctive lack of O_2 pathway conservancy. Given that each globin simulation required on average 40 distinct job submission and file transfer steps, the computation of the O_2 PMF maps for the 12 globins of **Fig. 23** would have taken many months without NAMD-G.

5.2. Outlook

The development of methods and tools for describing gas migration pathways and for automating large numbers of simulations has many immediate applications beyond hydrogenase. Many important families of proteins, such as oxygenases, oxidases, and globins, for example, must interact with O_2 or other gas ligands to perform their function. The ability to map O_2 pathways in these proteins is of great value for an understanding of how proteins function.

An important goal of this research is to provide scientists who design new nanodevices with ways to quickly screen and characterize many potential structures to identify the ones with the most promise of exhibiting some desired behavior. The design of an O_2-tolerant hydrogenase is a prime example for such an approach as it requires the extensive computational testing of mutations, each of which needs its own set of simulations and analyses; if the simulations and analyses were to be performed for each mutant individually by the researcher, the total number of testable mutations would be severely limited. Because the gas migration profiles can be predicted in a relatively automated manner, the integration of gas migration analysis tools with the NAMD-G simulation automation tool will also result in the ability to characterize how these rates are affected by specific amino acid substitutions. All these developments will tremendously speed up the process of optimizing or blocking the gas migration process inside proteins for biotechnological purposes.

Although the methodological advances have been dramatic, additional methods are still needed to better describe gas permeation in proteins. In particular, kinetic rates of gas transport to or from the active site of a protein are not readily measured by simulation. Although it is possible to simulate a minuscule number of gas migration trajectories inside a protein and infer diffusion times from them, a protein-wide volumetric approach similar to what has been done for locating the pathways ***(212,213,220,221)*** would provide a much more reliable estimate for gas transport rates, which could then be directly compared to experiment and used for the design of hydrogenases.

6. Conclusion

Just as cars and planes today are designed and simulated by the computer, going directly from simulation to manufacture, biomolecules and nanodevices are starting to be designed and tested efficiently within the computer. In the life sciences, computational biology already has played a role in many major

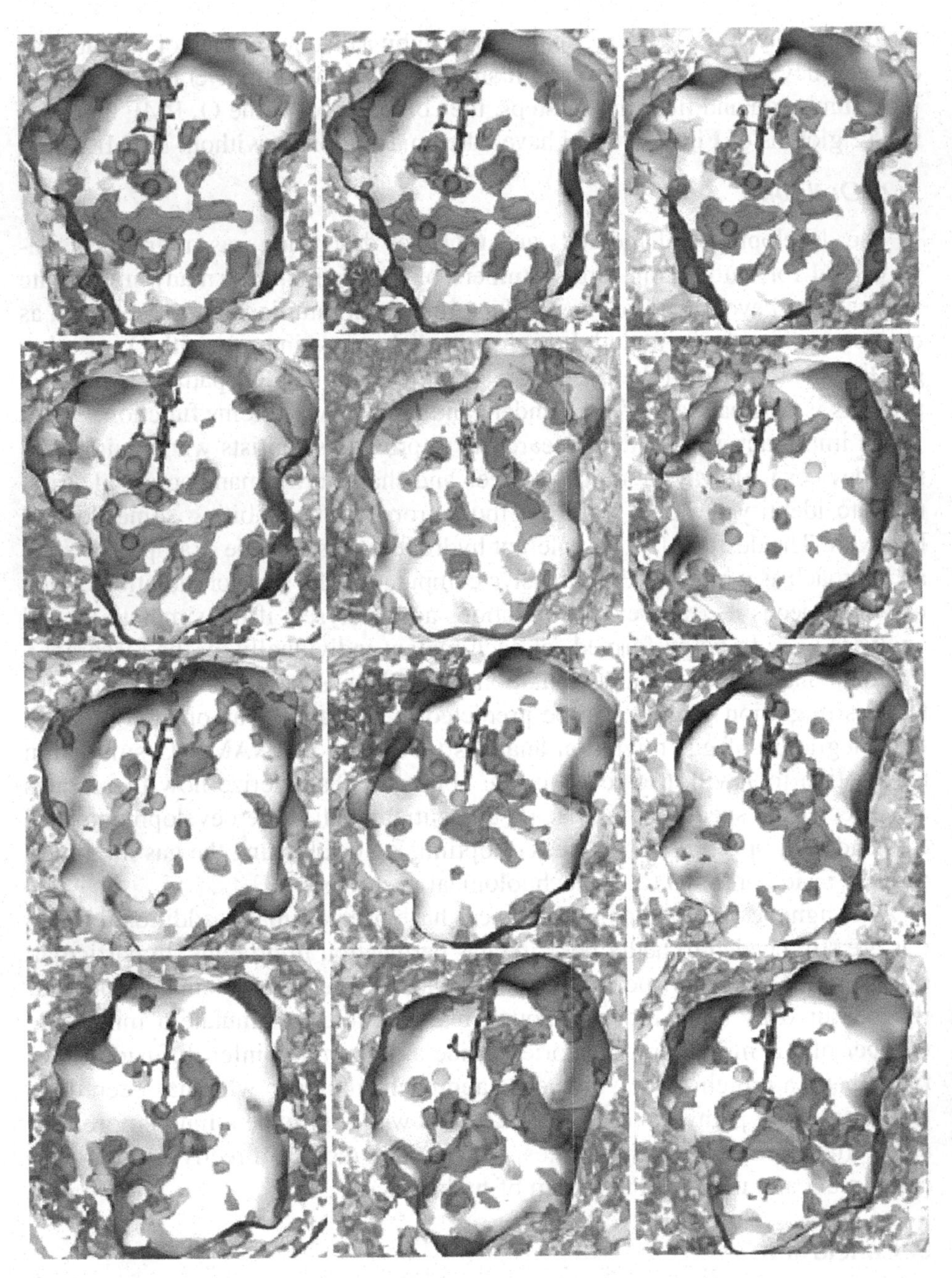

Fig. 23. O_2 **potential of mean force** (PMF) maps for various monomeric globins. Shown are the O_2 pathways for 12 monomeric globins (sperm whale oxy-myoglobin [Mb], deoxy-Mb, and YQR mutant Mb; horse and sea hare Mb; soy and lupin leghemoglobins; roundworm, trematode, bloodworm, clam, and midge hemoglobin).

advances, having brought about genome sequencing and sequence analysis, and having played a crucial role in discovering the physical mechanisms underlying cell function. This review illustrates in four cases the role that computer simulations can play in biotechnology.

In biotechnology, computational modeling offers a much-needed imaging tool covering the nanometer scale, for which no other tool offers a view of devices operating under physiological conditions. The "computational microscope" revealed how DNA threads itself through synthetic nanopores; how nanotubes surround themselves with water and ions and even DNA; how nanodiscs made of lipids and reengineered lipoproteins aggregate; and how proteins conduct gasses and what modifications optimize gas conduction. The qualitative views provided are of tremendous help in the design of new devices and methods in bionanotechnology, but the computational microscope also yields quantitative descriptions that permit testing of the computational micrographs and detailed interpretation of observational data.

The application of modeling to biotechnology demands a number of new methods beyond the modeling techniques used in traditional computational biology. The main challenge is that biotechnological devices combine inorganic and organic materials (e.g., silicon membranes or CNTs) with physiological solution and biomolecule. Modeling such devices requires force fields that are sufficiently accurate for both the inorganic and organic components, a development effort that will need to be pursued for years to come. Other challenges are posed by the large sizes and long time scales of devices and processes, involving millions of atoms and millisecond to second duration. Advanced computational technology and new concepts in simplifying models using "coarse graining" allow these challenges to be addressed. Finally, extensive computational testing, even when technically feasible, easily surpasses the work capacity of device engineers in sheer numbers of samples to be examined; only automation of modeling calculations as made possible through the NAMD-G program described offers a solution.

Many hurdles must yet be overcome to hone the computational approach to bionanotechnology, but the examples given show how computational modeling can already be a useful partner in nanodevice engineering, much like it is in macrodevice development. Just as we fly in planes that are designed and tested with computers, soon we will entrust medical diagnosis and treatment to devices engineered in the computer.

References

1. Humphrey W, Dalke A, Schulten A. (1996) VMD — Visual Molecular Dynamics. *J. Mol. Graphics* **14**, 33–38.
2. Phillips JC, Braun R, Wang W, et al. (2005) Scalable molecular dynamics with NAMD. *J. Comp. Chem.* **26**, 1781–1802.

3. Adcock SA, McCammon JA. (2006) Molecular dynamics: survey of methods for simulating the activity of proteins. *Chem. Rev.* **106**, 1589–1615.
4. Allen MP, Tildesley DJ. (1987) *Computer Simulation of Liquids*. Oxford University Press, New York.
5. Schlick T. (2002) *Molecular Modeling and Simulation: An Interdisciplinary Guide*. Springer-Verlag, New York.
6. Marsh S, Van Booven D, McLeod H. (2006) Global pharmacogenetics: giving the genome to the masses. *Pharmacogenomics* 7, 625–631.
7. Waltz E. (2006) After criticism, more modest cancer genome project takes shape. *Nat. Med.* **12**, 259.
8. Robertson JA. (2003) The $1000 genome: ethical and legal issues in whole genome sequencing of individuals. *Am. J. Bioethics* **3**, W35–W42.
9. Service RF. (2006) The race for the $1000 genome. *Science* **311**, 1544–1546.
10. Kobayashi S, Imaeda M, Matsumoto S. (2006) Single electron transistor fabricated with SOI wafer. *Mater. Sci. Eng. C* **26**, 889–892.
11. Brenning H, Kubatkin S, Erts D, Kafanov S, Bauch T, Delsing P. (2006) A single electron transistor on an atomic force microscope probe. *Nano Lett.* **6**, 937–941.
12. Heng JB, Aksimentiev A, Ho C, et al. (2005) Beyond the gene chip. *Bell Labs Tech. J.* **10**, 5–22.
13. Gracheva ME, Xiong A, Leburton JP, Aksimentiev A, Schulten K, Timp G. (2006) Simulation of the electric response of DNA translocation through a semiconductor nanopore-capacitor. *Nanotechnology* **17**, 622–633.
14. Aksimentiev A, Schulten K. (2004) Extending the molecular modeling methodology to study insertion of membrane nanopores. *Proc. Natl. Acad. Sci. U. S. A.* **101**, 4337–4338.
15. Heng JB, Ho C, Kim T, et al. (2004) Sizing DNA using a nanometer-diameter pore. *Biophys. J.* **87**, 2905–2911.
16. Heng JB, Aksimentiev A, Ho C, et al. (2005) Stretching DNA using an electric field in a synthetic nanopore. *Nano Lett.* **5**, 1883–1888.
17. Aksimentiev A, Heng JB, Timp G, Schulten K. (2004) Microscopic kinetics of DNA translocation through synthetic nanopores. *Biophys. J.* **87**, 2086–2097.
18. Heng JB, Aksimentiev A, Ho C, et al. (2006) The electromechanics of DNA in a synthetic nanopore. *Biophys. J.* **90**, 1098–1106.
19. Gracheva ME, Aksimentiev A, Leburton JP. (2006) Electrical signatures of single-stranded DNA with single base mutations in a nanopore capacitor. *Nanotechnology* **17**, 3160–3165.
20. Cornell WD, Cieplak P, Bayly CI, et al. (1995) A second generation force field for the simulation of proteins, nucleic acids, and organic molecules. *J. Am. Chem. Soc.* **117**, 5179–5197.
21. MacKerell AD Jr, Bashford D, Bellott M, et al. (1992) Self-consistent parameterization of biomolecules for molecular modeling and condensed phase simulations. *FASEB J.* **6**(1), A143.
22. MacKerell A Jr, Bashford D, Bellott M, et al. (1998) All-atom empirical potential for molecular modeling and dynamics studies of proteins. *J. Phys. Chem. B* **102**, 3586–3616.

23. Foloppe N, MacKerrell AD Jr. (2000) All-atom empirical force field for nucleic acids: I. parameter optimization based on small molecule and condensed phase macromolecular target data. *J. Comp. Chem.* **21**, 86–104.
24. Cruz-Chu ER, Aksimentiev A, Schulten K. (2006) Water-silica force field for simulating nanodevices. *J. Phys. Chem. B* **110**, 21497–21508.
25. Huff NT, Demiralp E, Cagin T, Goddard WA III. (1999) Factors affecting molecular dynamics simulated vitreous silica structures. *J. Non-Cryst. Solids* **253**, 133–142.
26. Patolsky F, Zheng G, Lieber CM. (2006) Nanowire-based biosensors. *Anal. Chem.* **78**, 4261–4269.
27. Hong JW, Quake SR. (2003) Integrated nanoliter systems. *Nat. Biotechnol.* **21**, 1179–1183.
28. Nawrocki J. (1997) The silanol group and its role in liquid chromatography. *J. Chromatogr. A* **779**, 29–71.
29. Heller M. (2002) DNA microarray technology: devices, systems and applications. *Annu. Rev. Biomed. Eng.* **4**, 129–153.
30. Wendel JA, Goddard WA III. (1992) The Hessian biased force-field for silicon nitride ceramics: predictions of the thermodynamic and mechanical properties for α- and β-Si_3N_4. *J. Chem. Phys.* **97**, 5048–5062.
31. Heng JB, Aksimentiev A, Ho C, et al. (2006) The electromechanics of DNA in a synthetic nanopore. *Biophys. J.* **90**, 1098–1106.
32. Werder T, Walther JH, Jaffe RL, Halicioglu T, Koumoutsakos P. (2003) On the water–carbon interaction for use in molecular dynamics simulations of graphite and carbon nanotubes. *J. Phys. Chem. B* **107**, 1345–1352.
33. Curthoys G, Davydov VY, Kiselev AK, Kiselev SA, Kuznetsov BV. (1974) Hydrogen bonding in adsorption on silica. *J. Colloid Interface Sci.* **48**, 58–72.
34. Palit D, Moulik S. (2001) Adsorption behaviors of L-histidine and DL-tryptophan on cholesterol, silica, alumina and graphite. *J. Colloid Interface Sci.* **239**, 20–26.
35. Cheng H, Zhang K, Libera J, de la Cruz M, Bedzyk M. (2006) Polynucleotide adsorption to negatively charged surfaces in divalent salt solutions. *Biophys. J.* **90**, 1164–1174.
36. Mathé J, Aksimentiev A, Nelson DR, Schulten K, Meller A. (2005) Orientation discrimination of single stranded DNA inside the α-hemolysin membrane channel. *Proc. Natl. Acad. Sci. U. S. A.* **102**, 12377–12382.
37. Aksimentiev A, Schulten K. (2005) Imaging alpha-hemolysin with molecular dynamics: Ionic conductance, osmotic permeability and the electrostatic potential map. *Biophys. J.* **88**, 3745–3761.
38. Sotomayor M, Vasquez V, Perozo E, Schulten K. (2007) Ion conduction through MscS as determined by electrophysiology and simulation. *Biophys. J.* **92**, 886–902.
39. Freites JA, Tobias DJ, von Heijne G, White SH. (2005) Interface connections of a transmembrane voltage sensor. *Proc. Natl. Acad. Sci. U. S. A.* **102**, 15059–15064.
40. Anishkin A, Sukharev S, Colombini M. (2006) Searching for the molecular arrangement of transmembrane ceramide channels. *Biophys. J.* **90**, 2414–2426.
41. Van Zeghbroeck B. (2004) *Principles of Semiconductor Devices*. Colorado University Press, Boulder, CO.

42. Hänsch W. (1991) *The Drift Diffusion Equation and Its Applications in MOSFET Modeling*. Springer, New York.
43. Lud S, Nikolaides M, Haase I, Fischer M, Bausch A. (2006) Field effect of screened charges: electrical detection of peptides and proteins by a thin-film resistor. *Chemphyschem* **7**, 379–384.
44. Dzyadevych SV, Soldatkin AP, El'skaya AV, Martelet C, Jaffrezic-Renault N. (2006) Enzyme biosensors based on ion-selective field-effect transistors. *Anal. Chim. Acta* **568**, 248–258.
45. Marrakchi M, Dzyadevych S, Biloivan O, Martelet C, Temple P, Jaffrezic-Renault N. (2006) Development of trypsin biosensor based on ion sensitive field-effect transistors for proteins determination. *Mater. Sci. Eng. C* **26**, 369–373.
46. Estrela P, Stewart A, Keighley S, Migliorato P. (2006) Biologically sensitive field-effect devices using polysilicon TFTs. *J. Korean Phys. Soc.* **48**, S22–S26.
47. Archer M, Fauchet P. (2003) Electrical sensing of DNA hybridization in porous silicon layers. *Physica Status Solidi A* **198**, 503–507.
48. Archer M, Christophersen M, Fauchet P. (2004) Macroporous silicon electrical sensor for DNA hybridization detection. *Biomed Microdevices* **6**, 203–211.
49. Chen H, Ilan B, Wu Y, Zhu F, Schulten K, Voth GA. (2007) Charge delocalization in proton channels. I. The aquaporin channels and proton blockage. *Biophys. J.* **92**, 46–60.
50. Guthold M, Falvo M, Matthews W, et al. (2000) Controlled manipulation of molecular samples with the NanoManipulator. *IEEEMT* **5**, 189–198.
51. Santos N, Castanho M. (2004) An overview of the biophysical applications of atomic force microscopy. *Biophys. Chem.* **107**, 133–149.
52. Li H, Cao E, Han B, Jin G. (2005) Stretching short single-stranded DNA adsorbed on gold surface by atomic force microscope. *Progr. Biochem. Biophys.* **32**, 1173–1177.
53. Wu P, Hogrebe P, Grainger D. (2006) DNA and protein microarray printing on silicon nitride waveguide surfaces *Biosens. Bioelectron.* **21**, 1252–1263.
54. Lyuksyutov S. (2005) Nano-patterning in polymeric materials and biological objects using atomic force microscopy electrostatic nanolithography. *Curr. Nanosci.* **1**, 245–251.
55. Liang X, Mao G, Ng K. (2004) Mechanical properties and stability measurement of cholesterol-containing liposome on mica by atomic force microscopy. *J. Colloid Interface Sci.* **278**, 53–62.
56. Bergveld P. (1991) A critical evaluation of direct electrical protein-detection methods. *Biosens. Bioelectron.* **6**, 55–72.
57. Kim D, Park J, Shin J, Kim P, Lim G, Shoji S. (2006) An extended gate FET-based biosensor integrated with a Si microfluidic channel for detection of protein complexes. *Sensor Actuat. B-Chem.* **117**, 488–494.
58. Pruneanu S, Ali Z, Watson G, et al. (2006) Investigation of electrochemical properties of carbon nanofibers prepared by CCVD method. *Particul. Sci. Technol.* **24**, 311–320.
59. Chen W, Yao H, Tzang C, Zhu J, Yang M, Lee S. (2006) Silicon nanowires for high-sensitivity glucose detection. *Appl. Phys. Lett.* **88**, 213104.

60. Weber J, Kumar A, Kumar A, Bhansali S. (2006) Novel lactate and pH biosensor for skin and sweat analysis based on single walled carbon nanotubes. *Sensor Actuat. B-Chem.* **117**, 308–313.
61. Iijima S. (1991) Helical microtubules of graphitic carbon. *Nature* **354**, 56–58.
62. Saito R, Dresselhaus G, Dresselhaus MS. (1998) *Physical Properties of Carbon Nanotubes*. Imperial College Press, London.
63. Kong J, Franklin NR, Zhou C, Chapline MG. (2000) Nanotube molecular wires as chemical sensors. *Science* **287**, 622–625.
64. Barone PW, Baik S, Heller DA, Strano MS. (2005) Near-infrared optical sensors based on single-walled carbon nanotubes. *Nat. Mater.* **4**, 86–92.
65. Heller DA, Jeng ES, Yeung T, et al. (2006) Optical detection of DNA conformational polymorphism on single-walled carbon nanotubes. *Science* **311**, 508–511.
66. Hummer G, Rasaiah JC, Noworyta JP. (2001) Water conduction through the hydrophobic channel of a carbon nanotube. *Nature* **414**, 188–190.
67. Noon WH, Ausman KD, Smalley RE, Ma J. (2002) Helical ice-sheets inside carbon nanotubes in the physiological condition. *Chem. Phys. Lett.* **355**, 445–448.
68. Mashl RJ, Joseph S, Aluru NR, Jakobsson E. (2003) Anomalously immobilized water: a new water phase induced by confinement in nanotubes. *Nano Lett.* **3**, 589–592.
69. Zhu F, Schulten K. (2003) Water and proton conduction through carbon nanotubes as models for biological channels. *Biophys. J.* **85**, 236–244.
70. Joseph S, Mashl RJ, Jakobsson E, Aluru NR. (2003) Electrolytic transport in modified carbon nanotubes. *Nano Lett.* **3**, 1399–1403.
71. Wei C, Srivastava D. (2003) Theory of transport of long polymer molecules through carbon nanotube channel. *Phys. Rev. Lett.* **91**, 235901.
72. Yeh IC, Hummer G. (2004) Diffusion and electrophoretic mobility of single-stranded RNA from molecular dynamics simulations. *Biophys. J.* **86**, 681–689.
73. Gao H, Kong Y. (2004) Simulation of DNA-nanotube interactions. *Annu. Rev. Mater. Res.* **34**, 129–152.
74. Lopez CF, Nielsen SO, Moore PB, Klein ML. (2004) Understanding nature's design for a nanosyringe. *Proc. Natl. Acad. Sci. U. S. A.* **101**, 4431–4434.
75. Lopez CF, Nielsen SO, Ensing B, Moore PB, Klein ML. Structure and dynamics of model pore insertion into a membrane. *Biophys. J.* **88**, 3083–3094.
76. Mann DJ, Halls MD. (2003) Water alignment and proton conduction inside carbon nanotubes. *Phys. Rev. Lett.* **90**, 195503.
77. Dellago C, Naor MM, Hummer G. (2003) Proton transport through water-filled carbon nanotubes. *Phys. Rev. Lett.* **90**, 105902.
78. Lu D, Li Y, Rotkin SV, Ravaioli U, Schulten K. (2004) Finite-size effect and wall polarization in a carbon nanotube channel. *Nano Lett.* **4**, 2383–2387.
79. Li Y, Lu D, Rotkin SV, Schulten K, Ravaioli U. (2005) Screening of water dipoles inside finite-length armchair carbon nanotubes. *J. Comp. Electron.* **4**, 161–165.
80. Lu D, Li Y, Ravaioli U, Schulten K. (2005) Empirical nanotube model for biological applications. *J. Phys. Chem. B* **109**, 11461–11467.
81. Lu D, Li Y, Ravaioli U, Schulten K. (2005) Ion-nanotube terahertz oscillator. *Phys. Rev. Lett.* **95**, 246801.

82. Lu D, Aksimentiev A, Shih AY, et al. (2006) The role of molecular modeling in bionanotechnology. *Phys. Biol.* **3**, S40–S53.
83. Sorin EJ, Pande VS. (2006) Nanotube confinement denatures protein helices. *J. Am. Chem. Soc.* **128**, 6316–6317.
84. Kolesnikov AI, Zanotti JM, Loong CK, et al. (2004) Anomalously soft dynamics of water in a nanotube: a revelation of nanoscale confinement. *Phys. Rev. Lett.* **93**, 035503.
85. Brünger, Schulten Z, Schulten K. (1983) A network thermodynamic investigation of stationary and non-stationary proton transport through proteins. *Z. Phys. Chem.* **NF136**, 1–63.
86. Schulten Z, Schulten K. (1985) Model for the resistance of the proton channel formed by the proteolipid of ATPase. *Eur. Biophys. J.* **11**, 149–155.
87. Schulten Z, Schulten K. (1986) Proton conduction through proteins: an overview of theoretical principles and applications. *Meth. Enzymol.* **127**, 419–438.
88. Lindahl E, Hess B, van der Spoel D. (2001) Gromacs 3.0: a package for molecular simulation and trajectory analysis. *J. Mol. Mod.* **7**(8), 306–317.
89. Bayly CI, Cieplak P, Cornell WD, Kollman PA. (1993) A well-behaved electrostatic potential based method using charge restraints for deriving atomic charges: the RESP model. *J. Phys. Chem.* **97**, 10268–10280.
90. Cornell WD, Cieplak P, Bayly CI, Kollman PA. (1993) Application of RESP charges to calculate conformational energies, hydrogen-bond energies, and free-energies of solvation. *J. Am. Chem. Soc.* **115**(21), 9620–9631.
91. Reich S, Maultzsch J, Thomsen C, Ordejón P. (2002) Tight-binding description of graphene. *Phys. Rev. B* **66**, 035412.
92. Lu D. (2005) Empirical nanotube model: applications to water channels and nano-oscillators. Ph.D. thesis, University of Illinois, Urbana-Champaign, Urbana, IL.
93. Pantarotto D, Briand JP, Prato M, Bianco A. (2004) Translocation of bioactive peptides across cell membranes by carbon nanotubes. *Chem. Commun.* **1**, 16–17.
94. Kam NWS, Jessop TC, Wender PA, Dai H. (2004) Nanotube molecular transporters: internalization of carbon nanotube-protein conjugates into mammalian cells. *J. Am. Chem. Soc.* **126**, 6850–6851.
95. Kam NWS, Dai H. (2005) Carbon nanotubes as intracellular protein transporters: generality and biological functionality. *J. Am. Chem. Soc.* **127**, 6021–6026.
96. Pantarotto D, Singh R, McCarthy D, et al. (2004) Functionalized carbon nanotubes for plasmid DNA gene delivery. *Angew. Chem. Int. Ed.* **43**, 5242–5246.
97. Cai D, Huang Z, Carnahan D, et al. (2005) Highly efficient molecular delivery into mammalian cells using carbon nanotube spearing. *Nat. Methods* **2**, 449–454.
98. Klumpp C, Kostarelosc K, Pratob M, Bianco A. (2006) Functionalized carbon nanotubes as emerging nanovectors for the delivery of therapeutics. *Biochim. Biophys. Acta* **1758**, 404–412.
99. Zheng M, Jagota A, Semke ED, Diner BA, et al. (2003) DNA-assisted dispersion and separation of carbon nanotubes. *Nat. Mater.* **2**, 338–342.

100. Chen RJ, Zhang Y, Wang D, Dai H. (2001) Noncovalent sidewall functionalization of single-walled carbon nanotubes for protein immobilization. *J. Am. Chem. Soc.* **123**, 3838–3839.
101. King GM, Golovchenko JA. (2005) Probing nanotube–nanopore interactions. *Phys. Rev. Lett.* **95**, 216103.
102. Woolley AT, Guillemette C, Cheung CL, Housman DE, Lieber CM. (2000) Direct haplotyping of kilobase-size DNA using carbon nanotube probes. *Nat. Biotechnol.* **18**, 760–763.
103. Wong SS, Harper JD, Lansbury PT Jr, Lieber CM. (1998) Carbon nanotube tips: high resolution probes for imaging biological systems. *J. Am. Chem. Soc.* **120**, 603–604.
104. Kam NWS, O'Connell M, Wisdom JA, Dai H. (2005) Carbon nanotubes as multifunctional biological transporters and near-infrared agents for selective cancer cell destruction. *Proc. Natl. Acad. Sci. U. S. A.* **102**, 11600–11605.
105. Sirdeshmukh R, Teker K, Panchapakesan B. (2004) Functionalization of carbon nanotubes with antibodies for breast cancer detection applications. *Proc. 2004 Int. Conf. MEMS NANO Smart Syst.* 48–53.
106. Ritz T, Hu X, Damjanovi A, Schulten K. (1998) Excitons and excitation transfer in the photosynthetic unit of purple bacteria. *J. Luminesc.* **76–77**, 310–321.
107. Damjanovi A, Ritz T, Schulten K. (2000) Excitation energy trapping by the reaction center of *Rhodobacter sphaeroides. Int. J. Quantum Chem.* **77**, 139–151.
108. Damjanovi A, Ritz T, Schulten K. (2000) Excitation transfer in the peridinin-chlorophyllprotein of *Amphidinium carterae. Biophys. J.* **79**, 1695–1705.
109. Sener MK, Park S, Lu D, et al. (2004) Excitation migration in trimeric cyanobacterial photosystem I. *J. Chem. Phys.* **120**, 11183–11195.
110. Thomsen W, Frazer J, Unett D. (2005) Functional assays for screening GPCR targets. *Curr. Opin. Biotechnol.* **16**, 655–665.
111. Klabunde T, Hessler G. (2002) Drug design strategies for targeting G-protein coupled receptors. *ChemBioChem* **3**, 928–944.
112. Flower DR. (1999) Modelling G-protein-coupled receptors for drug design. *Biochim. Biophys. Acta* **1422**, 207–234.
113. Gudermann T, Nurnberg N, Schultz G. (1995) Receptors and G proteins as primary components of transmembrane signal transduction. *J. Mol. Med.* **73**, 51–63.
114. Seddon AM, Curnow P, Booth PJ. (2004) Membrane proteins, lipids and detergents: not just a soap opera. *Biochim. Biophys. Acta* **1666**, 105–117.
115. Wang M, Briggs MR. (2004) HDL: The metabolism, function, and therapeutic importance. *Chem. Rev.* **104**, 119–137.
116. Bayburt TH, Grinkova YV, Sligar SG. (2002) Self-assembly of discoidal phospholipid bilayer nanoparticles with membrane scaffold proteins. *Nano Lett.* **2**, 853–856.
117. Sligar SG. (2003) Finding a single-molecule solution for membrane proteins. *Biochem. Biophys. Res. Commun.* **312**, 115–119.
118. Service RF. (2004) Sushi-like discs give inside view of elusive membrane protein. *Science* **304**, 674.

119. Nath A, Atkins WM, Sligar SG. (2007) Applications of phospholipid bilayer nanodiscs in the study of membrane and membrane proteins. *Biochemistry* **46**, 2059–2069.
120. Bayburt TH, Sligar SG. (2002) Single-molecule height measurements on microsomal cytochrome P450 in nanometer-scale phospholipid bilayer disks. *Proc. Natl. Acad. Sci. U. S. A.* **99**, 6725–6730.
121. Bayburt TH, Sligar SG. (2003) Self-assembly of single integral membrane proteins into soluble nanoscale phospholipid bilayers. *Protein Sci.* **12**, 2476–2481.
122. Civjan NR, Bayburt TH, Schuler MA, Sligar SG. (2003) Direct solubilization of heterologously expressed membrane proteins by incorporation into nanoscale lipid bilayers. *Biotechniques* **35**, 556–560, 562–563.
123. Baas BJ, Denisov IG, Sligar SG. (2004) Homotropic cooperativity of monomeric cytochrome P450 3A4 in a nanoscale native bilayer environment. *Arch. Biochem. Biophys.* **430**, 218–228.
124. Duan H, Civjan NR, Sligar SG, Schuler MA. (2004) Co-incorporation of heterologously expressed *Arabidopsis* cytochrome P450 and P450 reductase into soluble nanoscale lipid bilayers. *Arch. Biochem. Biophys.* **424**, 141–153.
125. Shaw AW, McLean MA, Sligar SG. (2004) Phospholipid phase transitions in homogeneous nanometer scale bilayer discs. *FEBS Lett.* **556**, 260–264.
126. Davydov DR, Fernando H, Baas BJ, Sligar SG, Halpert JR. (2005) Kinetics of dithionite-dependent reduction of cytochrome P450 3A4: heterogeneity of the enzyme caused by its oligomerization. *Biochemistry* **44**, 13902–13913.
127. Denisov IG, McLean MA, Shaw AW, Grinkova YV, Sligar SG. (2005) Thermotropic phase transition in soluble nanoscale lipid bilayers. *J. Phys. Chem. B* **109**, 15580–15588.
128. Denisov IG, Grinkova YV, Baas BJ, Sligar SG. (2006) The ferrous-dioxygen intermediate in human cytochrome P450 3A4: Substrate dependence of formation and decay kinetics. *J. Biol. Chem.* **281**, 23313–23318.
129. Li L, Wetzel S, Pluckthun A, Fernandez JM. (2006) Stepwise unfolding of ankyrin repeats in a single protein revealed by atomic force microscopy. *Biophys. J.* **90**, L30–L32.
130. Bayburt TH, Grinkova YV, Sligar SG. (2006) Assembly of single bacteriorhodopsin trimers in bilayer nanodiscs. *Arch. Biochem. Biophys.* **450**, 215–222.
131. Boldog T, Grimme S, Li M, Sligar SG, Hazelbauer GL. (2006) Nanodiscs separate chemoreceptor oligomeric states and reveal their signaling properties. *Proc. Natl. Acad. Sci. U. S. A.* **103**, 11509–11514.
132. Leitz AJ, Bayburt TH, Barnakov AN, Springer BA, Sligar SG. (2006) Functional reconstitution of Beta2-adrenergic receptors utilizing self-assembling nanodisc technology. *Biotechniques* **40**, 601–612.
133. Shaw AW, Pureza VS, Sligar SG, Morrissey JH. (2007) The local phospholipid environment modulates the activation of blood clotting. *J. Biol. Chem.* **282**, 6556–6563.
134. Denisov IG, Baas BJ, Grinkova YV, Sligar SG. (2007) Cooperativity in P450 CYP3A4: linkages in substrate binding, spin state, uncoupling and product formation. *J. Biol. Chem.* **282**, 7066–7076.

135. Borhani DW, Rogers DP, Engler JA, Brouillette CG. (1997) Crystal structure of truncated human apolipoprotein A-I suggests a lipid-bound conformation. *Proc. Natl. Acad. Sci. U. S. A.* **94**, 12291–12296.
136. Ajees AA, Anantharamaiah GM, Mishra VK, Hussain MM, Murthy HMK. (2006) Crystal structure of human apolipoprotein A-I: insights into its protective effect against cardiovascular diseases. *Proc. Natl. Acad. Sci. U. S. A.* **103**, 2126–2131.
137. Koppaka V, Silvestro L, Engler JA, Brouillette CG, Axelsen PH. (1999) The structure of human lipoprotein A-I. Evidence for the "belt" model. *J. Biol. Chem.* **274**, 14541–14544.
138. Panagotopulos SE, Horace EM, Maiorano JN, Davidson WS. (2001) Apolipoprotein A-I adopts a belt-like orientation in reconstituted high density lipoproteins. *J. Biol. Chem.* **276**, 42965–42970.
139. Li H, Lyles DS, Thomas MJ, Pan W, Sorci-Thomas MG. (2000) Structural determination of lipid-bound ApoA-I using fluorescence resonance energy transfer. *J. Biol. Chem.* **275**, 37048–37054.
140. Tricerri MA, Behling Agree AK, Sanchez SA, Bronski JA, Jonas A. (2001) Arrangement of apolipoprotein A-I in reconstituted high-density lipoprotein disks: an alternative model based on fluorescence resonance energy transfer experiments. *Biochemistry* **40**, 5065–5074.
141. Silva RA, Hilliard GM, Li L, Segrest JP, Davidson WS. (2005) A mass spectrometric determination of the conformation of dimeric apolipoprotein A-I in discoidal high density lipoproteins. *Biochemistry* **44**, 8600–8607.
142. Li Y, Kijac AZ, Sligar SG, Rienstra CM. (2006) Structural analysis of nanoscale self-assembled discoidal lipid bilayers by solid-state NMR spectroscopy. *Biophys. J.* **91**, 3819–3828.
143. Gorshkova IN, Liu T, Kan HY, Chroni A, Zannis VI, Atkinson D. (2006) Structure and stability of apolipoprotein a-I in solution and in discoidal high-density lipoprotein probed by double charge ablation and deletion mutation. *Biochemistry* **45**, 1242–1254.
144. Phillips JC, Wriggers W, Li Z, Jonas A, Schulten K. (1997) Predicting the structure of apolipoprotein A-I in reconstituted high density lipoprotein disks. *Biophys. J.* **73**, 2337–2346.
145. Nichols AV, Gong EL, Blanche PJ, Forte TM, Shore VG. (1984) Interaction of model discoidal complexes of phosphatidylcholine and apolipoprotein A-I with plasma components. Physical and chemical properties of the transformed complexes. *Biochim. Biophys. Acta* **793**, 325–337.
146. McGuire KA, Davidson WS, Jonas A. (1996) High yield overexpression and characterization of human recombinant proapolipoprotein A-I. *J. Lipid Res.* **37**, 1519–1528.
147. Davidson WS, Sparks DL, Lund-Katz S, Phillips MC. (1994) The molecular basis for the difference in charge between pre-beta- and alpha-migrating high density lipoproteins. *J. Biol. Chem.* **269**, 8959–8965.
148. Jonas A. (1986) Reconstitution of high-density lipoproteins. *Methods Enzymol.* **128**, 553–582.

149. Shih AY, Denisov IG, Phillips JC, Sligar SG, Schulten K. (2005) Molecular dynamics simulations of discoidal bilayers assembled from truncated human lipoproteins. *Biophys. J.* **88**, 548–556.
150. Klon AE, Segrest JP, Harvey SC. (2002) Molecular dynamics simulations on discoidal HDL particles suggest a mechanism for rotation in the apo A-I belt model. *J. Mol. Biol.* **324**, 703–721.
151. Segrest JP, Jones MK, Klon AE, et al. (1999) A detailed molecular belt model for apolipoprotein A-I in discoidal high density lipoprotein. *J. Biol. Chem.* **274**, 31755–31758.
152. Sheldahl CJ, Harvey SC. (1999) Molecular dynamics on a model for nascent discoidal high-density lipoprotein: role of salt-bridges. *Biophys. J.* **76**, 1190–1198.
153. Denisov IG, Grinkova YV, Lazarides AV, Sligar SG. (2004) Directed self-assembly of monodisperse phospholipid bilayer nanodiscs with controlled size. *J. Am. Chem. Soc.* **126**, 3477–3487.
154. Davidson WS, Silva RAGD. (2005) Apolipoprotein structural organization in high density lipoproteins: belts, bundles, hinges and hairpins. *Curr. Opin. Lipidol.* **16**, 295–300.
155. Shih AY, Arkhipov A, Freddolino PL, Schulten K. (2006) Coarse grained protein-lipid model with application to lipoprotein particles. *J. Phys. Chem. B* **110**, 3674–3684.
156. Shih AY, Freddolino PL, Arkhipov A, Schulten K. (2007) Assembly of lipoprotein particles revealed by coarse-grained molecular dynamics simulations. *J. Struct. Biol.* **157**, 579–592.
157. Marrink SJ, de Vries AH, Mark AE. (2004) Coarse grained model for semiquantitative lipid simulations. *J. Phys. Chem. B* **108**, 750–760.
158. Shelley JC, Shelley MY, Reeder RC, Bandyopadhyay S, Moore PB, Klein ML. (2001) Simulations of phospholipids using a coarse grain model. *J. Phys. Chem. B* **105**, 9785–9792.
159. Shelley JC, Shelley MY, Reeder RC, Bandyopadhyay S, Klein ML. (2001) A coarse grain model for phospholipid simulations. *J. Phys. Chem. B* **105**, 4464–4470.
160. Stevens MJ, Hoh JH, Woolf TB. (2003) Insights into the molecular mechanism of membrane fusion from simulations: evidence for the association of splayer tails. *Phys. Rev. Lett.* **91**, 188102.
161. Izvekov S, Voth GA. (2005) Multiscale coarse graining of liquid-state systems. *J. Chem. Phys.* **123**, 134105.
162. Markvoort AJ, Pieterse K, Steijaert MN, Spijjker P, Hilbers PAJ. (2005) The bilayer–vesicle transition is entropy driven. *J. Phys. Chem. B* **109**, 22649–22654.
163. Stevens MJ. (2004) Coarse-grained simulations of lipid bilayers. *J. Chem. Phys.* **121**, 11942–11948.
164. Nielsen SO, Lopez CF, Srinivas G, Klein ML. (2004) Coarse grain models and the computer simulation of soft materials. *J. Phys. Condens. Matter* **16**, R481–R512.

165. Nielsen SO, Ensing B, Ortiz V, Moore PB, Klein ML. (2005) Lipid bilayer perturbations around a transmembrane nanotube: a coarse grain molecular dynamics study. *Biophys. J.* **88**, 3822–3828.
166. Pickholz M, Saiz L, Klein ML. (2005) Concentration effects of volatile anesthetics on the properties of model membranes: a coarse-grain approach. *Biophys. J.* **88**, 1524–1534.
167. Faller R, Marrink SJ. (2004) Simulation of domain formation in DLPC-DSPC mixed bilayers. Langmuir **20**, 7686–7693.
168. Marrink SJ, Mark AE. (2003) Molecular dynamics simulation of the formation, structure, and dynamics of small phospholipid vesicles. *J. Am. Chem. Soc.* **125**, 15233–15242.
169. Marrink SJ, Mark AE. (2003) The mechanism of vesicle fusion as revealed by molecular dynamics simulations. *J. Am. Chem. Soc.* **125**, 11144–11145.
170. Marrink SJ, Risselada J, Mark AE. (2005) Simulation of gel phase formation and melting in lipid bilayers using a coarse grained model. *Chem. Phys. Lipids* **135**(2), 223–244.
171. de Vries AH, Yefimov S, Mark AE, Marrink SJ. (2005) Molecular structure of the lecithin ripple phase. *Proc. Natl. Acad. Sci. U. S. A.* **102**, 5392–5396.
172. Smeijers AF, Pieterse K, Markvoort AJ, Hilbers PAJ. (2006) Coarse-grained transmembrane proteins: hydrophobic matching, aggregation, and their effect on fusion, *J. Phys. Chem. B* **110**, 13614–13623.
173. van der Spoel D, Lindahl E, Hess B, Groenhof G, Mark AE, Berendsen HJC. (2005) Gromacs: Fast, flexible, and free. *J. Comp. Chem.* **26**, 1701–1718.
174. Jonas A, Wald JH, Toohill KLH, Krul ES, Kézdy KE. (1990) Apolipoprotein AI structure and lipid properties in homogeneous reconstituted spherical and discoidal high density lipoproteins. *J. Biol. Chem.* **265**, 22123–22129.
175. Sparks DL, Davidson WS, Lund-Katz S, Phillips MC. (1995) Effects of the neutral lipid content of high density lipoprotein on apolipoprotein A-I structure and particle stability. *J. Biol. Chem.* **270**, 26910–26917.
176. Reith D, Pütz M, Müller-Plathe F. (2003) Deriving effective mesoscale potentials from atomistic simulations. *J. Comp. Chem.* **24**, 1624–1636.
177. Bahar I, Kaplan M, Jernigan RL. (1997) Short-range conformational energies, secondary structure propensities, and recognition of correct sequence-structure matches. *Proteins* **29**, 292–308.
178. Tozzini V, McCammon A. (2005) A coarse grained model for the dynamics of flap opening in HIV-1 protease. *Chem. Phys. Lett.* **413**, 123–128.
179. Zhang L, Skolnick J. (1998) How do potentials derived from structural databases relate to "true" potentials? *Protein Sci.* **7**, 112–122.
180. Sun Q, Faller R. (2005) Systematic coarse-graining of atomistic models for simulation of polymeric systems. *Comp. Chem. Eng.* **29**, 2380–2385.
181. Shih AY, Freddolino PL, Sligar SG, Schulten K. (2007) Disassembly of nanodiscs with cholate. *Nano Lett.* **7**, 1692–1696.
182. Aliabadi HM, Lavasanifar A. (2006) Polymeric micelles for drug delivery. *Expert Opin. Drug Deliv.* **3**, 139–162.

183. Gaucher G, Dufresne M-H, Sant VP, Kang N, Maysinger D, Leroux J-C. (2005) Block copolymer micelles: preparation, characterization and application in drug delivery. *J. Control. Release* **109**, 169–188.
184. Kshirsagar NA, Pandya SK, Kirodian GB, Sanath S. (2005) Liposomal drug delivery system from laboratory to clinic. *J. Postgrad. Med.* **51**(suppl 1), S5–S15.
185. Nishiyama N, Kataoka K. (2006) Current state, achievements, and future prospects of polymeric micelles as nanocarriers for drug and gene delivery. *Pharmacol. Ther.* **112**, 630–648.
186. Torchilin VP. (2006) Recent approaches to intracellular delivery of drugs and DNA and organelle targeting. *Annu. Rev. Biomed. Eng.* **8**, 343–375.
187. Torchilin VP. Lipid-core micelles for targeted drug delivery. *Curr. Drug Deliv.* **2**, 319–327.
188. Anrather D, Smetazko M, Saba M, Alguel Y, Schalkhammer T. (2004) Supported membrane nanodevices. *J. Nanosci. Nanotechnol.* **4**, 1–22.
189. Yeung ES. (2004) Dynamics of single biomolecules in free solution. *Annu. Rev. Phys. Chem.* **55**, 97–126.
190. Yoo KH, Ha DH, Lee JO, et al. (2001) Electrical conduction through poly(dA)-poly(dT) and poly(dG)-poly(dC) DNA molecules. *Phys. Rev. Lett.* **87**, 198102-1–198102-4.
191. Mirkin CA, Taton TA. (2000) Semiconductors meet biology. *Science* **405**, 626–627.
192. Peter BJ, Kent HM, Mills IG, et al. (2004) BAR domains as sensors of membrane curvature: the amphiphysin BAR structure. *Science* **303**, 495–499.
193. Vignais PM, Billoud B, Meyer J. (2001) Classification and phylogeny of hydrogenases. *FEMS Microbiol. Rev.* **25**, 455–501.
194. Nicolet Y, Cavazza C, Fontecilla-Camps JC. (2002) Fe-only hydrogenases: structure, function and evolution. *J. Inorg. Biochem.* **91**, 1–8.
195. Boichenko VA, Greenbaum E, Seibert M. (2004) Hydrogen production by photosynthetic microorganisms. In: Archer MD, Barber J, eds. *Photoconversion of Solar Energy: Molecular to Global Photosynthesis*. London: Imperial College Press, pp. 397–452.
196. Mertens R, Liese A. (2004) Biotechnological applications of hydrogenases. *Curr. Opin. Biotechnol.* **15**, 343–348.
197. Ghirardi ML, Zhang L, Lee JW, et al. (2000) Microalgae: a green source of renewable H_2. *Trends Biotechnol.* **18**, 506–511.
198. Peters JW, Lanzilotta WN, Lemon BJ, Seefeldt LC. (1998) X-ray crystal structure of the Fe-only hydrogenase (CpI) from *Clostridium pasteurianum* to 1.8 angstrom resolution. *Science* **282**, 1853–1858.
199. Salomonsson L, Lee A, Gennis RB, Brzezinski P. (2004) A single-amino-acid lid renders a gas-tight compartment within a membrane-bound transporter. *Proc. Natl. Acad. Sci. U. S. A.* **101**(32), 11617–11621.
200. Buhrke T, Lenz O, Krauss N, Friedrich B. (2005) Oxygen tolerance of the H2-sensing [NiFe] hydrogenase from *Ralstonia eutropha* H16 is based on limited access of oxygen to the active site. *J. Biol. Chem.* **280**(25), 23791–23796.

201. Schotte F, Lim M, Jackson TA, et al. (2003) Watching a protein as it functions with 150-ps time-resolved X-ray crystallography. *Science* **300**, 1944–1947.
202. Scott EE, Gibson QH, Olson JS. (2001) Mapping the pathways for O_2 entry into and exit from myoglobin. *J. Biol. Chem.* **276**(7), 5177–5188.
203. Brunori M, Vallone B, Cutruzzola B, et al. (2000) The role of cavities in protein dynamics: crystal structure of a photolytic intermediate of a mutant myoglobin. *Proc. Natl. Acad. Sci. U. S. A.* **97**, 2058–2063.
204. Gibson QH, Regan R, Elber R, Olson JS, Carver TE. (1992) Distal pocket residues affect picosecond ligand recombination in myoglobin. *J. Biol. Chem.* **267**,22022–22034.
205. Rohlfs RJ, Olson JS, Gibson QH. (1988) A comparison of the geminate recombination kinetics of several monomeric heme proteins. *J. Biol. Chem.* **263**(4), 1803–1813.
206. Bossa C, Amadei A, Daidone I, et al. (2005) Molecular dynamics simulation of sperm whale myoglobin: effects of mutations and trapped CO on the structure and dynamics of cavities. *Biophys. J.* **89**, 465–474.
207. Hummer G, Schotte F, Anfinrud PA. (2004) Unveiling functional protein motions with picosecond x-ray crystallography and molecular dynamics simulations. *Proc. Natl. Acad. Sci. U. S. A.* **101**, 15330–15334.
208. Amara P, Andreoletti P, Jouve HM, Field MJ. (2001) Ligand diffusion in the catalase from *Proteus mirabilis*: a molecular dynamics study. *Protein Sci.* **10**, 1927–1935.
209. Elber R, Karplus M. (1990) Enhanced sampling in molecular dynamics: use of the time-dependent Hartree approximation for a simulation of carbon monoxide diffusion through myoglobin. *J. Am. Chem. Soc.* **112**(25), 9161–9175.
210. Hofacker I, Schulten K. (1998) Oxygen and proton pathways in cytochrome *c* oxidase. *Proteins* **30**(1), 100–107.
211. Cohen J, Kim K, Posewitz M, et al. (2005) Molecular dynamics and experimental investigation of H_2 and O_2 diffusion in [Fe]- hydrogenase. *Biochem. Soc. Trans.* **33**, 80–82.
212. Cohen J, Kim K, King P, Seibert M, Schulten K. (2005) Finding gas diffusion pathways in proteins: application to O_2 and H_2 transport in CpI [FeFe]-hydrogenase and the role of packing defects. *Structure* **13**, 1321–1329.
213. Cohen J, Arkhipov A, Braun R, Schulten K. (2006) Imaging the migration pathways for O_2, CO, NO, and Xe inside myoglobin. *Biophys. J.* **91**, 1844–1857.
214. Straub JE, Karplus M. (1991) Energy equipartitioning in the classical time-dependent Hartree approximation. *J. Chem. Phys.* **94**(10), 6737–6739.
215. Ghirardi ML, King PW, Posewitz MC, et al. (2005) Approaches to developing biological H_2-photoproducing organisms and processes. *Biochem. Soc. Trans.* **33**, 70–72.
216. Beveridge DL, DiCapua FM. (1989) Free energy via molecular simulation: applications to chemical and biological systems. *Annu. Rev. Biophys. Biophys. Chem.* **18**, 431–492.

217. Kollman P. (1993) Free energy calculations: applications to chemical and biochemical phenomena. *Chem. Rev.* **93**, 2395–2417.
218. Johnson BJ, Cohen J, Welford RW, et al. (2007) Exploring molecular oxygen pathways in *Hanseluna Polymorpha* copper-containing amine oxidase. *J. Biol. Chem.* **282**, 17767–17776.
219. Cohen J, Schulten K. (2007) O_2 migration pathways in monomeric globins are determined by residue composition, not tertiary structure. *Biophys. J.* **93**, 3591–3600.
220. Wang Y, Cohen J, Boron WF, Schulten K, Tajkhorshid E. (2007) Exploring gas permeability of cellular membranes and membrane channels with molecular dynamics. *J. Struct. Biol.* **157**, 534–544.
221. Ghirardi ML, Cohen J, King P, Schulten K, Kim K, Seibert M. (2006) [FeFe]-hydrogenases and photobiological hydrogen production. *Proc. SPIE* **6340**, 253–258.
222. King PW, Svedruzic D, Cohen J, Schulten K, Seibert M, Ghirardi ML. (2006) Structural and functional investigations of biological catalysts for optimization of solar-driven, H_2 production systems. *Proc. SPIE* **6340**, 259–267.
223. Gower M, Cohen J, Phillips J, Kufrin R, Schulten K. (2006) Managing biomolecular simulations in a grid environment with NAMD-G. In: Proceedings of the 2006 TeraGrid Conference.

12

What Can We Learn From Highly Connected β-Rich Structures for Structural Interface Design?

Ugur Emekli, K. Gunasekaran, Ruth Nussinov, and Turkan Haliloglu

Summary

Most hubs' binding sites are able to transiently interact with numerous proteins. We focus on β-rich hubs with the goal of inferring features toward design. Since they are able to interact with many partners and association of β-conformations may lead to amyloid fibrils, we ask whether there is some property that distinguishes them from low-connectivity β-rich proteins, which may be more interaction specific. Identification of such features should be useful as they can be incorporated in interface design while avoiding polymerization into fibrils. We classify the proteins in the yeast interaction map according to the types of their secondary structures. The small number of the obtained β-rich protein structures in the Protein Data Bank likely reflects their low occurrence in the proteome. Analysis of the obtained structures indicates that highly connected β-rich proteins tend to have clusters of conserved residues in their cores, unlike β-rich structures with low connectivity, suggesting that the highly packed conserved cores are important to the stability of proteins, which have residue composition and sequence prone to β-structure and amyloid formation. The enhanced stability may hinder partial unfolding, which, depending on the conditions, is more likely to lead to polymerization of these sequences.

Key Words: β-Structures; connectivity; hubs; interface design; network; yeast network.

1. Introduction

What can we learn from protein interaction networks that would assist us in design of protein–protein interfaces? Protein interaction networks are of crucial importance for figuring out processes in the living cell. The interactions involved in such networks provide hints to cellular functions, dysfunctions, and diseases *(1)*. The proteome has been shown to be organized as a scale-free network characterized by an uneven distribution of connectedness. Rather than having a random pattern of connections, some nodes in the networks are "very connected" hubs, whereas others have fewer edges connecting them to other

From: *Methods in Molecular Biology, vol. 474: Nanostructure Design: Methods and Protocols*
Edited by: E. Gazit and R. Nussinov © Humana Press, Totowa, NJ

proteins. Such an organization leads to system robustness since disrupting the functions of most proteins is unlikely to affect the system in a major way; yet, it also leads to sensitivity if a key protein in the network dysfunctions. This organization influences dramatically the way the network functions.

The central proteins that interact with several partners are expected to be vital for the survival of the cell since many pathways funnel through them ***(2)***. Since the interactivity of proteins ultimately rests on physical properties such as the sequence and structure of the protein surface and the protein's conformational attributes, understanding the molecular basis of protein structures and their interactions in the network is of fundamental importance. Proteins can have very different number of partners with which they interact. Current data suggest that the numbers range from practically (almost) none to several tens or even hundreds. While such numbers are likely to be overestimates, based on some overexpression experimental screens, nevertheless it does suggest a large number of biological partners.

Here, we study the system with the goal of abstracting information useful for design. Our premise is that evolution has perfected the system to avoid unwanted consequences. Since hub proteins interact with many partners while at the same time maintaining selectivity and high affinity, clues regarding which features can be used and which to avoid can be obtained.

Previous analysis of the yeast protein interaction network has suggested a correlation between the connectivity and evolutionary constraints ***(3)***. Evolutionary constraints have been proposed to be due to protein–protein interactions and to be correlated with the number of protein interaction partners ***(4)***. Converse arguments also exist for the latter, suggesting that the correlation may arise due to biases ***(5,6)***. A better understanding would require the complete data sets and the incorporation of an integrated approach that will combine the individual and the contextual properties of all constituents in the network.

We have observed that protein-binding sites that interact with many partners preferentially belong to helical proteins, and the multipartner binding sites largely consist of α-helices ***(7)***. This is likely to be the outcome of the variable ways in which α-helices can associate. Yet, β-rich proteins also serve as hubs. The β-structures hairpins ***(8,9)*** and sheets ***(10)*** have been studied extensively. To date, no correspondence has been observed between disease-associated amyloidogenic proteins such as Alzheimer's amyloid-β (Aβ), islet amyloid polypeptide (IAPP; causative agent in type 2 diabetes), and others that under physiological conditions form amyloids and proteins consisting largely of β-conformations ***(11)***. Thus, either the β-structures of native proteins possess inherent features, particularly at their surfaces, that are not amenable to amyloidogenicity ***(12)*** or in β-structures, like in other secondary structural types, amyloidogenicity involves at least partial unfolding prior to aggregation ***(13–15)***. It is quite likely that both factors are at play. Hence, since centrally connected proteins interact with many partners, and thus the same binding site is likely to associate with multiple proteins, and since β proteins are highly likely to associate via β-strand addition across the interfaces ***(16,17)***, here we ask whether structures largely consisting of β-conformations that serve as hubs have features distinguishing them from β-rich structures with low connectivity. Since there is no information about the location of the binding sites, we can only assume that in β-structure-rich proteins, binding will involve the β-strand addition to extend the sheet via strand insertion, β-sheet augmentation, or fold complementation ***(18)***. A binding site interacting with many partners through such mechanisms might conceivably be more sensitive to amyloid formation compared to binding sites in low-connectivity proteins, binding to one or few partners.

To address these questions in a systematic way, we chose an organism for which both comprehensive protein–protein interaction screens are available and enough proteins have been crystallized in the structural database. Yeast is unique in this respect; high-throughput experiments ***(18–21)*** yielded the most extensive protein–protein interaction map to date, despite a presumed large number of false positives. *Human, mouse*, and *Escherichia coli* have more proteins crystallized, but their protein–protein interaction coverage is far less satisfying; on the other hand, *fly* and *worm* have large protein–protein interaction datasets available but unsatisfactory coverage in the Protein Data Bank (PDB). Further, among the protein networks, the yeast network is the one that has been widely explored in terms of interactions ***(22)***. The structural details of a very high percentage of its proteins in the network are well identified ***(23)***.

Interestingly, we observe that among the properties we study (e.g., residue conservation, composition, distribution of residue types, exposure/burial, location of bulges on the β-strands) with respect to the connectivity of the proteins, highly connected β-rich proteins tend to have clusters of conserved residues, and these are largely buried in the protein cores. These well-packed clusters are likely to lend stability to the proteins, thus lowering the chance of partial unfolding and amyloid formation through the β-prone residues populating these structures. Consequently, to use β-structures in design, one strategy to avoid polymerization is to stabilize the protein cores.

This work is within the nanostructure design framework carried out by our group to employ naturally occurring proteins and their building blocks in such design efforts ***(24–30)***.

2. Materials

We outline the strategy we have used to study the high-connectivity β-rich structures in the yeast proteome. A similar strategy can be followed in studies of other features or species.

2.1. Dataset Determination

Since we are interested in structural analysis of the yeast proteome *(23)*, we first determined the list of yeast proteins with structures that are available. The list of yeast PDB files *(31)* was extracted from the National Center for Biotechnology Information (NCBI) taxonomic table of PDB proteins, totaling 775 files and 346 yeast proteins. Some proteins have been repeatedly crystallized, leading to more PDB files than proteins. For example, there are more than 80 available structures of cytochrome c peroxidase.

To determine the nonredundant list of proteins in the PDB, we have cross referenced the PDB with SwissProt *(32)*. SwissProt references from PDB files were checked, and missing SwissProt references were determined by BLAST (Basic Local Alignment Search Tool) *(33)*. Only chains longer than 10 residues were considered. Each protein was represented by that chain in the PDB that had the highest sequence identity with the entire protein sequence as obtained from the SwissProt database. Proteins determined by cryoelectron microscopy were excluded since the resolution of these structures is too low. This left 303 proteins, 3 of which were redundant due to PDB listings with both primary and secondary SwissProt accession numbers. Preference was given to X-ray structures, if available.

A protein was considered to have a *good structural coverage* if its best PDB representative covered at least 80% of the entire protein sequence. Proteins without sufficient structural coverage were excluded from our analysis since important interfaces or domains might be missing from the structure.

2.2. Database of Interacting Proteins Dataset

The interactivity data was taken from DIP, the Database of Interacting Proteins *(34,35)*, which contains the entire yeast dataset and core yeast dataset. The complete yeast dataset as of July 4, 2004, consisted of 4772 proteins and 15,461 interactions. The core dataset consisted of 2415 proteins and 6574 interactions. We used the SwissProt ID to get the mapping between DIP accession number and the PDB file. Of our 300 proteins represented in the PDB, 172 have good structural coverage in the PDB, 263 have connectivity data, and 152 have both. We further eliminated proteins with some missing structural information. We finally obtained 146 unique chains from proteins of different families with known connectivity and structural data, and this constituted our main database, set 1. This set is given in **Table 1**.

2.3. Connectivity

Connectivity is defined as the number of nodes (proteins) with which a given node (protein chain) interacts. Nodes with 5 or fewer partners are considered low connected, and nodes with more than 10 partners are considered high connected.

2.4. Secondary Structure

We used backbone dihedral angles (ϕ, ψ) criteria to identify secondary structures—extended conformation: ϕ = −180 to −30, ψ = 60 to 180 and −180 to −150; right-handed helical conformation: ϕ = −140 to −30, ψ = −90 to 45; left-handed helical conformation: ϕ = 20 to 125, ψ = −45 to 90. A segment is identified as a β-strand or α-helical if at least four consecutive residues are present in an extended conformation or α-helical conformation, respectively *(36)*. We divided set 1 into subset datasets according to the secondary structure content to identify proteins as β-rich or α-rich structures. We used both the percentage (of the total) and the ratio (β residues/α residues) values as criteria. The sets are listed in **Table 1**.

Sets 2 and are composed of β-rich and α-rich proteins, respectively, based on the ratio of β-residues to α-residues. Set 2 was our primary set, whereas sets 1 and 3 and sets A to D, which are based on the percentages described in this subheading, were used for comparison. We tested different ratio values of the β-strand to α-helix residues in the construction of sets 2 and 3, respectively. Our goal was to obtain datasets with a reasonable number of members, yet distinguishing β- from α-rich structures. For set 2, we settled on 1.4 since higher values and smaller values of this ratio led to either too small a dataset or a misleading β-rich behavior, respectively. For set 3, 0.7 (~1/1.4) was used. We had 14 structures in set 2 with a ratio of at least 1.4 β-residues to α residues and 80 structures in set 3 with a ratio of at most 0.7 β-residues to α-residues. This led to only 14 distinct β-rich structures to analyze as compared with 80 distinct α-rich structures, reflecting the abundance of α-structures as compared to β-structures.

Given this disparity, sets A–D were also compiled. Here, we have used the percentage of residues with a specific secondary structure with respect to the total number of residues. Set A was composed of structures with a percentage of β-strand residues higher than 30 (25 structures). Set B consisted of structures with the percentage of α-helix residues less than 20 (9 structures). Set C consisted of structures with a percentage of α-helix residues less than 30 (23 structures). Set D, which was constructed by removing elements of set 2 from set C (leading to a set of β- and α-poor structures) was also employed to further test whether the set 2 results originated from the β-rich behavior of the structures and not from their low α-helix content.

2.5. Conservation

We computed the conservation data for each residue using the ConSurf-HSSP database *(37)*, the bulge and secondary structure data using PDBsum[21]. The sequence data were obtained from the PDB structure files. ConSurf-HSSP employs multiple-sequence alignment to calculate and normalize the

Table 1
Datasets with the structures and the number of structures, N, in each

Datasets (N)	Structures
Set 1 (142)	**All Structures:** 1a4eD, 1afwB, 1afwB, 1ejbE1, f5mB, 1f89B, 1fa0B, 1fntS, 1gpuB, 1gy7B, 1hr6F, 1i1dD, 1ig0B, 1jd1C, 1jd2W, 1jd21, 1jd22, 1jd2P, 1k39B, 1kyoM, 1kyqB, 1n0vD, 1n9pD, 1ox4B, 1pguB, 1pvdB, 1qhfB, 1qsmD, 1qvvB, 1rypY, 1tlbW, 1u4cB, 1yaaD, 1ypiB, 1yprB, 1auaZ, 1edzA, 1eq6A, 1b54Z, 1cofZ, 1sxjB, 1qcqA, 1ofrD, 1i3qE, 1r5uH, 1wgiA, 1h7rA, 1g62A, 1jk0A, 1f7vA, 1qpgZ, 1jk9B, 1jd2C, 1plqZ, 1g6q1, 1ci0A, 1id3C, 1dqwA, 1hr6A, 1q0wB, 2uczZ, 1kb9H, 1a3wA, 1ek0A, 1ry2A, 1g7cA, 1qsdA, 1qqeA, 1i3ql, 1jatB, 1jbbA, 1lbqA, 1sxjE, 1m2oA, 1yatZ, 1asyA, 1ig8A, 1g65L, 1istA, 1jehA, 1p1kA, 1ky2A, 1kbiA, 1m2vB, 1kl7A, 1i50C, 1hv2A, 1cmxA, 1kb9E, 1gcbZ, 1mqsA, 1jd2E, 1bobZ, 1ebfA, 1nh2D, 1fpwA, 1n6zA, 1rklA, 1un0A, 1lxjA, 1m2oB, 1n0hA, 1lkjA, 1nynA, 1g2qA, 1ho8A, 1dp5A, 1d1qA, 1ap8Z, 1jihA, 1dl2A, 1ox7A, 1ka1A, 1fi4A, 1ukzZ, 1yagA, 1j70A, 1ixvA, 1sxjD, 1rjdA, 1i4wA, 1akyZ, 1rypl, 1sq9A, 1m46A, 1mr3F, 1kyoG, 2pccB, 1odfA, 1g65M, 2oneA, 1kokA, 1fuuB, 1t0iA, 1ayzA, 2jcwZ, 1i6hK, 1kb9A, 1jd2A, 1h4pA, 1ps0A, 1i6hJ
Set 2 (14)	**β Strand Residues / α Helix Residues >1.4**: 1eq6A, 1jk9B, 1u4cB, 1pguB, 1g7cA, 1sq9A, 1g6q1, 2jcwZ, 1gy7B, 1wgiA, 1plqZ, 1yatZ, 1istA, 1q0wB
Set 3 (80)	**β Strand Residues / α Helix Residues <0.7**: 1a4eD, 1afwB, 1ejbE, 1f5mB, 1fntS, 1gpuB, 1hr6F, 1jd2W, 1jd21, 1jd22, 1k39B, 1n8pD, 1pvdB, 1qhfB, 1qvvB, 1tlbW, 1yaaD, 1ypiB, 1auaZ, 1edzA, 1b54Z, 1sxjB, 1qcqA, 1ofrD, 1id3qE, 1h7rA, 1jk0A, 1f7vA, 1qpgZ, 1jd2C, 1id3C, 1dqwA, 1hr6A, 2uczZ, 1a3wA, 1ry2A, 1qsdA, 1qqeA, 1jbbA, 1ibqA, 1sxjE, 1m2oA, 1ig8A, 1kbiA, 1m2vB, 1kl7A, 1hv2A, 1cmxA, 1gcbZ, 1mqsA, 1ebfA, 1fpwA, 1un0A, 1n0hA, 1lkjA, 1ho8A, 1d1qA, 1dl2A, 1ox7A, 1ka1A, 1fi4A, 1ukzZ, 1yagA, 1ixvA, 1sxjD, 1rjdA, 1i4wA, 1akyZ, 1m46A, 2pccB, 1odfA, 2oneA1kokA, 1fuuB, 1t0iA, 1ayzA, 1h4pA, 1r5tA, 1ps0A, 1i6hJ
Set A (25)	**>30% β strand**: 1pguB, 1sq9A, 1gy7B, 1u4cB, 1plqZ, 1nh2D, 1jk9B, 1eq6A, 1yatZ, 1g6q1, 1ci0A, 2jcwZ, 1rypl, 1g7cA, 1yprB, 1q0wB, 1rypY, 1i1dD, 1cofZ, 1istA, 1qsmD, 1lxjA, 1g2qA, 1g65L, 1g65M
Set B (9)	**<20%α helix**: 1jk9B, 1u4cB, 1pguB, 1sq9A, 2jcwZ, 1wgiA, 1plqZ, 1yatZ, 1istA
Set C (23)	**<30% β helix (Encloses Set 2)**: 1eq6A, 1ci0A, 1ps0A, 1ry2A, 1jk9B, 1jatB, 1asyA, 1u4cB, 1pguB, 1g7cA, 1g65L, 1sq9A, 1g6q1, 1rypl, 2jcwZ, 1gy7B, 1wgiA, 1plqZ, 1i50C, 1yatZ, 1istA, 1q0wB, 1ap8Z
Set D (9)	**<30% α helix and <1.4 β/α (Set C – Set 2)**: 1g6q1, 1r5tA, 1ek0A, 1i3ql, 1yatZ, 1ig8A, 1akyZ, 1kl7A, 1d1qA

conservation scores for residues and the HSSP database to collect homologs of proteins to take the three-dimensional (3D) structure into account ***(37)***. The HSSP homolog database appeared to provide better results than the UniProt homolog database ***(38,39)***, especially in identifying functional regions in proteins ***(37)***. For the two low-connectivity cases in set 1, only a few homologs were available in the HSSP database (13 homologs for 1eq6 [*Saccharomyces cerevisiae* ran-binding protein]; 7 homologs for 1jk9 [*Saccharomyces cerevisiae* superoxide dismutase]).

A residue is considered as highly conserved if the conservation score is 8 or 9 on a 1–9 scale and conserved if the conservation score is higher than 6 on the 1–9 scale in the ConSurf-HSSP database ***(37)***. We focused mainly on highly conserved residues with a conservation score of 8 or 9. The conservation score of a residue is a normalized value with respect to other residues in a given protein and can be compared only within the protein but not with the scores of other proteins. It is used only to determine and record the conserved residues in each protein.

2.6. Residue Type and Accessibility

Residues are labeled as polar if they are one of the following amino acids: Arg, Asn, Asp, Csy, Gln, Glu, His, Lys, Ser, Thr, and Tyr; Arg, His, and Lys are considered as plus charged, Asp and Glu as minus charged, and Cys, Ile, Leu, Met, Phe, Pro, Trp, Tyr, Val, and Ala as hydrophobic; His, Phe, Trp, and Tyr are considered aromatic. To determine the surface-exposed and buried residues in a protein, the accessible surface areas (ASAs) of the residues were calculated using the Lee and Richards algorithm ***(40)***. A probe radius of 1.4 Å was used in the calculation. The percentage of ASA of an individual residue in a protein was calculated using the standard value of maximum ASA of a residue *X* placed in a Gly-*X*-Gly motif. Residues with 20% or less ASA were identified as buried and those with greater than 20% ASA as exposed.

We carried out statistical and visual analysis to extract information using connectivity as the independent variable and changing the dependent variables for the three different datasets: all structures (set 1), β-rich structures (set 2), and α-rich structures (set 3). Datasets A, B, C, and D served as control sets and were subjected to the same analysis. In the statistical analysis of the datasets, we have searched for the distributions of the highly conserved and almost conserved residues, bulges, charged and polar residues, hydrophobic and aromatic residues over different secondary structural regions, and these residue's respective positions in the structure as core (buried) or surface residues. For the distribution of the conserved residues, a clustering algorithm has been used to search for the extent of association between the conserved residues ***(41)***. The function of the protein has also been considered in the analysis as the interaction with other partners and the functions of the proteins in the cell are intimately related ***(2)***.

3. Results and Discussion

We analyzed the 142 proteins in set 1. The distribution of the conserved residues in the secondary structural units and the conserved residues' preference to be on the surface or in the core of the protein showed no correlation with the connectivity. Similarly, no correlation was observed between the distribution of the hydrophobic, charged, and polar residues and the connectivity.

3.1. Percentages of Conserved Residues

Figure 1 displays the percentages of the conserved residues as a function of the connectivity for each structure in set 1 and for the 14 β-rich structures in set 2. As the connectivity increased, the percentage of the conserved residues decreased for β-rich structures but shows an almost random distribution for the structures of set 1. This behavior suggests that in β-rich structures the number of conserved residues is likely to be higher for low-connected proteins. On the other hand, this number decreased as the protein connected to more partners in the network. Since the size of the dataset was small, to verify that this decrease was not incidental we analyzed five additional datasets with 14 proteins randomly drawn from the 142 proteins. A trend with a nonzero slope was not observed for any of these new datasets.

3.2. Percentages of β-Strand Residues

Figure 2 displays the percentage of the β-strand residues, surface residues, and conserved residues versus the connectivity for each β-rich structure in set 2. No significant correlation has been observed for the dependence of the percentages of the secondary structures and of the surface residues on the connectivity. This implies that the correlation with the conservation observed in **Fig. 1** is not related to the changes in the secondary structure or the surface residue content that might be associated with the change in the connectivity.

3.3. Location and Distribution of Conserved Residues

We analyzed the 14 cases to determine the location and the distribution of the conserved residues in 3D space. We observed that the highly conserved residues were distributed over the protein structure for low-connected proteins. However, the highly conserved residues were more likely to be localized at certain regions in β-rich proteins with high connectivity (**Fig. 3a,b**). The clusters of the highly conserved residues might provide higher stabilities for the structures. This was not the case in highly connected α-proteins. The distribution of the conserved residues and their clustering in the structure are further elaborated.

Figure 4a–d displays examples of conserved β-strand residues (dots) for structures with β-rich proteins with low and high connectivity. Surface residues are presented as a layer. In the two low-connectivity cases, the surface layer is comprised of overlapping dots (dark regions), implying that conserved β-strand

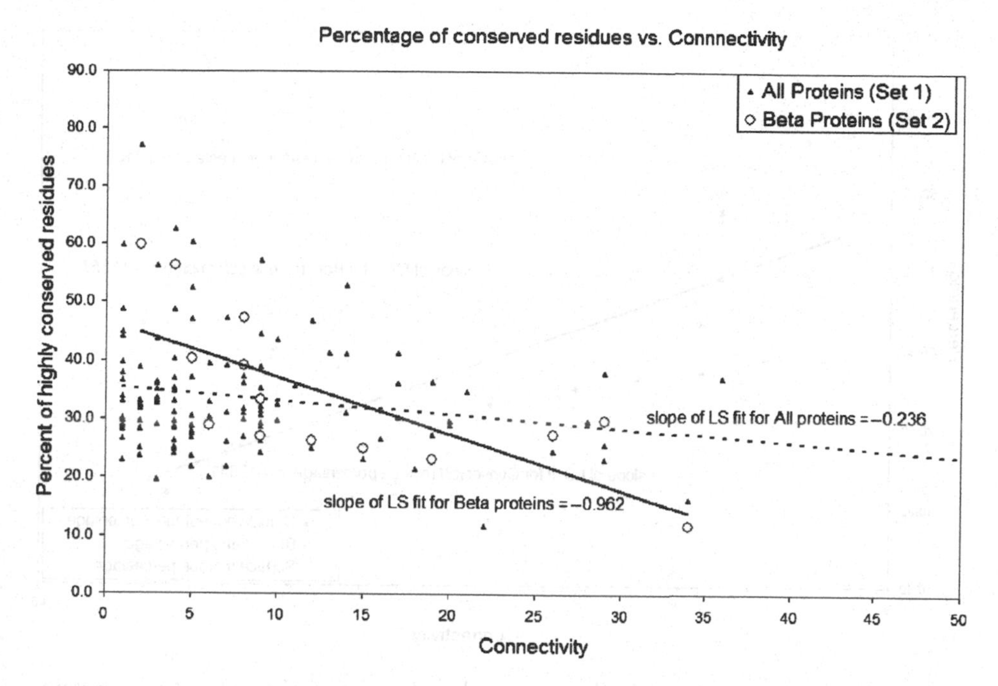

Fig. 1. The percentage of highly conserved residues for sets 1 and 2 versus the connectivity. Dispersion of triangles (set 1) and almost zero slope of fitted line show that connectivity was uncorrelated with conserved residue percentages. However, circles (β-rich structures, set 1) show decreasing trend with increasing connectivity.

residues were on the surface. In the two low-connectivity cases, the surface is comprised of darker regions, implying that conserved β-strand residues were on the surface. In structures with high connectivity, none of the conserved β-strand residues were on the surface, although the overall β-strand percentages on the surface did not show a change with connectivity (**Fig. 2**). Further, as discussed next, the results of the statistical analysis indicated that in low-connectivity β-rich structures the distribution of the conserved residues is more homogeneous, whereas in high-connectivity conserved residues tend to cluster in the protein core.

3.4. Using a Clustering Algorithm for Determination of the Degree of Localization

To determine the degree of localization, we used a simple clustering algorithm ***(41)*** to cluster the conserved β-strand residues based on the distance between the conserved residues. Conserved β-strand residues were collected into a pool, and the distances between each pair of residues were calculated. For each residue, the number of neighbors within an 8-Å distance was calculated. The residue with the highest number of neighbors was considered the center of the first cluster. All members of the cluster were removed from the pool. The

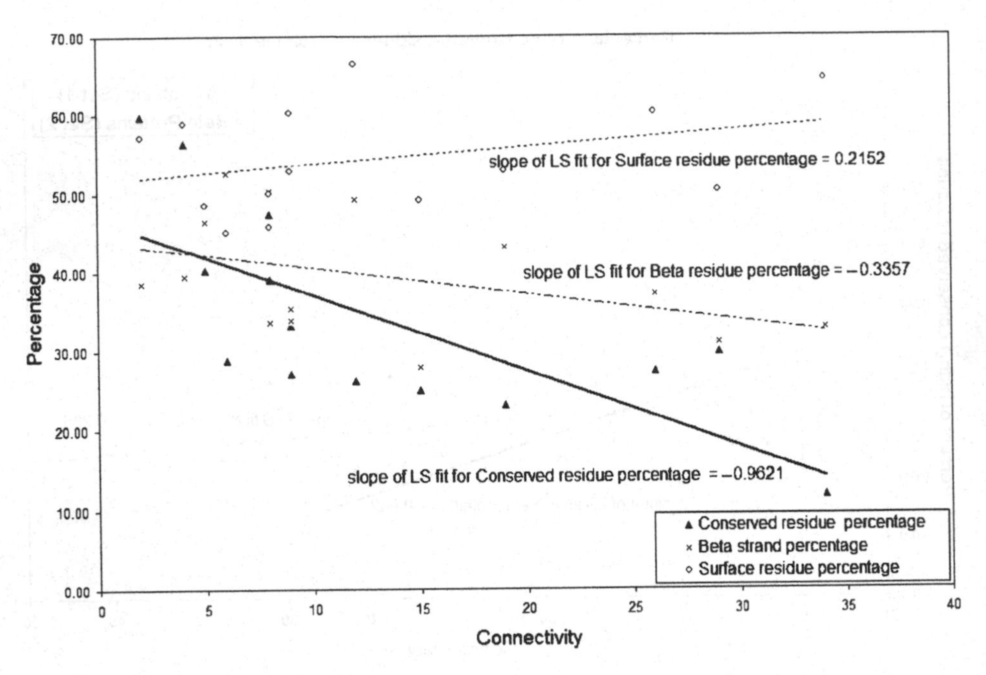

Fig. 2. The percentages of the β-, conserved, and surface residues for β-rich proteins (set 2) versus connectivity. Dashed line (triangles) shows a decreasing trend in percentage of the conserved residues with increasing connectivity. The surface residue (circles) and β-strand percentages (crosses) show random distribution.

center of the second cluster was similarly determined using the new pool of the conserved β-strand residues. The procedure was repeated until each conserved residue in the pool was assigned to a cluster. As the total number of residues and the secondary structure content varied from one structure to another, to normalize the results the percentage of the residues in the largest two clusters with respect to the total number of conserved β-strand residues was determined.

The clustering results are given in **Fig. 3**. In **Fig. 3a**, the bars represent the percentage of the two largest cluster sizes from the total number of clustered residues. **Figure 3b** gives the number of clusters versus connectivity. For

→

Fig. 3. (**a**) Degree of clustering: the ratio of the number of the residues in the largest two clusters to total number of residues clustered, which are the conserved β-strand residues versus connectivity for β-rich structures; *x*-axis represents the Protein Data Bank (PDB) code and connectivity information, and *y*axis represents the percentage. Apparent increase after sixth structure for β-rich proteins shows that for highly connected cases conserved β-strand residues are more likely to be in the clusters.

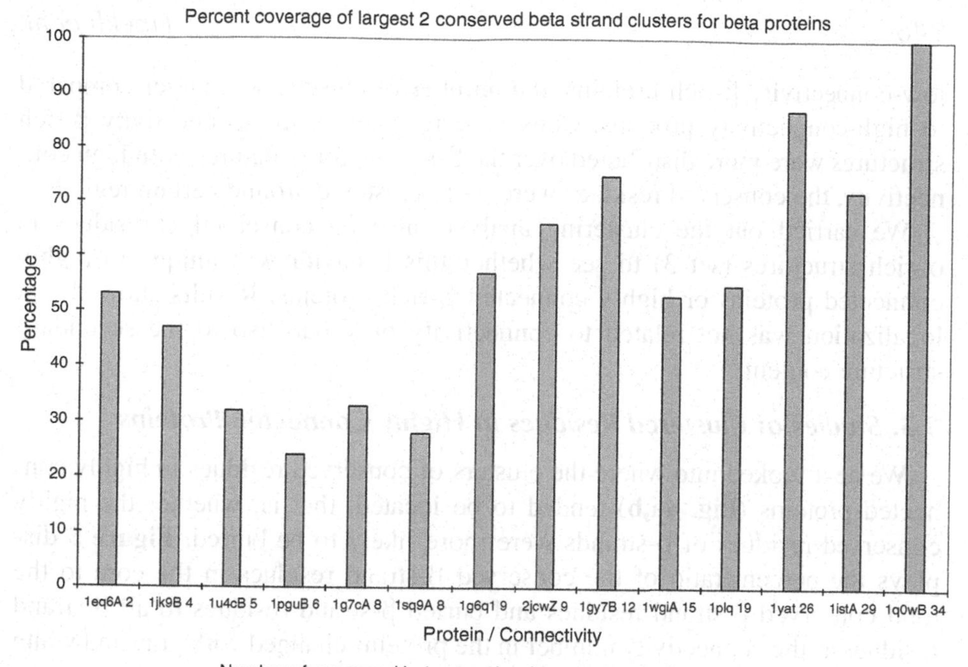

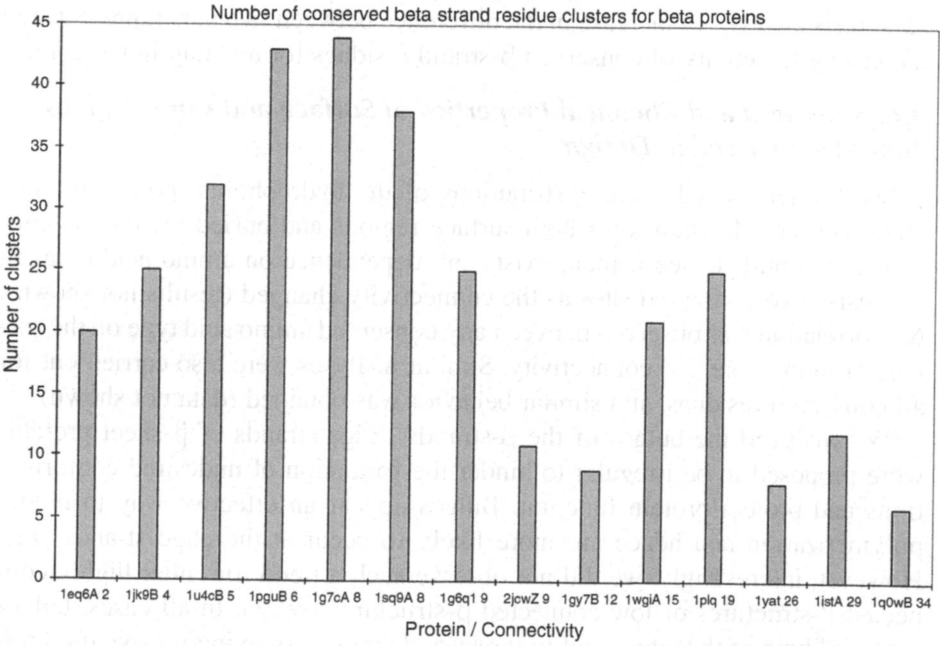

Fig. 3. (*continued*) Localization is not apparent for the α-rich structures. (**b**) Number of clusters versus connectivity. The *x*-axis represents the PDB code and connectivity information, and the *y*-axis represents the number of clusters. Clustering includes all conserved β-strand residues. There is an apparent decrease after the sixth protein, which shows fewer clusters and more localization for the highly connected structures, which could not be observed at α-helix-rich structures.

low-connectivity β-rich proteins, the number of clusters was larger compared to high-connectivity proteins. Conserved residues in low-connectivity β-rich structures were more distributed over the β-sheets; for structures with high connectivity, the conserved residues were more clustered around certain regions.

We carried out the clustering analysis also on conserved α-residues in α-rich structures (set 3) to see whether this behavior was unique to highly connected proteins or highly connected β-rich proteins. Results showed that localization was not related to connectivity only but also to the secondary structure content.

3.5. Studies of Clustered Residues in Highly Connected Proteins

We next looked into where the clusters of conserved residues in highly connected proteins (**Fig. 4a,b**) tended to be located, that is, whether the highly conserved residues of β-strands were more likely to be buried. **Figure 5** displays the percent ratio of the conserved β-strand residues in the core to the total conserved β-strand residues and buried β-strand residues to all β-strand residues as the connectivity number in the proteins changed with gray and white bars, respectively. An increase in the difference between two bars represents the increasing selectivity of conserved β-strand residues for residing in the core.

3.6. Structural and Chemical Properties of Surface and Core Regions That May Be Used in Design

We further looked at the distributions of the hydrophobic, polar, and aromatic conserved residues for both surface regions and buried regions on conserved β-strands to see if there exists any dependence on amino acid types in the respective conserved sites as the connectivity changed (results not shown). No correlation was observed between any conserved amino acid type on the surface or in the core and connectivity. Similar analyses were also carried out for all conserved residues, and similar behavior was obtained (data not shown).

We analyzed the bulges of the β-strands. Edge strands of β-sheet proteins were proposed to be irregular to hinder the formation of undesired conformations and protect protein function. Bulges appear an effective way to hinder polymerization and hence are more likely to occur at the edge strands *(12)*. However, interestingly, we did not observe such a trend for either highly connected β-structures or low-connected β-structures (set 1). In all cases, bulges occurred both at the edges and in the buried strands, showing no specific preference for being at the edge or at the highly conserved regions. For example, for propeller geometries, there were many β-bulges distributed all over the sheets. This geometry has been observed in many families, especially for low-connectivity cases (proteins with connectivities of 3 to 6). In those cases, highly conserved residues were distributed rather than clustered.

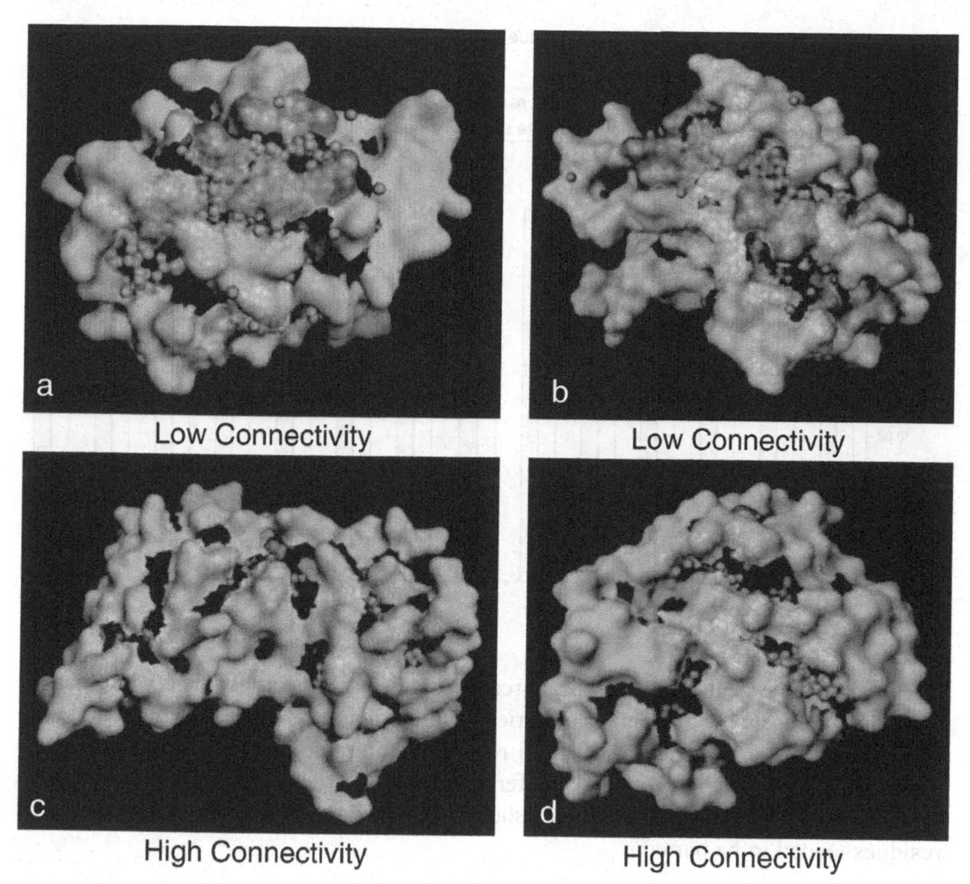

Fig. 4. Two structures with low connectivity: 1eq6 chain A (connectivity = 2; involved in nuclear protein import) (**a**), 1u4c chain B (connectivity = 5; required for cell cycle arrest in response to loss of microtubule function) (**b**). Surface residues are shown as a layer, and conserved β-strand residues are shown as dots. Darker parts of the layer represent the conserved β-strand residues on the surface. Many conserved β-strand residues are available on the surface. Buried conserved β-strand residues can be observed as spots in the inner parts. Two structures with high connectivity: 1plq (connectivity = 19; an auxiliary protein of DNA polymerase delta) (**c**) and 1wgi chain A (connectivity = 15; catalytic activity, converts diphosphate and H_2O to two phosphate) (**d**). Nonconserved surface residues (no darker region on the outer layer) cover the inner conserved β-strands (spots inside the core). Conserved regions can only be seen at the inner part.

We analyzed set A, which contained proteins with α-helix residue percentage less than 30%, that is, α-helix-poor structures (**Fig. 6**). This dataset consisted of 26 proteins, including the 14 β-rich structures (set 2). The dependence of the conserved residues on the connectivity showed a trend similar to that of the

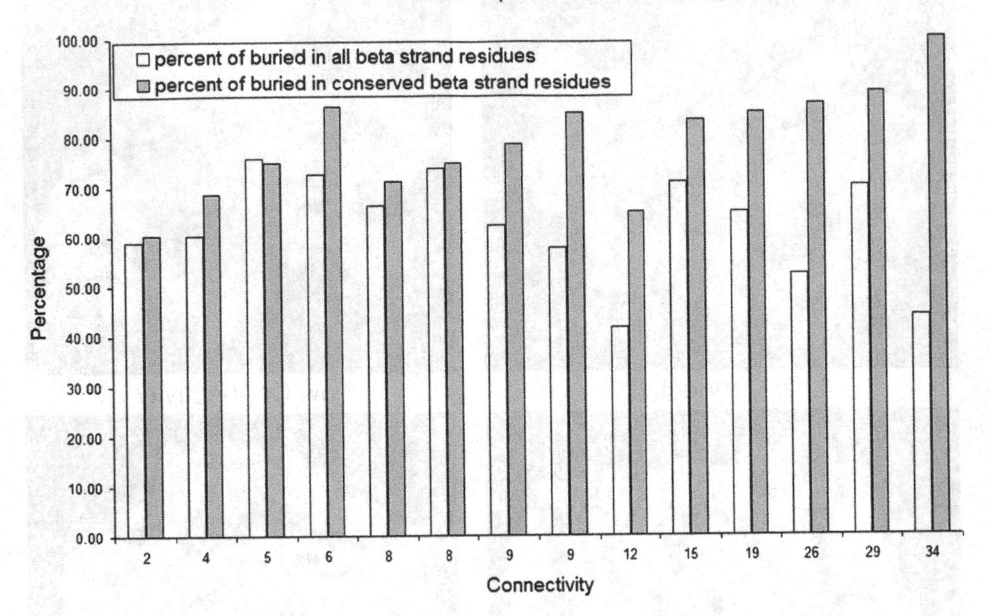

Fig. 5. The percentage of the buried residues versus connectivity. The white bars in the graph represent the percentage of buried residues among β-strand residues, and gray bars represent the percentage of buried residues among conserved β-strand residues. The difference of the bars gives the preference of conserved residues to be buried. The increase in the difference between bars shows that as connectivity increased, conserved residues tended to be buried.

β-rich-only structures, however, with a smaller slope value. Next, we excluded the 14 β-rich structures, leaving only the 12 α-poor (set D) ones. Those 12 proteins showed no trend in the behavior of their conserved residue percentage with respect to the connectivity. The percentage of the highly conserved residues versus connectivity plot displays a random distribution. This result shows that the decreasing trend in the percentage of highly conserved residues was the outcome of β-richness rather than α-poorness.

We also inspected the homolog structures for 5 of 14 β-rich proteins in our primary set 2: 1b64, 1or8(A), 1srd(B), 1fkk, and 1awq were analyzed similarly as the homolog structures of 1g7c(A), 1g6q(1), 2jcw, 1yat, and 1ist(A), respectively. The homology between the structures varied from 50% to 70% with e-values well below 10^{-10}. The aim in this analysis was to determine whether it is possible to draw the same conclusions with homologous structures. The results for these proteins verified that for high-connectivity cases the conserved residues decreased in percentage and tended to be buried, while this was not the case for low-connectivity proteins.

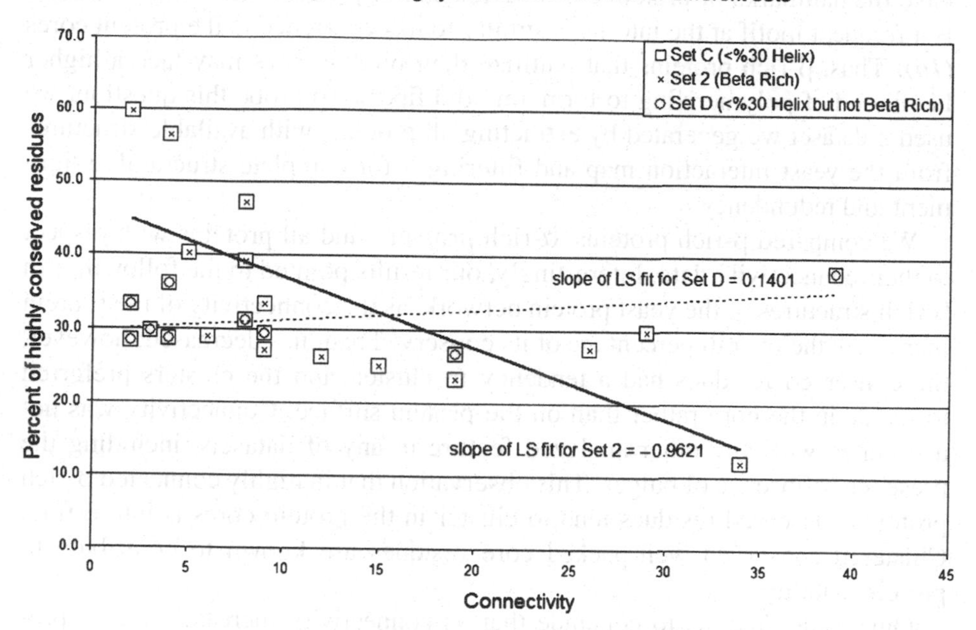

Fig. 6. The percentage of conserved residues versus connectivity for sets 2, C, and D (which is set C minus set 2). There was an obvious decrease in sets 2 and C, and in contrast the slight increase in set D, which is formed by removing set 2 elements in set C, showed that the decrease in the trend was caused by set 2 structures specifically. This means the decreasing trend in the percentage of highly conserved residues originated from β-richness of the structures rather than α-poorness.

4. Conclusions

We asked whether β-rich proteins that interact with many partners have features distinguishing them from β-rich proteins that interact with few partners. Availability of such features can be used in interface design. Our premise was that since evolution has perfected the structures of multiply interacting protein hubs, we may obtain useful features from such an analysis. In particular, we focused on β-rich proteins since in principle such proteins may lead to polymeric fibrils. While knowledge of the binding sites for most proteins in the yeast interaction map is lacking, here we assumed that β-rich proteins interact via β-strand addition. Binding sites of central hub proteins that interact with a large number of partners are likely to be multiply used. For binding sites consisting of β-structures, the ability to interact with multiple partners may suggest that they might be more sensitive to amyloid formation. Protein–protein interfaces have the hallmarks of protein cores. Extension of β-sheets across the interface is a frequent motif at the interface, similar to its occurrence in the protein cores *(16)*. Thus, β-rich proteins that reutilize their binding sites may face a higher likelihood of polymerizing to form amyloid fibers. To probe this question, we used a dataset we generated by extracting all proteins with available structures from the yeast interaction map and filtering it for complete structural assignment and redundancy.

We compared β-rich proteins, α-rich proteins, and all proteins with respect to their connectivity data. Interestingly, our results pointed to the following: In β-rich structures of the yeast protein network, as the connectivity of the protein increased, the overall percentage of its conserved residues decreased; however, the conserved residues had a tendency to cluster, and the clusters preferred to reside in the core rather than on the protein surface. Connectivity was not associated with any other analyzed feature in any of datasets, including the presence or absence of bulges. This observation that in highly connected β-rich proteins conserved residues tend to cluster in the protein cores is interesting. Clustered conserved well-packed core residues are known to contribute to protein stability.

Our results lead us to conclude that as connectivity increases, β-rich proteins bury and pack their conserved residues and tend to have more sequence variations on the surface. The conserved residues serve as clustered hot spots in the core to stabilize the fold, hindering the (partial) unfolding and amyloid formation of the β-structure-prone sequences. A similar mechanism of stabilization has been observed in protein–protein interfaces *(42)*. Consequently, these are the features that might be employed when using multi-interacting proteins as building blocks in design.

Acknowledgments

The dataset of proteins used in this study was generated by Kristina Rogale. We thank Isil Ulug in the analysis for bulges. This research was supported in part by the Intramural Research Program of the National Institutes of Health, National Cancer Institute, Center for Cancer Research and was funded in whole or in part with federal funds from the National Cancer Institute, National Institutes of Health, under contract NO1-CO-12400. The content of this publication does not necessarily reflect the views or policies of the Department of Health and Human Services, and mention of trade names, commercial products, or organizations does not imply endorsement by the U.S. government. T. Haliloglu acknowledges the Turkish Academy of Sciences in the framework of the Young Scientist Award Program (EA-TUBA-GEBIP/2001-1-1), State Planning Organization grants 03K120250 and EU_FP6-2004-ACC-SSA-2: Project 517991.

References

1. Pellegrini M, Haynor D, Johnson JM. (2004) Protein interaction networks. *Expert Rev. Proteomics* **1**(2), 239–249.
2. Jeong H, Mason SP, Barabasi AL, Oltvai ZN. (2001) Lethality and centrality in protein networks. *Nature* **411**(6833), 41–42.
3. Hahn MW, Conant GC, Wagner A. (2004) Molecular evolution in large genetic networks: does connectivity equal constraint? *J. Mol. Evol.* **58**(2), 203–211.
4. Fraser HB, Hirsh AE. (2004) Evolutionary rate depends on number of protein-protein interactions independently of gene expression level. *BMC Evol. Biol.* **4**, 13.
5. Bloom JD, Adami C. (2003) Apparent dependence of protein evolutionary rate on number of interactions is linked to biases in protein–protein interactions data sets. *BMC Evol. Biol.* **3**, 21.
6. Bloom JD, Adami C. (2003) Evolutionary rate depends on number of protein–protein interactions independently of gene expression level: response. *BMC Evol. Biol.* **4**, 14.
7. Keskin O, Nussinov R. (2007) Similar binding sites and different partners: implications to shared proteins in cellular pathways. *Structure* **15**, 341–354.
8. Gunasekaran K, Ramakrishnan C, Balaram P. (1997) Beta-Hairpins in proteins revisited: lessons for de novo design. *Protein Eng.* **10**(10), 1131–1141.
9. de la Cruz X, Hutchinson EG, Shepherd A, Thornton JM. (2002) Toward predicting protein topology: an approach to identifying beta hairpins. *Proc. Natl. Acad. Sci. U. S. A.* **99**(17), 11157–11162.
10. Fooks HM, Martin AC, Woolfson DN, Sessions RB, Hutchinson EG. (2006) Amino acid pairing preferences in parallel beta-sheets in proteins. *J. Mol. Biol.* **356**(1), 32–44.
11. Serpell LC, Sunde M, Blake CC. (1997) The molecular basis of amyloidosis. *Cell. Mol. Life Sci.* **53**(11–12), 871–887.
12. Richardson JS, Richardson DC. (2002) Natural β-sheet proteins use negative design to avoid edge-to-edge aggregation. *Proc. Natl. Acad. Sci. U. S. A.* **99**(5), 2754–2759.
13. Mousseau N, Derreumaux P. (2005) Exploring the early steps of amyloid peptide aggregation by computers. *Acc. Chem. Res.* **38**(11), 885–891.
14. Santini S, Mousseau N, Derreumaux P. (2004) In silico assembly of Alzheimer's $Abeta_{16-22}$ peptide into beta-sheets. *J. Am. Chem. Soc.* **126**(37), 11509–11516.
15. Langedijk JP, Fuentes G, Boshuizen R, Bonvin AM. (2006) Two-rung model of a left-handed beta-helix for prions explains species barrier and strain variation in transmissible spongiform encephalopathies. *J. Mol. Biol.* **360**(4), 907–920.
16. Tsai CJ, Xu D, Nussinov R. (1997) Structural motifs at protein–protein interfaces: protein cores versus two-state and three-state model complexes. *Protein Sci.* **6**, 1793–1805.
17. Remaut H, Waksman G. (2006) Protein–protein interaction through β-strand addition. *Trends Biochem. Sci.* **31**, 436–444.
18. Ito T, Chiba T, Ozawa R, Yoshida M, Hattori M, Sakaki Y. (2001) A comprehensive two-hybrid analysis to explore the yeast protein interactome. *Proc. Natl. Acad. Sci. U. S. A.* **98**(8), 4569–4574.

19. Uetz P, Giot L, Cagney G, et al. (2000) A comprehensive analysis of protein-protein interactions in *Saccharomyces cerevisiae. Nature* **403(6770)**, 623–27.
20. Ho Y, Gruhler A, Heilbut A, et al. (2002) Systematic identification of protein complexes in *Saccharomyces cerevisiae* by mass spectrometry. *Nature* **415**(6868), 180–183.
21. Gavin AC, Bosche M, Krause R, et al. (2002) Functional organization of the yeast proteome by systematic analysis of protein complexes. *Nature* **415**(6868), 141–147.
22. Laskowski RA. (2001) PDBsum: summaries and analyses of PDB structures. *Nucleic Acids Res*. **29**(1), 221–222.
23. Schwikowski B, Uetz P, Fields S. (2000) A network of protein–protein interactions in yeast. *Nat. Biotechnol.* **18**(12), 1257–1261.
24. Tsai CJ, Zheng J, Zanuy D, et al. (2007) Principles of nanostructure design with protein building blocks. *Proteins* **68**(1), 1–12.
25. Haspel N, Zanuy D, Zheng J, Aleman C, Wolfson H, Nussinov R. (2007) Changing the charge distribution of {beta}-helical based nanostructures can provide the conditions for charge transfer. *Biophys. J.* **93**(1), 245–253.
26. Tsai CJ, Zheng J, Alemán C, Nussinov R. (2006) Structure by design: from single proteins and their building blocks to nanostructures. *Trends Biotechnol.* **24**, 449–454.
27. Zanuy D, Nussinov R, Alemán C. (2006) From peptide-based material science to protein fibrils: discipline convergence in nanobiology. *Phys. Biol.* **3**, S80–S90.
28. Alemán C, Zanuy D, Jiménez AI, et al. (2006) Concepts and schemes for the re-engineering of physical protein modules: generating nanodevices via targeted replacements with constrained amino acids. *Phys. Biol.* **3**, S54–S62.
29. Tsai CJ, Zheng J, Nussinov R. (2006) Designing a nanotube using naturally occurring protein building blocks. *PLoS Comput. Biol.* **2**, e42.
30. Zheng J, Zanuy D, Haspel N, Tsai CJ, Aleman C, Nussinov R. (2007) Nanostructure design using protein building blocks enhanced by conformationally constrained synthetic residues. *Biochemistry* **46**, 1205–1218.
31. Berman HM, Battistuz T, Bhat TN, et al. (2000) The Protein Data Bank. *Nucleic Acids Res*. **28**, 235–242.
32. Boeckmann B, Bairoch A, Apweiler R, et al. (2003) The SWISS-PROT protein knowledgebase and its supplement TrEMBL in 2003. *Nucleic Acids Res*. **31**, 365–370.
33. Altschul SF, Madden TL, Schaffer AA, et al. (1997) Gapped BLAST and PSI-BLAST: a new generation of protein database search programs. *Nucleic Acids Res*. **25**, 3389–3402.
34. Marcotte EM, Xenarios I, Eisenberg D. (2001) Mining literature for protein-protein interactions. *Bioinformatics* **17**, 359–363.
35. Salwinski L, Miller CS, Smith AJ, Pettit FK, Bowie JU, Eisenberg D. (2004) The Database of Interacting Proteins: update. *Nucleic Acids Res*. **32**, 449–451.
36. Hamelryck T, Manderick, B. (2003) PDB file parser and structure class implemented in Python. *Bioinformatics* **19**(17), 2308–2310.

37. Glaser F, Rosenberg Y, Kessel A, Pupko T, Ben-Tal N. (2005) The ConSurf-HSSP database: the mapping of evolutionary conservation among homologs onto PDB structures. *Proteins* **58**, 610–617.
38. Apweiler R, Bairoch A, Wu CH, et al. (2004) UniProt: the Universal Protein Knowledgebase. *Nucleic Acids Res.* **32**, D115–D119.
39. Glaser F, Pupko T, Paz I, et al. (2003) ConSurf: identification of functional regions in proteins by surface-mapping of phylogenetic information. *Bioinformatics* **19**, 163–164.
40. Lee B, Richards FM. (1971) The interpretation of protein structures: estimation of static accessibility. *J. Mol. Biol* **55**, 379–400.
41. Haliloglu T, Keskin O, Ma B, Nussinov R. (2005) How similar are protein folding and protein binding nuclei? Examination of vibrational motions of energy hot spots and conserved residues. *Biophys. J.* **88**, 1552–1559.
42. Keskin O, Ma B, Nussinov R. (2005) Hot regions in protein–protein interactions: the organization and contribution of structurally conserved hot spot residues. *J. Mol. Biol.* **345**, 1281–1294.

Index

B

C